Atlas of Sand Grain Surface Textures and Applications

WILLIAM C. MAHANEY

ATLAS OF SAND GRAIN SURFACE TEXTURES AND APPLICATIONS

OXFORD
UNIVERSITY PRESS
2002

OXFORD
UNIVERSITY PRESS

Oxford New York
Auckland Bangkok Buenos Aires Cape Town Chennai
Dar es Salaam Delhi Hong Kong Istanbul Karachi Kolkata
Kuala Lumpur Madrid Melbourne Mexico City Mumbai Nairobi
São Paulo Shanghai Singapore Taipei Tokyo Toronto

Published by Oxford University Press, Inc.
198 Madison Avenue, New York, New York 10016

www.oup.com

Library of Congress Cataloging-in-Publication Data
Mahaney, W.C.
Atlas of sand grain surface textures and applications / by William C. Mahaney.
p. cm.
Includes bibliographical references and index.
ISBN 0-19-513812-0
1. Quartz—Surfaces—Pictorial works. 2. Scanning electron microscopy. I. Title.
QE391.Q2 M25 2002
549′.68—dc21 2001054851

9 8 7 6 5 4 3 2 1

Printed in the United States of America
on acid-free paper

Acknowledgments

I am deeply indebted to the late Klaus Fecher of Philips Universität, Marburg, Germany. Until his death in 1990, Klaus was the SEM technician in the Geology Faculty; he introduced me to the scanning electron microscope (SEM) and energy-dispersive spectrometer (EDS). He was a principal player in the Mount Kenya Quaternary History Project along with Professor Walter Vortisch of the same institution (presently at Leoben University, Institute of Prospektion und Angenwandt, Leoben, Austria). Many project publications benefited immensely from Klaus and the imagery that he produced with his CAMSCAN-4 SEM. Nothing of importance was too small to escape his scrutiny and go unnoticed in the various analytical sessions that took place in his lab before and after numerous visits to Mount Kenya. His enthusiasm over the power of the SEM and EDS, in the study of mineral composition and micro-features, and various scientific problems that could benefit from SEM–EDS applications, were a major inspiration to many colleagues and students who worked with him at Marburg. His microphotography could only be described as superb, and many examples of it, especially his high magnification imagery, are found on several pages in this book.

This research was funded in large part by Quaternary Surveys of Toronto and Deutscheforschungsgemeinschaft (DFG), with minor support from the Natural Sciences and Engineering Research Council of Canada (NSERC), the NSERC-DFG exchange program, and the Minor Research Grants Program of Atkinson College, York University.

I am indebted to a number of colleagues for criticism of various chapters, stimulating discussion of various geologic processes, and for the loan of unpublished and/or previously published material. In particular, I thank the following individuals for reviews of various chapters: Randy Dirszowsky (1–5 and 7), Edwin Larson and Richard Mount (3), Andrew Stewart (4), Ron Dorn (1–5), Krister Lindé (1–5 and 9) James Dohm (6), Volli Kalm, John Menzies and Ian Smalley (9), Kalle Kirismäe (11), Peter Birkeland (12), Dennis Netoff (13), Tiia Kurvits (13), Michael W. Milner (3, 10–11, 13–15), Helmuth Schloessin, Patrick Julig, Andrew Stewart and Ron Dorn (15). I also appreciate the comments and criticism received from numerous graduate students, especially the Quaternary stratigraphy group at York University (1999) and Walter Vortisch at Leoben University, Austria. John Dawson prepared all the illustrations.

The author and publisher wish to thank the following for permission to reprint or modify copyright material:

Figure 9.1, adapted from Mahaney, 1990c, *Boreas*, **19:** 81–89, by permission of the publisher, Taylor & Francis AS.

Figures 9.3A, 9.3B, 9.4, 9.5, and 9.6, reprinted from Mahaney & Kalm, 2000, *Boreas*, **29:** 35–51, by permission of the publisher, Taylor & Francis AS.

Figures 10.1, 10.2, and 10.3, reprinted from Mahaney & Sjöberg, 1993, *Zeitschrift für Geomorphologie*, **37:** 337–348, by permission of the publisher.

Figures 12.1 and 12.2, reprinted from Vortisch, Mahaney, & Fecher, 1987, *Geologica et Paleontologica*, **21:** 245–255, by permission of the editor.

Figure 13.2, reprinted from Kurvits, Mahaney, & Kalm, 2000, *Proc. Estonian Acad. Sci. Geol.*, **49:** 17–27, by permission.

Figures 14.1A, 14.1B, 14.1C, 14.1D, 14.1E, 14.1F, 14.2A, and 14.6C, reprinted from Mahaney & Milner, 1998, *Boreas*, **27:** 140–152, by permission of the publisher, Taylor & Francis AS.

Figure 15.5 and 15.6, imagery of samples provided by M.W. Milner, geological consultant, Toronto, Canada.

Figure 15.8. imagery provided by Ron Dorn, Arizona State University, Tempe, Arizona.

Figure 15.9, reprinted figures 4 and 6 from Endlicher & Tillman, 1997, *Archaeometry*, **39:** 333–342, by permission of the authors and *Archaeometry*.

Contents

Atlas of Sand Grain Surface Textures and Applications

1 Introduction

The SEM atlas of Krinsley & Doornkamp (1973), which introduced the subject of sand grain surface microtextures to sedimentology using electron microscopy, has been widely used by a number of researchers since the late 1970s. As a repository of sand grain imagery derived from electron microscopy the original atlas provided the first attempt to collate microfeatures on sands and link them to specific geologic agents and/or sedimentary environments. After two and half decades of research on sand grain morphometry it is now known that only a few microtextures are uniquely indicative of a specific environment or geologic process, and that one needs the power of a representative population of grains to definitively identify the conditions under which grains were formed and/or modified. In theory, at least, SEM microtextures record the sojourn of quartz and other lithologies from time of genesis through different sedimentary/weathering environments to a final resting place.

The SEM analysis of sediments, pioneered by Biederman (1962), Porter (1962), and Krinsley & Takahashi (1962), has undergone several modifications over the last four decades. Since its inception in the early 1960s, the emphasis of most research presented in scientific journals has been on sedimentological applications. However, increasingly, the technique has been applied to other fields where SEM–EDS methods can be employed to build a database of microtextures, microstructures, and chemical spectra that help to answer existing questions as well as generate new ones. For example, a standardized database on precipitates and coatings, which was lacking in the Krinsley & Doornkamp (1973) volume, is now available which can clarify problems related to soil formation (pedology), weathering–diagenesis, soil water and groundwater movement (hydrology) and relative dating of sediments. For the first time, new information on mass wasting (chapter 8), including fast-gravitational effects on grains in rockfall and imagery from grain impact and cohesion in talus piles are presented. Grains released from source rock in such areas as the Caucasus, Venezuelan Andes, coastal mountains of Oregon, and Cape Breton, Nova Scotia are compared and contrasted, with the objective of finding a common set of microtextures that might prove useful in identifying bedrock release. Weathering studies and soil morphogenesis (chapter 12) are research areas where both the SEM and EDS can be used to advantage; however, relatively few studies have been undertaken that hinge on these techniques (Roaldset, 1984; Mahaney, 1990a, 1992a; Robinson & Williams, 2000).

The 1973 volume on surface microtextures is long out of date, but has remained the basic reference for researchers interested in the application of the technique since that time. Requests are still received from various parts of the world asking where copies can be obtained. Additional books such as those of Whalley (1978) and Marshall (1987) have contributed a number of useful case studies in the field since the appearance of the 1973 atlas. This new atlas contains additional case studies intended to complement and expand upon these earlier reference works; however, the main purpose is to bring the reader up to date on the techniques used and the range of microtextures now recognized from different sedimentary environments. The intent is to explore the range of known microtextures and illustrate how they can be used to solve scientific problems, not just in geology, but also in allied fields such as those mentioned above. Copies of the original atlas, where available, may be profitably used along with this volume to help design research projects dealing with the discrimination of sedimentary environments.

This new book includes a discussion (chapter 3) of the most common SEM in the secondary mode and EDS techniques used to image and identify sand-sized particles. Newer digital SEM and cathodoluminescence equipment are not specifically discussed as they are not yet widely available to most researchers. The volume also provides a discussion of sample selection and preparation (chapter 4) and a comprehensive microtexture index with descriptions (chapter 5) before dealing with the possibilities and problems of relating microtextures to specific geologic environments, processes and materials (chapters 6–14) and in other applications (chapter 15). As mentioned, case studies are interspersed throughout the volume in chapters familiar to users of the original atlas (i.e., eolian, fluvial, and glacial environments), plus several new ones. Thus, great pains have been taken to include a microtexture index and to closely scrutinize the range of each microtexture as it appears in different paleoenvironmental settings.

The identification of microtextures peculiar to individual environments has been a prime motivation in searching my own files and logs and those of colleagues engaged in similar quests. Along with this, the main objective is to determine whether different microtextures appear with a characteristic periodic frequency of occurrence in different sedimentary environments and thus may be used for discrimination. From the evidence it is possible, for example, to separate till from glaciolacustrine and glaciofluvial grains within the glacial sedimentary environment (chapter 9). Similarly, while it has proved difficult to separate tills genetically, it is possible to generate information on glacier erosion processes and thickness, as well as the relative amount of water transport within the glacier system.

In general terms, the formation of glacial grains from bedrock release (plucking and quarrying) and the eventual production of microstriations, the one uniquely glacial microtexture, can only be assessed from individual samples of till if the ice responsible attained a minimum thickness and applied a minimum force (Mahaney et al., 1988b). Indeed, SEM microtexture imagery of grains grown in an ice column and subjected to uniaxial stress under controlled conditions has indicated the importance of greatly increased glacial crushing that occurs along a transition from cold-based to warm-based ice with a minimum thickness of at least 500 m. At temperatures close to p-melting, and with cryostatic pressures equal to 1- and 2-km-thick ice sheets, stick–slip motions appear to induce an array of microtextures on quartz sand grains that closely resemble those microtextures observed in tills deposited by the thickest Pleistocene glaciers of Canada, Europe, and the Antarctic. Aside from striated surfaces, it is not the presence of individual microfeatures that identifies the mechanism of glaciation, but the range of microtextures associated with varying degrees of glacial crushing. With decreasing ice thickness the range of microfeatures diminishes until, with thin cirque ice, little damage appears on grains transported from source to end moraines.

In the eolian environment (chapter 6), rounding and frosting on larger grains due to abrasion along with occasional upturned plates and the presence of numerous craters of variable geometry, apparently due to turbulent collisions during transport, are the main criteria used for identification. The very fact that eolian and glacial grains could be differentiated in an analysis of Saharan ergs (Mahaney & Andres, 1996), without any previous knowledge on the part of the SEM operator, should convince skeptics that there is, indeed, an advantage in using SEM imagery in sedimentological analyses of various kinds. Eolian microtextures are generally limited to superficial "skin" damage to individual grains which distinguishes them from the deeper, more oriented "Wallner lines" produced by glaciation.

The fluvial environment (chapter 7), like the eolian one, produces an array of superficial fractures and abrasion microfeatures. However, for the most part, fluvial transport tends to produce more severe abrasion that is correlated with the discharge and turbulence of river flow. More forceful grain collisions also produce v-shaped percussion scars, the one microtexture exclusive to the fluvial environment. Interpretive difficulties arise because similar microfeatures are also produced by glaciers, as for instance, in moulin systems or in fast-moving, turbulent subglacial and englacial meltwater channels. If such grains are fluvially overprinted once they leave the glacial system, the origin of the grain may be difficult to decipher (Mahaney & Kalm, 1996). Of course, in nonglacial areas, streams leave an indelible record of water transport on quartz and other mineral fragments (Mahaney & Kalm, 2000).

Mass wasting (chapter 8), a neglected geologic agent in microtextural studies (Bull et al., 1987), initially leaves a record of bedrock release on grains; however, fast or slow movement in gravitationally induced systems leads to the generation of new microtextures that involve grain packing and slickensides. In some talus deposits grain packing appears to be associated with the presence or absence of different species of expandable clay minerals. The most distinctive microtexture in mass wasted sediments is the fracture face, a relatively clean break on the grain surface, often surrounded by varying degrees of chemical dissolution. Fracture faces are occasionally found on grains originating in other sedimentary environments, but not with the frequency of occurrence and high ratio of fracture face to grain surface that occurs with mass wasting. Thus, it seems SEM investigations have much to offer in understanding mass wasting.

Shock-melted quartz (chapter 11), though relatively little studied, possesses an easily recognizable microtexture that occurs along with various radial fractures from bolide impacts. Bolide (asteroid) impact is considered by some researchers (Oberbeck et al., 1993) to provide a plausible explanation for the occurrence of some tillites (ancient glacial deposits) characterized mainly by fragmented and fractured quartz particles. The few tillites that have been examined, however, are lacking in shock-melted quartz, which is the main product of extreme high-energy impacts. Additional investigations of tillites are needed to determine if shock-melted quartz can be located in sufficient quantity to support the theory.

Quartz sand grains subjected to pressure in shear zones on either side of a fault (chapter 10) are rounded and stretched, producing a unique "rolled" microtexture (Kanaori et al., 1991) that can be studied by SEM. Such grains have a distinctive character that distinguishes them from grains in all other sedimentary environments. The wrenching motion produced along main fault lines produces a range of rolled, flattened and cracked grains. Reworking of these grains by eolian and lacustrine processes may lead to their further modification in small-scale fault displacement systems in periglacial (proglacial) sediments, which tend to produce cleaved surfaces with thin "wafer-like" particles of quartz clinging to larger rolled quartz.

This new monograph should be useful to those who still find the 1973 *Atlas* of Krinsley and Doornkamp valuable.

2 Literature Review

The scanning electron microscope (SEM) is used in the secondary electron (SE) mode to image sand grain surface topography (Krinsley & Doornkamp, 1973), and occasionally in the cathodoluminescent (CL) mode (Goldstein et al., 1981; Krinsley & Trusty, 1986) to define and source environments. The SE mode is thus used to separate ancient depositional environments associated with wind, water, ice (glacial environments), and weathering–diagenesis (Krinsley & Doornkamp, 1973). Certain subenvironments have also been recognized. The technique has been applied to grains as old as Lower Paleozoic and Precambrian (Kurvits et al., 2000), but in these sediments, the original mechanical textures (assuming they were present) have usually been eliminated by diagenesis.

Henry Clifton Sorby, who pioneered the application of the petrographic microscope and thin section to sedimentary petrology (Sorby, 1880), was the first person to comment on sand grain surface textures, and clearly is the "father of the discipline." Sorby discussed the rounding of sand grains in various sedimentary environments and described certain corrosion and mechanical textures he observed using the light microscope. The classic treatise on size analysis, preparation of samples, orientation, shape, and roundness was published by Krumbein & Pettijohn (1938), thus establishing principles for the optical study of grains. In addition they focused on the modification of grain shape that occurs with transport by different geological agents.

This earlier work was refined by Shepard & Young (1961), Kuenen (1960), and Kuenen & Perdok (1962) who introduced the concept of "pivotability" or the tendency of a grain to roll on an inclined plane. Pivotability depends on shape and roundness, with angular grains having the greatest stability. Grain shape and methods of determining roundness and angularity were discussed by Mueller (1967). Pettijohn (1975) focused on the "surface character" of sedimentary grains and suggested that quantitative definition and measurement of microtextures had been lacking up to that point. He also believed that interpretation based on microtextures was less than satisfactory and that more experimental and observational data were needed before a satisfactory understanding of the subject could be achieved.

The history of surface sand grain textural analysis from the beginning of the twentieth century to the late 1940s is summarized in an excellent Ph.D. thesis by

Lou Williams (1947) prepared under the direction of Francis Pettijohn. The little work that had been conducted up to that time, had all been done using only the light microscope, which, of course, lacks the high resolution and stereo imagery of the SEM (Reed, 1997). It is difficult to see much textural detail when examining individual grains, as resolution with the light microscope is relatively low (–2000 Å). Depth of field is also restricted making it impossible to view even a very fine sand of approximately 100 μm in diameter at reasonably high magnification with more than 40 percent of the grain in focus at any one time (Bond, 1954). Many of the microtextures now known are on the order of several micrometers in diameter and thus could not be observed with reasonable clarity by these early workers.

The above difficulties were overcome to some extent when geologists began to use the transmission electron microscope (TEM) to study surface microfeatures on sand grains. Biederman (1962), Porter (1962), and Krinsley & Takahashi (1962) were the first workers to describe sand grain surfaces at high magnification and with the resolution permitted by the TEM. The latter authors were the first to relate groups of specific environmental surface microtextures to particular environments of deposition, such as eolian, subaqueous, and glacial. However, a number of problems limited the use of the TEM. For example, it proved necessary to construct replicas of individual sand grains, a difficult and time-consuming process that limited the number of samples that could be analyzed. Also, since the original surfaces were not examined, many features were misidentified, or not recognized at all. Specifically, the replica technique did not duplicate features that contained a great deal of microrelief. Thus, certain microfeatures were not observed until the scanning electron microscope became available.

The SEM did not become available at many universities until the late 1960s, and it was not used to describe surfaces of sand grains until 1969 (Krinsley & Margolis, 1969). All subsequent work of any importance in this area was carried out with the SEM in the secondary mode. Not until the EDS became available as an accessory in the latter part of the 1970s could individual grains be examined for their chemical composition at the same time. Coatings and precipitates on grains could be analyzed to determine their thickness which proved important in weathering and diagenesis studies (Mahaney, 1990a).

The early TEM and SEM microtextural analysis involved both observation and experimentation. Initially, quartz sand grains from specific environments such as eolian, beach, glacial, and river were selected, imaged and classified for characteristic surface topography or texture. Laboratory experiments involving the crushing of fresh grains, for example, were then performed in an attempt to duplicate the textures observed. It was hoped that the experience gained by producing the textures artificially would provide insights into the natural mechanisms involved in grain evolution. In comparison with independent data from known environments, experimental results (Krinsley & Takahashi, 1962; Kaldi et al., 1978; Lindé & Mycielska-Dowgiallo, 1980) proved to be useful only in a general way, but needed (and still need) to be refined.

The method was then used to identify ancient environments (Krinsley & Doornkamp, 1973) and checked against other, independent data from known environments. The results were usually satisfactory in unconsolidated, semi-consolidated, or consolidated sediments, only if diagenesis had not been too intense. It was recognized that only in younger sediments are the original surface microtextures likely to be preserved. In light of the difficulties, most workers acknowledged that the technique should not be used alone, but in conjunction with other methods of determining environment; if all else failed, the method could be used alone to support hypotheses, but with reservations. The basic technique, and early ideas on environmental analysis along with representative photomicrographs, were summarized in the atlas (Krinsley & Doornkamp, 1973). The book is still in use and the representative micrographs it contains are an excellent complement to the present work.

What is the basis for using the surface microtextures of sand grains to distinguish ancient sedimentary environments? Miller & Olson (1955) divided all properties of contemporary environments of sedimentation into three groups. The first includes properties that exist only in contemporary sediments and not in the lithology. The second includes properties that could be analyzed both in modern and ancient sediments, but which change over time with diagenetic processes. The third group encompasses properties that are the same in both modern sediments and lithology.

Surface microtextures fall into the second and third groups; probably most of the characteristic surface microtextures described in this volume are third group properties, but a considerable number may belong to the second group. Some microtextures result from weathering processes such as solution; other microtextures appear to increase in frequency of occurrence where there is evidence of alteration or chemical mobilization. However, the correspondence between modern and ancient microtextures may be very close depending upon the original textural imprint (imprint energy), grain age, and all the diagenetic processes that operated over post-depositional time. If the origin of particular microtextures, including those due to diagenesis, are well understood, environments can be delineated with a fair degree of certainty. It is important, however, to study sand grains or groups of grains in great detail, since the effects of more than one stage of development may be present. Knowledge of the physical and chemical parameters that influenced or created characteristic surface microtextures can lead to more exact environmental discrimination.

Long-term post-depositional solution and precipitation as well as intense weathering in the modern environment may eliminate evidence of previous mechanical action. However, if a large number of sand grains are studied, it is often possible to find relatively restricted areas on some grains where chemical action has not been sufficiently strong to eliminate or overprint earlier microtextures. The reason for this is that diagenetic chemical action and/or physical changes are highly variable so that, if the search is long enough, one or several mechanical episodes may be discernible. Relative age relations also may possibly be determined by studying and comparing dissolution or etching features on groups of grains.

In some cases where grains possess a diagenetic coating, such as precipitated silica, it may be possible to etch the surface down to its original state. Assuming the original surface has not been removed by weathering, or abrasion, the original environmental information would be revealed. The technique has not been systematically investigated, and methods are not yet well developed. Cathodoluminescence (CL) is not used in my laboratory to investigate grain surfaces below the top few nanometers or so, although other workers indicate it has great potential (D. Krinsley, personal communications, 1998, 1999). This technique provides subsurface information down to depths of at least a micrometer (10,000 Å), as compared to secondary electron depths of several hundred angstroms. However, the type of information available with CL is somewhat different than that obtained in the secondary electron mode, so that correspondence between the two may be difficult to determine (Krinsley & Hyde, 1971; Krinsley & Tovey, 1978; Tovey & Krinsley, 1980; Seyedolali et al., 1997). Nevertheless, Seyedolali et al. (1997) were able to show that source areas for quartz sand grains examined in CL using the SEM, could be distinguished in cross-section.

Another problem is the distinction between modern surface chemical action and diagenetic alteration. This is difficult as the "veil of diagenesis" discussed by Birkeland (1999) for the field description of paleosols could be applied here. Unfortunately, criteria have not yet been established that will reliably distinguish between pedogenesis and diagenesis. A number of differing early stage solution and precipitation features have been identified (Waugh, 1965, 1970; Pye & Sperling, 1983) but so far little work has been done on pre- and post-depositional groupings of microtextures (Mahaney & Kalm, 1995). However, a later stage, in which quartz crystal terminations are formed, only seems to occur in diagenesis. Initially, terminations form on the edges of cleavages and abrasion fractures; these then enlarge, and finally coalesce. Presumably the overgrowths or new growths that form individual quartz grains would eventually grow together as the terminations interlocked and a dense quartzite would be produced (Mahaney, 1995a). The stages described have not been studied in detail, but if carefully done, would enhance our understanding of diagenesis.

This review, though brief, indicates that most applications of SEM-based microtextural analysis have had the objective of distinguishing various depositional and weathering–diagenetic environments. A number of other uses have been attempted, or considered, and are outlined below. The following list is by no means exhaustive and is intended to convey the vast potential of the method.

1. Studies of how diagenesis occurs at the micro level; for example, cementation and pressure solution processes (Waugh, 1965, 1970; Pye & Mazzullo, 1994; Kurvits et al., 2000), or relation to time spans when magnetism is acquired (Larson & Walker, 1975).
2. Comparison of source area grains as compared to basin samples. Studies of multiple sources (Krinsley, 1981; Mazzullo & Ritter, 1991; de Boer and Crosby, 1995; Mahaney et al., 2001c), including statistical analysis (see, e.g., Bull, 1978).
3. Tracing of particular environments within a sedimentary column by comparison of specific microtextures in a series of progressively younger samples (Krinsley, 1981; Barlusseau & Kounkou, 1989; Cremer & Legigan, 1989; Kenig, 1995).

4. Classification of fault gouges in terms of structure and origin (Kanaori, 1980; Kanaori et al., 1980, 1985, 1991).
5. Determination of terrestrial environments at specific times in Earth history from sediment extracted in ocean cores (Krinsley & McCoy, 1977; Bull, 1981; Helland & Holmes, 1995, 1997).
6. Study of ancient wind velocities (Krinsley & Wellendorf, 1980; Wellendorf & Krinsley, 1980; also Krinsley et al., 1979 by comparison with experimental abrasion of eolian sand grains).
7. Studies of eolian and other environmental fracture–abrasion mechanisms in the laboratory (e.g., Lindé, 1978, 1984; Krinsley et al., 1979; Marshall, 1979; McKee et al., 1979; Lindé & Mycielska-Dowgiallo, 1980; Osborne et al., 1993). Wind tunnel studies of ripple formation (Lindé, 1978).
8. Studies of the origin of fine silt and clay particles in the geological column (Krinsley & Smalley, 1973; Krinsley & McCoy, 1978; McKee & Krinsley, 1978; Assallay et al., 1998).
9. CL in quartz grains indicating previous fracture histories (Krinsley & Hyde, 1971; Krinsley & Tovey, 1978; Tovey & Krinsley, 1980; Seyedolali et al., 1997).
10. Simulation studies of artificial Martian eolian sands (Krinsley et al., 1979).
11. Forensic geology in relation to sand grain surface textures (Murray & Tedrow, 1975, 1992).
12. The $^{87}Rb/^{87}Sr$ dating of quartz grains and its relation to surface microtextures (Powers et al., 1979).
13. Micro-correlation of geographic areas with respect to individual sedimentary layers. This has not been done as yet.
14. Origin of brittle materials, crack propagation, and plastic deformation of quartz grains in relation to surface microtextures (Marshall, 1979, 1987; Middleton & Kassera, 1987; Mahaney, 1995b).
15. Relation of quartz sands to various soil horizons and types of soils or paleosols (Mahaney, 1990a, 1992a; Mahaney & Hancock, 1993; Ghosh, 1997; Mahaney et al., 1997a; Earl-Goulet et al., 1998).
16. Sand grain shape as related to tropical weathering (Pye, 1983; Borger, 1993a; Pye & Mazzullo, 1994) and subtropical weathering (Little et al., 1978).
17. Detailed interpretation of Pleistocene and Recent glacial histories (Mahaney et al., 1988a, 1996a; Mahaney, 1995a,b; Mahaney & Kalm, 1995, 2000; Mahaney & Andres, 1996).
18. Analogue studies of grains under polar climates in relation to weathering on Mars (in progress at York University's Geomorphology and Pedology Laboratory; Mahaney et al., 2001a).
19. Studies of periglacial environments, especially rock glaciers (Lewis & Hanvey, 1993) and patterned ground processes.
20. Quartz surface microtextures and grain size characteristics (Rogerson & Hudson, 1983; Mahaney, 1990a; Assallay et al., 1998).
21. Micromorphology as an indicator of paleoenvironments (Mahaney, 1990a; Earl-Goulet et al., 1998; Borger, 2000).

The statistical approach has been used to great advantage in recent years (Mahaney et al., 2001b), and is absolutely necessary since large numbers of microtextures, grains, and samples must be related to environment. This topic is discussed in some detail later in this book (see, e.g., chapter 4; also see Mahaney & Kalm, 2000; Mahaney et al., 2001b).

Finally, I hope that this book will provide the reader with better ways of simulating quartz grain surface modifications and of relating surface microtextures to environment.

SEM and Energy-Dispersive Spectrometer

3

This book is based on the use of the scanning electron microscope (SEM), in the secondary electron imaging mode (SE), to image the surface microtextures of mainly quartz sand grains. In the simplest light microscope, in use since the mid-fifteenth century, a single lens is used to magnify an object. The compound microscope uses an additional lens, or objective, to enlarge the image produced by the primary lens. In the 1920s, Louis de Broglie demonstrated that an electron beam constituted wave motion with a wavelength shorter than light. Shortly thereafter, the discovery that magnetic fields could be used as lenses to focus electrons and other charged particles (Passchier & Trouw, 1998) set the stage for the development of the electron microscope, the instrument that would eventually make possible the study of microtextures and microstructures.

To examine a specimen in the SEM, a sample is placed into a chamber on a X–Y–Z tilt rotation stage. Above the chamber is a column containing electromagnetic lenses and at the top of the column an electron source (typically a tungsten filament or LaB_6 source) emits electrons with energies up to 40 keV (electron volts). After placing the sample in the chamber both the chamber and column are pumped down to high vacuum and power is supplied to the electron source. Electrons are collimated into a beam which is scanned in an ordered x–y system (tilt and shift on the control panel) across a specimen surface by means of electromagnetic deflectors in the microscope column. The sample is scanned raster line by line by an electron beam. The beam is swept in successive horizontal lines, each line a fixed distance from the others, at variable speeds up to tens of milliseconds per line. The distance between scanning lines is sufficiently small to include most of the surface topography on the specimen. As electrons impinge on the specimen surface (in this case, the rough surface of a sand grain), they create a number of phenomena, the most significant in this discussion being the emission of secondary electrons with energies up to a few tens of eV and X-rays. Other types of emissions include backscattered (reflected primary) electrons, "Auger" electrons, photons, continuum X-rays and heat (see Goldstein et al. (1981) and Slayter & Slayter (1992) for a detailed description of the SEM).

Secondary electrons are used to produce conventional images of surface topography in the SEM due to their ability to image fine surface detail. Secondary electrons are "resident" electrons of the sample that are "knocked out" of their orbit by impinging primary (source) electrons and can be segregated from other emissions by their energy state. As the electron beam moves across the specimen the secondary electrons that are emitted are sequentially collected by a crystal scintillator, amplified via a photomultiplier tube, and scanned across a cathode ray tube synchronously with the scanned electron beam to produce a television-like image of the sample surface. In order to prevent a build-up of electrical charge in the sample, non-metallic specimens must be coated with a continuous layer of carbon, gold, or gold–palladium several nanometers in thickness that is continuous with the metal stub the specimen is mounted on. This coating provides a constant "ground" for the specimen and prevents an unpredictable rapid discharge of electrical current which causes image artifacts.

The primary electrons, upon hitting the sample, undergo elastic and inelastic scattering. During elastic scattering, the incident electrons interact with the nuclei of atoms in the specimen, causing trajectory deflation, but with little loss of energy. Two types of elastic scattering may occur. Rutherford scattering occurs in the Coulomb field of the nucleus and is a single scattering event that results in a change in trajectory, sometimes exceeding 90°. During inelastic scattering, incident electrons may interact with the nuclei of atoms in the specimen, thus losing energy in the Coulomb field and emitting white or continuum X-ray radiation, or they may collide with loosely bound electrons in the specimen, causing the latter to be ejected as secondary electrons (Watt, 1985). Because secondary electrons have low energies (typically less than 50 eV), they are unlikely to escape from the specimen if produced more than about 50 Å below the surface. Due to the scattering of electrons within the specimen, secondary electrons are emitted from a pear-shaped volume beneath the focus point of the beam. The actual volume of the specimen that produces secondary electrons is dependent on both the energy per unit area of the electron beam and the atomic number (Z) of the specimen.

The emission of secondary electrons and their intensity is sensitive to the angle at which the electron beam strikes the specimen surface, and, therefore, to the topographic features on that surface. The signals generated at any spot on the specimen are produced from a specific volume just below the surface of the specimen, the size of which depends upon the area and intensity of the primary electron beam (E_0) and the average atomic number (Z) of the specimen. The output is composed of electrons and photons of variable energy emitted from each point on the specimen. With electronic amplification this signal may be run through a synchronously scanned oscilloscope to form an image that can be photographed. Variations in the resulting signal strength as the electron beam is scanned across the specimen are used to vary the brightness of the trace on a cathode ray tube which is being scanned in synchronization with the beam.

The magnification produced by the SEM is the ratio between the dimension of the final image display and the field scanned on the specimen. In other words, if the electron beam is scanned across an area 1 mm^2 on the specimen, and the output is displayed on a screen 100 mm^2, the linear magnification is 100. Changes in magnification are brought about by altering the extent of the scan on the specimen, while keeping the size of the display constant. In practical terms, the magnification control on the microscope actually varies the angle through which the electron beam is deflected, the linear extent of the scan depending on the working distance of the specimen from the final lens. In a typical microscope, the magnification range is from $\times 20$ to $\times 100{,}000$ at a working distance of 30 mm. The magnification of the SEM instrument is 100 times greater than that of a normal optical microscope.

The wide range of magnifications available coupled with a large depth of field is the greatest advantage of secondary electron scanning microscopy. However, on the minus side the imagery is always monochromatic; therefore, recognition of mineral species by interference colors is not possible with the SEM. However, as will be seen later with the EDS, mineral identification with SEM imagery is possible by means of the gray tone on the monitor (atomic-number contrast; as explained by Hall & Lloyd, 1981). In any case, material irradiated by the electron beam emits (or back scatters) an electron stream that is proportional to the atomic number, in most cases beginning with Na (AN = 11).

In the SEM, object detail is determined point by point, the resolving power being directly influenced by the smallest area that can be sampled (i.e., by the diameter of the electron beam). Thus, high resolution can be obtained by a lens system with small aberrations and high electrical stability. In practice, the smallest beam that can be formed with enough electrons to produce a decent picture depends on the focal length or working distance of the final beam-forming lens, and the strength of the electron gun. For a conventional gun, working at a distance of about 10 mm, the smallest attainable probe size is about 4 nm, and the secondary electron resolving power is between 4 and 7 nm. Better resolving power may be obtained with a stronger lense or a more intense electron source (Goldstein et al., 1981; Watt, 1985; Slayter & Slayter, 1992; Krinsley et al., 1998).

Image Recording

The SEM image is recorded sequentially; the face of a cathode-ray tube is photographed as a time exposure, while the image is drawn on it line by line for one complete frame. Commonly, the cathode-ray tube is coated with a blue phosphor (P11) with a short persistence time. With older analog SEMs the operator must take care to allow phosphorescence to "burn" off the screen before taking a photograph, a process that takes only a few seconds. Failure to "burn off" the phosphorescence results in horizontal streaks or "ghost images" of the previous view being superimposed, which reduces the photographic resolution and produces substandard imagery. The recording tube has its own brightness and contrast controls which can be set to suit the characteristics of the photographic material being used for recording. The image is scanned with more lines per picture than the screen can resolve, so that no scanning lines are seen in the photographic record. The record scan is produced more slowly than the visual scan used for selection and focusing, in order to give a smooth, grain-free photograph. The frame time may range from 10 to 200 s, depending on the noise of the image. High-magnification images made with small, focused beams require longer exposure times, since the signal-to-noise ratio is low (Watt, 1985; see also Smart & Tovey, 1988).

The photographic record may be made on 35 mm, or larger, format film; some researchers prefer to use Polaroid film if an instant photographic print is required. Brightness and contrast settings are different, however, for Polaroid and 35 mm or 120 mm film. Many current SEMs produce an image that is rectangular and compatible with the 1:1.25 aspect of 4″ × 5″ photographs (Watt, 1985).

Newer SEMs usually have a digital recording capability. In this case, the image is scanned and digitally stored in the memory of a computer disk, and displayed as needed. Since no photographic record is necessary, a considerable amount of time and space may be saved. In addition, the image can be printed out immediately and/or manipulated and improved with programs such as Adobe Photoshop and Corel Photohouse.

Energy Dispersive Spectrometry (EDS)

On the basis of shape alone, it may be difficult to distinguish quartz sands from other grains, such as feldspar. This is particularly the case when the grains are not sectioned, but simply attached to a glass slide or aluminum stub (see chapter 4). It is of importance in many cases to be able to identify clay minerals, other adhering substances, and coatings on sand grains. To do this, X-rays released from the specimen during electron microscopy may be used in energy dispersive spectrometry (EDS). It should be noted that EDS (also called energy dispersive X-ray microanalysis or EDX) supplies chemical elemental data only, and not mineralogical information. For instance, if one analyzes potassium feldspar, the information obtained would indicate only that the specimen possesses various concentrations of K, Al, and Si (and possibly O if a "windowless" CNO detector is used). With experience, it is possible to identify K-feldspar with a fair amount of certainty.

In EDS, a semiconductor detector is used to classify X-radiation in terms of its energy rather than its wavelength. This type of detector usually consists of a single-crystal disc of lithium-drifted silicon [Si(Li)], 3–5 mm thick, with an active area of between 10–30 mm. The energy of X-ray photons coming from the specimen after bombardment by an electron beam and impinging on the crystal are converted into pulses of current proportional to the energy of the photons. The pulses are amplified, counted, digitized, and fed into a multichannel analyzer or a computer memory where they are stored in a location appropriate to the pulse height.

The memory may have up to 1000 channels, each 10, 20, 40, or 80 eV wide. The ongoing counting process and the final spectrum can be displayed on a color TV monitor. The entire energy dispersive spectrum is displayed simultaneously, as is shown in illustrations later in this book.

Computer software enables the displayed spectrum to be expanded horizontally and vertically, and analyzed by means of data stored in the computer memory. This can be done very rapidly; for instance, one may wish to identify each sand grain studied, which takes just a few seconds for an experienced operator. The energy range 1–10 keV is very useful, containing $K\alpha$ lines of Z from 30 to 80 and M lines from $Z = 62$ upward (Watt, 1985). It is useful, however, to note that EDS cannot be used to identify small grains, such as those less than 5 μm, as the X-rays are produced in a "pear shaped volume... where the electron beam hits the surface" (Passchier & Trouw, 1998).

The detecting crystal and the first stage of the built-in pre-amplifier are cooled with liquid nitrogen to optimize the electronic signal-to-noise ratio of the spectrometer. A cooled crystal, even within a high vacuum, would rapidly be covered with ice and other condensed vapors from the SEM column. To avoid this, the crystal is screened from the vacuum by a thick (7.5 μm) beryllium "window." The window presents an analysis problem for low-atomic-number chemical elements. Thus, the normal EDS or EDX system is usable only for atomic numbers from 11 (sodium) upward. Thinner windows (so-called "windowless detectors") can be used to detect elements as low as oxygen and carbon, but these thin windows are expensive; other problems ensue from the use of carbon coatings (Watt, 1985). In order to prevent electronic charging, specimens are normally coated with carbon, gold, or gold–palladium. Obviously, if one is analyzing for gold, or for any chemical elements in the central critical palladium range, the specimens should not be coated with gold–palladium. Some of the newer SEMs, however, operate at very low voltages, and coating of grains may not be necessary. Resolution tends to drop at low voltages.

The use of polished specimens mounted at a predetermined angle allows quantitative point analyses (Fecher, 1985) with the electron beam and X-ray spectrometer. This quantitative work requires a known penetration depth–absorption path length ratio of the beam, which is obtained from the relative positions of the detector and specimen and from the surface orientation of the specimen on the stage. As long as the intensity of X-ray generation and absorption are homogeneous and the surface plane between point of incidence and exit area are known, the same procedure can be applied to rough surfaces on a single grain to increase the accuracy of the chemical analysis. Various other stereoscopic techniques for X-ray microanalysis of rough specimens may be used to provide information about the homogeneity in depth of a particular microfeature (Bomback, 1973).

As pointed out by Xu (1985), the SEM is one of the most versatile instruments for the analysis of microtextures and microstructures in solid materials of various kinds. In most cases one can learn to use and operate the SEM–EDS within a few hours. Taking high-resolution and high-magnification images, however, with film or Polaroid equipment requires a greater time commitment. Also, the operator must realize that brightness and contrast settings differ from Polaroid to film. The amount of data that can be generated in a few hours on an SEM–EDS, however, makes the time spent well worthwhile. For most projects, sands can be studied by light microscopy first at relatively low cost, and then, individual grains may be subjected to high-resolution imagery on the SEM. Combined with rapid delivery of chemical information from the EDS, this makes the instrument well worth the rental cost. Normally a six-hour stint on the SEM–EDS, assuming a photographic rate of 10–15 frames per hour, will cost from $200 to $400 (US) (including rental fees and photographic processing). Newer digital imaging systems cost less than older analog equipment because there are no photographic costs involved. On the other hand, commercial SEM facilities that must bear the full cost of operation might charge as much as $1000 (US) per day.

4 Sample Selection and Preparation

Microtextural and microstructural analyses, by their very nature, require investigators to consider the importance of analyzing exceedingly small samples that will produce as much information as possible about particular stratigraphic units and paleoenvironments. The larger mass unit under investigation is called the *population* or *universe*; for any population there may be several microtextural properties that describe or characterize it. These properties are described in detail in chapter 5. The value for each property is called a parameter and is usually in the form of a ratio or percentage value representing its frequency of occurrence (see Mahaney & Kalm, 2000; Mahaney et al., 2001b). It is important to realize that, in addition to *ratio* numbers, or scales of measurement, there are also *ordinal* or *nominal* scales which are of greater relevance to the description of sand surfaces and microtextures used in paleoenvironmental reconstruction. For instance, etching can be quantified along an ordinal scale from high to low values and coatings can be characterized on a nominal scale as present or absent. The scale one chooses reflects the degree of quantification desired and often the complexity of the microtexture.

As explained later in this chapter, one might use frequency of occurrence for individual microtextures where for *x* number of grains studied, the number of times (frequency) an individual microtexture is encountered is tabulated. Undoubtedly, some microtextures will not be seen, while others will vary in the degree to which they appear on the surfaces of observed grains (see, e.g., Mahaney & Kalm, 2000).

Sampling Method

Since sampling involves the study of the variability of earth materials, a major concern is whether the material under investigation is homogeneous or heterogeneous. While one is sometimes interested in uniformity, the greatest interest is in showing how one population of grains varies from another population; that is, how the A horizon of a soil differs from the B horizon, or how a fine sand bed differs from a coarse sand bed in a lacustrine deposit. When the time to sample arrives, *research design* depends on the exact problem under investigation. The strength of the SEM–EDS analyses discussed later in this book rests almost entirely on the precision and logic employed in field sampling. It is important to remember that high-precision laboratory analyses will not make up for poor field sampling.

Sampling is ideally carried out to obtain a representative sample of the population usually by sampling randomly (Fisher & Yates, 1967) and to estimate microtextural characteristics with high accuracy. This is balanced by practical limits on the number of samples or grains that may be examined. Usually subsamples of a sand fraction are selected, by counting out and selecting grains under the light microscope or hand lens. Alternatively, fine sands may be sprinkled onto a stub. If the population is homogeneous, two or three samples may provide the information needed, but often it is necessary to provide a statistical assessment, with sufficient replication, to insure that the microtextural evidence is not a chance occurrence. Earth materials are complex and often highly variable, even within a single outcrop, section, and/or profile. There are cases, however, where geological successions are exceedingly homogeneous (Mahaney et al., 1992), with little variation in key parameters up and down section. Homogeneity is useful to establish, as it often implies that materials all have a common source and/or were transported by one geologic agent; it also indicates the absence of any diagenetic or pedogenic processes following deposition.

If earth materials under investigation are heterogeneous, it is imperative to consider the inferences one wishes to make regarding the population under investigation. Two questions must be considered carefully when collecting field samples: *where* to collect and with what *frequency*? The greater the heterogeneity of the material, the greater the number of samples required to characterize it. In some cases, the variation about the mean is so high, even within one population, that it is not possible to distinguish it from another. With sands, for example, interhorizon or interbed parameter variations are often higher in soils or paleosols than in underlying parent material, so much so that sampling intensity may necessarily be higher in the soil than in the fresh, undifferentiated material from which it is formed. Some thought must be given to the number of samples that might be subjected to SEM–EDS analyses, as this will determine the number of samples to be collected from each unit. Identification of individual units must be based on stratigraphic criteria such as, for example, different processes where one might wish to characterize different sedimentary environments, or deposits suspected of having different source areas.

At this point the investigator may wish to consider the number of samples needed and the sampling error involved. A small sample may have less error than a random sample. With an increase in sample size, error within a random sample decreases and the strength of inferences constructed increases; with judgment samples (i.e., grains taken after examination with the light microscope), the error may decrease slowly unless the investigator is skilled in selecting "representative" sites.

Stratified random samples, where the population is broken into a number of subpopulations and a random sample taken from each subpopulation, allow inferences about each of the separate subpopulations and increase the reliability of inferences made for the entire population. Within a lithostratigraphic population, every stratum might be considered a subpopulation which could be analyzed for several parameters. However, there is seldom any reason to sample with this kind of intensity considering that the cost for SEM–EDS analysis may become prohibitive.

Sampling errors arise during collection from selected sites and are in addition to measurement errors. Together they amount to the total error. Sampling error occurs when the sample does not reflect the entire population. Selection error occurs when the sample is collected with a bias for some individual characteristic such as accessibility to that part of the section or perhaps coarse texture, for example. Measurement error arises when the observed measurement is not the true measurement of the unit to be collected.

Sample Collection

A sample collected in the field is only a representation of a population that may or may not be homogeneous. The sampling plan dictates the number of samples required to estimate sampling error and to characterize a population. Normally in the field, a single sample is taken from each horizon or stratigraphic unit (deposit or rock formation); however, even within a deposit or formation there are subunits (termed variously members, lentils, tongues, beds, etc.) that the investigator might be interested in sampling in order to characterize unusual or extreme events. Such sampling is said to be "systematic," and the sampling grid is considered to provide the requisite number of samples to form an adequate database from which interrelationships can be established and interpretations made that will help solve the problem at hand. The problem and degree of precision required will dictate the sampling grid to be established, but the investigator must decide on the interunit and intraunit differences to be established later on. The physical energy required to collect a "systematic sample" in the field, and to see a great number of samples through various laboratory analyses, is daunting to say the least, and may even on occasion be life-threatening (especially with mountain research). Thus, not all samples selected may be collected or used for subsequent microscopic work.

Representative Sample

After deciding where to collect and with what intensity, the researcher normally considers some of his or her samples as "representative" for the purpose of microscopic analysis. Thus, we have the concept of the most "typical" site and sample, which in the the collector's mind provides reliable information on the population. Because sampling in this way may introduce bias, some investigators are cautious and either collect the extremes of the population in order to show the variability within the population, or else analyze the extremes but exclude them from the population (Petersen & Calvin, 1986). In the latter case, the extremes are then considered to be "unrepresentative" of the population. In the majority of cases representative samples are collected on the basis of color, mineral content, particle size, or some other easily determined property in the field, and may or may not represent the population. Cross-bedded sediments are not well described by general samples, requiring the investigator to sample each stratum. A single sample of anything is regarded with extreme caution. Nevertheless, it is these representative samples that form the database for microscopic analysis.

Subsampling

In almost all types of sampling there is a need for subsampling in the laboratory. When this is necessary, the sample is divided into a smaller portion to carry out a particular test, such as particle size, moisture content determinations, or pH. The test is then carried out, presumably with a "representative" portion of the bulk sample (i.e., with splits or subsamples of the sand fractions), or with various fractions of sand recovered from wet sieving following particle size analysis. If samples are to be selected from the bulk samples, great care must be taken to check representative grains under the light microscope. For example, they may have dust coatings that need to be washed with distilled water, so that individual grain surfaces can be microscopically studied. In any case, it is necessary to wash samples prior to selection for SEM–EDS analysis.

The various fractions to be analyzed may be recovered following particle size analysis where the sands are wet sieved and dried. Individual sand fractions may be analyzed, from very coarse (1–2 mm), coarse (1 mm–500 μm), medium (250–500 μm) to the fine (250–125 μm), and very fine (125–63 μm) grade sizes. Dried grains, cleaned of all loosely bound particles and any organic detritus, provide excellent surfaces on which to study the imprint of different transport mechanisms, diagenesis, and weathering. The smaller the particles, the flatter they become. Fracture mechanisms tend to depend on a statistical distribution of defects. Thus, as particle size is reduced, fracture becomes progressively more difficult (Krinsley & Smalley, 1973). In addition, failure occurs more easily by cleavage (crossover point about 50–75 μm). As well as various microtextures and their relative frequency of occurrence, it is often a good idea to compare grain coatings observed with XRD analysis of the <2 μm fraction recovered following particle size analysis. More will be said about this in chapter 9.

A representative subsample can be prepared with a sample splitter, or by spreading the sample out on weighing paper, and then using a microspatula to select small amounts at more or less random intervals to produce the desired weight. For all practical purposes sands sampled for microscopic analyses are selected from dry sieved separates recovered from particle size analysis or from samples sieved directly from bulk down to subsample size of interest without any particle size analysis. From the subsample used for light microscopic analysis, individual grains may be selected and mounted on aluminum stubs (figure 4.1) using double-sided sticky tape or graphite colloidal cement as outlined by Vortisch et al. (1987) and Mahaney (1990a).

In rare cases composite samples, formed from a mix of several samples, can save time and produce cost-effective results. A composite sample is prepared by taking a number of field samples that are considered to represent the population and thoroughly mixing the samples to form one sample. This approach is based on the assumption that the mean of some characteristic of the population may be determined from a single analysis. Composite samples are only rarely studied

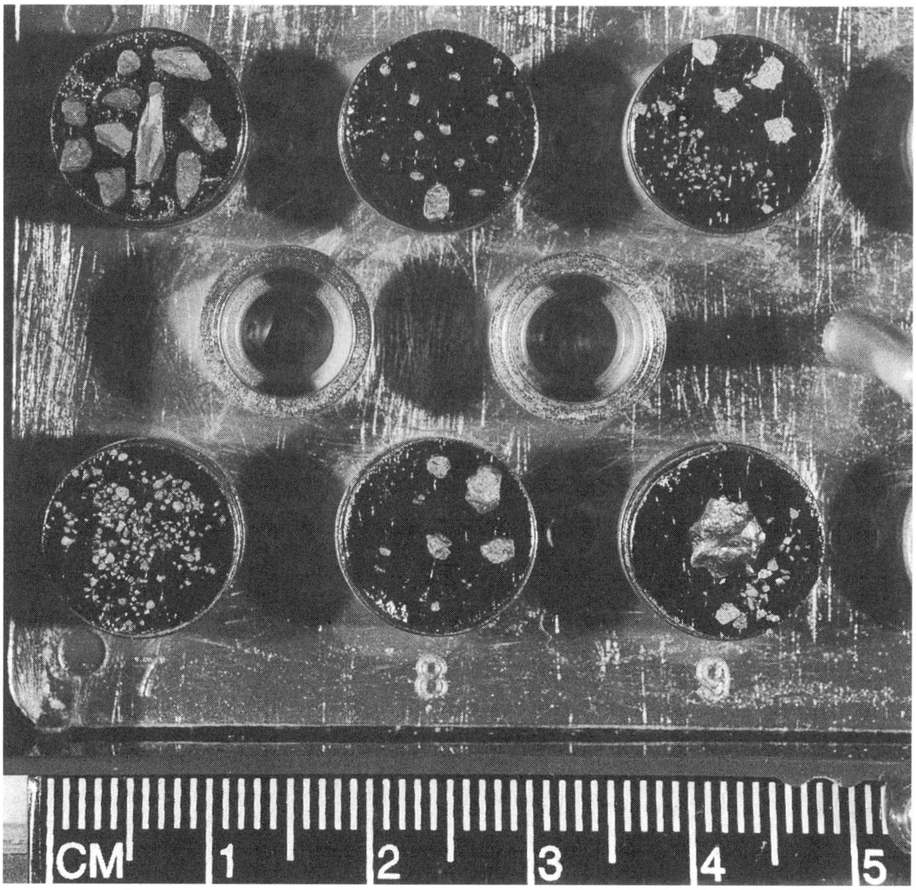

Figure 4.1 Representative array of stubs with coarse and fine grains. The grains in upper left are >2 mm; grains in lower left are medium sands (250–500 μm). The grains in upper right are a mix of >2 mm and fine sands; grains in lower right are >2 mm.

in the Earth sciences. They provide only a mean of the population from which the samples used to make the composite were drawn. This is not a recommended procedure for SEM–EDS analysis.

Sampling Scales

Sampling is intended to extract information about a particular earth material. Only occasionally does the sample consist of the entire mass of the material (e.g., a small meteorite). Samples are collected to provide information on the entire population, which may be large or small. Obviously there are vast ranges of scale in a population of Earth materials of one kind or another. The sample may represent a shale several hundred square kilometers in area and several dozens or hundreds of meters in thickness, or it may involve a soil horizon of only several meters lateral extent and several centimeters thickness. SEM analysis is done on an inordinately small, but hopefully representative, number of grains from that population.

Statistical Tests

The most basic means of summarizing microtextural data is to plot histograms or bar graphs showing the frequency of occurrence of individual microtextures. The frequency of occurrence of individual microtextures can also be subjected to chi-square or Student's T-tests to establish that observed differences are not a chance occurrence. In so doing, significance levels or confidence intervals are attached to statements of differences between two populations of grains as well as strengthening the interpretations arrived at by different investigators. The chi-square test is used for nominal or ordinal (count/frequency) data whereas Student's T-test is applicable when ratio data are available.

One means of analyzing bar graph summaries (frequency distributions) of microtexture observations on sediment grains is to quantify similarities or differences among samples in order to compare them. Specifically, one may compare samples from three distinct environments of deposition. Figure 9.7 shows summary results from a test site in Estonia using Devonian fluvial sands, Weichselian glaciofluvial sands, and Weichselian tills as discussed by Mahaney et al. (2001b). The purpose of this analysis is to determine if sedimentary environments can be differentiated based on microtexture observations. Vertical axes in the graphs on figure 9.7 represent the percentage of observations (in this case the percentage of grains observed in each environment) that exhibit a specific microtexture. Microtextures are arranged in the same order along the horizontal axis in each graph.

The bar graphs have different shapes, suggesting immediately that there are distinct signatures for each environment. More precisely, the two water-transported sands (fluvial and glaciofluvial) seem to resemble each other more than either resemble tills. This is first apparent in the range of microtextures observed: sands are represented by fewer (15) and tills by more (23) microtextures. The specific microtextures present and absent on waterlain sands are the same, except for "sharp angular features", which are present on glaciofluvial, but absent on fluvial sands. This first approximation of grouping—sands distinct from tills—can be tested and refined by application of a distance or similarity coefficient.

The idea that fluvial and glaciofluvial sands are more similar to each other than they are to tills can be tested by ordering these environments by a coefficient that reflects the distance (or similarity) among them based on percentage occurrence of microtextures. The Euclidean distances, shown in table 4.1, reflect straight-line distances between all possible pairs of environments based on the following equation:

Table 4.1 Distance values between sands from fluvial and glacial environments.

	Devonian Sands	Weichselian Tills	Weichselian Glaciofluvial Sands
Devonian Sands	0		
Weichselian Tills	145	0	
Weichselian Glaciofluvial Sands	77	111	0

$$d_{ij} = \sqrt{\sum_{k=1}^{p}(x_{ik} - x_{jk})^2}$$

where d is the distance between two cases (environments), i and j, measured on pairs of individual microtexture percentage values, x, for p pairs of variables beginning with the first ($k = 1$) and ending with the last ($k = p$) pair.

Usually, two assumptions are made when interpreting results (Shennan, 1988). First, the range in values for all variables should be similar; here, all variables are percentages and so values range from 0 to 100. Second, variables should be independent. Here, it is likely that some variables are dependent on others but with only three cases, it is not possible to demonstrate this one way or the other. Two-way plots suggest that some variables are probably correlated, among which are the following: edge-rounding and sharp angular features (negatively); preweathered surfaces and dissolution etching (positively); and overprinted grains, conchoidal fractures, and adhering particles (all positively related with each other). Quantitative analysis of observations on individual grains from all three environments would be required to show these relationships, which is something that would require principal components analysis or classical cluster analysis.

Distance values confirm the visual impression that fluvial and glaciofluvial sands are distinct from tills. They provide additional information, which allows the three environments to be ordered in a way that makes intuitive sense: Weichselian tills and Devonian sands are most distant (145), occurring at opposite ends of a spectrum; glaciofluvial sands are intermediate, but more closely related to fluvial sands (77) and more distantly related to tills (111). To review the imagery used in this analysis, see chapter 9 and Mahaney et al. (2001b).

Other statistical tests that rely on dendrograms drawn from cluster plots are described by Sheridan & Kortemeier (1987); comparative imagery charts for clast differentiation are discussed by Sheridan & Marshall (1987). The use of discriminant analysis and factor analysis in sorting grains by source area and in relating microtextures to different environments is discussed by Passchier et al. (1997). Fourier analysis of grain shape to identify cold sedimentary environments was discussed by Dowdeswell (1982).

A Sampling Strategy

The following is a typical sampling strategy. Given the above considerations we can outline a typical sampling strategy of a soil weathering or morphogenesis study where the specific objective may be to search for degrees of etching or coating microtextures. To understand the genesis of the soil profile, one might collect single samples from the A and B horizons, as well as the weathered part of the C (Cox = oxidized C horizon; see Birkeland, 1999), and the unweathered C (Cu = unweathered C horizon). It is important to analyze the C horizon because identification of the weathered and unweathered portions are a prime objective in pedology. If additional information is required on leaching and/or eolian influx, it might be necessary to collect two additional subsamples from the A horizon and two from the B horizon (at the top and bottom of each horizon) and concentrate on analyzing the range of minerals present and their weathered state. If wetting depth in the soil profile is well defined along a horizontal contact, it might be necessary to focus on the weathered portion above, and unweathered portion below. In other cases, the boundary may be diffuse or gradational, which could make sample collection more difficult, but the transition zone itself might be studied by SEM–EDS with significant results.

The collection of the parent material in the Cu horizon below the pedogenic profile is of paramount importance since it represents the material from which the soil formed. It is a *control* and its prime importance is as a standard in calibrating the soil to determine how much physical and chemical alteration has taken place during morphogenesis. The SEM is the ideal instrument to image chemical etching, and EDS can assist in providing chemical spectra of associated coatings of various kinds.

If it is desirable to assess the degree to which clay is produced in the soil, it may be necessary to study clay coatings on sands and compare and contrast them throughout the profile. Often the composition of clay coatings on sand grains, as determined by EDS, is supported by XRD analysis of clay minerals in the $<2\,\mu m$ fraction and particle size analysis. A certain species of clay mineral that is present in the B horizon but absent in the Cu and A horizons provides proof of a pedogenic origin. The SEM at high magnification can provide superb imagery showing different weathering products. The more intense the frequency of sampling down profile, coupled with SEM–EDS data, the higher the resolution of the database.

The same sampling strategy may be employed when dealing with geological sections once the sampling frequency is decided upon. In some cases one may find that one sample per lithic unit is sufficient; in others it may be necessary to increase the sampling frequency so that color, particle size and/or soil structural changes can be studied. As with soils and paleosols, the SEM–EDS is the ideal instrument to look for microstructures to help solve scientific problems.

5 Microtextures

A complete list of microtextures found on quartz and other related minerals is shown in figure 5.1. Photomicrographs illustrating the various microtextures originating in different sedimentary environments are displayed in figures 5.2 to 5.72. This imagery is representative of the microforms observed on both fresh and weathered grains of quartz and other light and heavy minerals. Since quartz is a ubiquitous mineral, widespread in nearly every environment, and relatively resistant to weathering (Bloss & Gibbs, 1963), it provides a recording surface with a long memory. It is, therefore, not uncommon to find grains overprinted by a succession of different environments. Because many grains are recycled in almost every sedimentary environment, and carry a record of events widely or narrowly separated in time, both simple and complex types were selected to illustrate all microtextural types. Often it is possible to reconstruct the entire environmental history of a grain as well as the relative time involved by paying close attention to preweathering and weathering features. A number of complex and/or weathered grains, including those with adhering particles and clay coatings, are provided here to show the degree of alteration that can be expected.

Sampling was covered in chapter 4, but suffice it to say that one must take great care to look at multiple samples and multiple sand fractions. The samples illustrated here are relatively ideal. Once the stratigraphic framework is established, it is necessary to decide on a suitable number of samples for analysis by SEM and EDS. The research problem should dictate the number of samples to be analyzed and the degree of precision required. An exploratory investigation might only require analysis of the fine and very fine sand fractions, whereas a high-resolution analysis would involve separate scans of the coarse, medium, and fine-to-very fine sands. One stub may hold several hundred fine-to-very fine sands and can provide a good cross-section of the mineralogy in a sample in addition to the range of microtextures present on individual surfaces.

One may use the SEM–EDS to establish sedimentary environmental records, to observe the state of weathering of a single mineral or multiple minerals, or even as a sourcing tool where provenance of grains is of interest. Whatever the problem involved, it is important to have an idea of the kinds of microtextures that exist in the sample suites. What follows is an index of representative microforms particularly common in the glacial, eolian, colluvial and fluvial environments. When observing microtextures, a log should be kept of their occurrence on each sample so that the frequency of occurrence or relative percentages of each microform can be tallied for each sample or for samples within a group.

With reference to subparallel linear and conchoidal fractures that are present on many grains, it is sometimes useful to document the relation between high and low frequency variations or spacings. For example, on glacial grains, it is often the case that frequency of fractures is directly proportional to ice thickness; or there may be a correlation between wind velocity and fracture frequency on eolian grains. The latter relationship still needs considerable testing with laboratory experiments to determine if the observed correlation is real. With glaciers, however, the cone of energy that accompanies stick–slip motion at the base of the ice may reach pressures of 200 atmospheres in thick continental ice sheets, producing the extreme vibrational effects that find release in fracture propagation through the mineral fabrics. The microtexture index can illustrate high versus low frequency microfeatures that may be related to ice thickness (Mahaney, 1995b). Examples of high and low frequency fractures are given in figures 5.59 and 5.60.

Microtexture Definition

The microtextures used herein are shown in figure 5.1 (page 30). Because various definitions for each microform have appeared in the literature over the last four decades, and to avoid confusion, they are defined below in the order in which they appear on the figure.

1. *Fresh surfaces*. Unweathered surfaces, lacking any dissolution etching or precipitation features.
2. *Craters*. Depressions of all shapes and sizes caused by impact.
3. *Sawtooth fractures*. Shaped like the teeth of a saw. Very rare, but occasionally occur in quartz. May occur under high compressional stress in glaciers, neotectonic release, fault plane gouging, and/or ejecta from asteroid impact.
4. *Rolled grains*. Quartz or garnet particles, grown in the shear zones of faults and subject to sufficient pressure and/or the action of hydrothermal fluids to undergo reshaping and crystallographic reorientation. They have smooth and flattened surfaces.
5. *Slickenside surfaces*. Polished and flat surfaces, often accompanied by grooved or striated microfeatures of variable size from the whole side of a grain to several micrometers in length and width. Occurring in shear zones adjacent to faults.
6. *Cracked grains*. Grains completely cracked or broken into fragments from a sharp, sudden impact. Fragments may become adhering particles.
7. *Abrasion fatigue*. Cracked, dislocated, and broken surface, partly broken precipitate, and epidermal grain surface.

8. *Parallel ridges*. Approximately 0.5–10 μm apart and placed between depressions. Relationships between microfeature length and grain size is unknown but length may be longer on larger diameter grains.
9. *Elongate depressions*. Situated between parallel ridges or may exist as singular features that reach 250 μm in diameter. They may be related to bulbous edges and are probably related to saltation.
10. *Smoothed-over depressions*. Where upturned plates are no longer sharp and depressions between plates have been filled in, a rolling topography is formed which encompasses the entire grain. Produced by saltation, these features are also related to bulbous edges.
11. *Bulbous edges*. Prominent, protruding and rounded grain edges in the shape of a parabolic curve.
12. *Shock lamellae*. Multiple sets of parallel, planar optical discontinuities or planes spaced from 0.05 to 10 μm apart. Some lamellae are thin platelets (DFs); others may be planar deformation features (PDFs); while still others, the decorated PDFs, may be decorated with cavities or vugs. PDFs are considered the hallmark of asteroid impact.
13. *Shock-melted grains*. Melting microfeatures including drip-forms and grain meltdown due to high-pressure shock waves and heat resulting from meteoric impact. Theoretically, shock melting may occur from eolian transport in rare instances (Lindé, 1978, and personal communication, 2001).
14. *Fracture faces*. Large, smooth, and clean fractures across at least 25 percent of the grain surface. These microfeatures may be flat or slightly curved but are well defined and not rounded. They are usually clean breaks but occasionally have minor subparallel fractures on margins.
15. *Radial fractures*. Linear fractures radiating from impact. The impact point may be highly focused on the grain surface, such as the apex of a cone of kinetic energy released on collision, or several micrometers in diameter if the released energy is low. The number of fractures and distances between them also relate to energy absorbed in collision.
16. *Subparallel linear fractures*. Also known as "Wallner lines." Shallow to deep linear fractures in the grain surface. They may be several micrometers long and grade into conchoidal fractures.
17. *Conchoidal fractures*. A smooth curved fracture, with a ribbed appearance, similar to the curve of a conch or seashell. It is characteristic of quartz, but also occurs occasionally on plagioclase and some heavy minerals.
18. *Curved grooves*. Deep elongate depressions or troughs. They are abraded microfeatures caused by a "wandering" stylus (engraving tool), commonly associated with glaciers where embedded grains score others of equal or less hardness. They may also occur on grains in mass wasted debris.
19. *Straight grooves*. Similar to curved grooves but with distinct linearity and sometimes more deeply inscribed in the grain surface. Straight grooves suggest greater force imparted by the "stylus" and less tension as it abrades in a straight direction.
20. *Sharp angular features*. Distinct sharp edges on grain surfaces.

21. *Low relief.* Nearly smooth surface without topographic irregularities.
22. *Medium relief.* Semi-smooth surface with topographic irregularities caused by impact with other grains and/or by weathering.
23. *High relief.* Topographically irregular surface with pronounced swells and swales. Relief is the difference between high and low points on the grain surface.
24. *Chattermarks.* Crescentic trails of parallel, oriented microfeatures of uniform spacing that occur randomly on garnets (Folk, 1975; Orr, 1981) and on quartz sands in lodgement till.
25. *High and low frequency fractures.* Identified by the spacing of conchoidal and subparallel fracture lines. Small spacings to the order of tenths of a micrometer or less constitute high frequency lines; Large spacings in excess of tenths of a micrometer are low frequency. The difference in frequency is thought to be related to vibrational energy released during grain-to-grain collisions in glacial ice (see Mahaney, 1995a,b).
26. *Crescentic gouges.* Crescent-shaped and often deep gouges, with convex and concave limbs, produced by the chiseling action of bladed fragments.
27. *Arc-shaped steps.* Deep tears or breaks in the mineral fabric caused by impact. Similar to conchoidal fractures but with depths greater than several micrometers and usually widely spaced (>5 μm).
28. *Linear steps.* Deep tears or breaks in the grain surface similar to subparallel linear fractures, with wide spacing (>5 μm).
29. *Deep troughs.* Grooves deeper than 10 μm.
30. *Mechanically upturned plates.* Impacted surfaces with small to large plates partially torn loose from the mineral surface. They are most common on quartz but are occasionally found on plagioclase.
31. *Lattice shattering.* Crystallographically disturbed or failed surfaces where crystal planes have collapsed.
32. *Abrasion features.* Rubbed or worn down grain surfaces produced by the mechanical grinding or scraping of rock particles in traction and moved by wind, water, ice, or gravity.
33. *Dissolution etching.* Surface with a felted net of cavities resulting from solution of mineral material. May be random or follow fractures or grain cleavage.
34. *Preweathered surfaces.* Weathered surfaces overprinted with fresh abrasion or fractures implying a weathered history followed by transport.
35. *Weathered surfaces.* Surface marked by etching. Minute pits produced by acid acting on a mineral surface may be distributed randomly or conform to grain crystallographic orientation.
36. *Precipitation features.* Thin to thick coatings of SiO_2, Fe_2O_3, and $CaCO_3$ precipitated out of solution onto grain surfaces, either during diagenesis, or as a result of pedogenesis.
37. *Adhering particles.* Fragments or particles adhering to mineral grains.
38. *V-shaped percussion cracks.* V-Shaped fractures of variable size and depth embedded in a grain surface.
39. *Edge rounding.* Rounded edges on grains usually found in conjunction with sharp edges. These are less prominent than bulbous-edged grains.

40. *Overprinted grains*. Grains with a variety of fracture–abrasion–weathering–precipitate microfeatures implying, for example, transport → weathering → transport → precipitation. Similar to preweathered grains, but with more complex histories.
41. *Clay coatings*. Coatings of various clay minerals with Si:Al ratios of 1:1 and 2:1, sometimes mixed with weathering products such as gibbsite or amorphous materials. The EDS is the instrument of choice for unraveling the composition of these materials and shedding light on the pedogenic–diagenetic history of the primary grains.

Microtexture Index

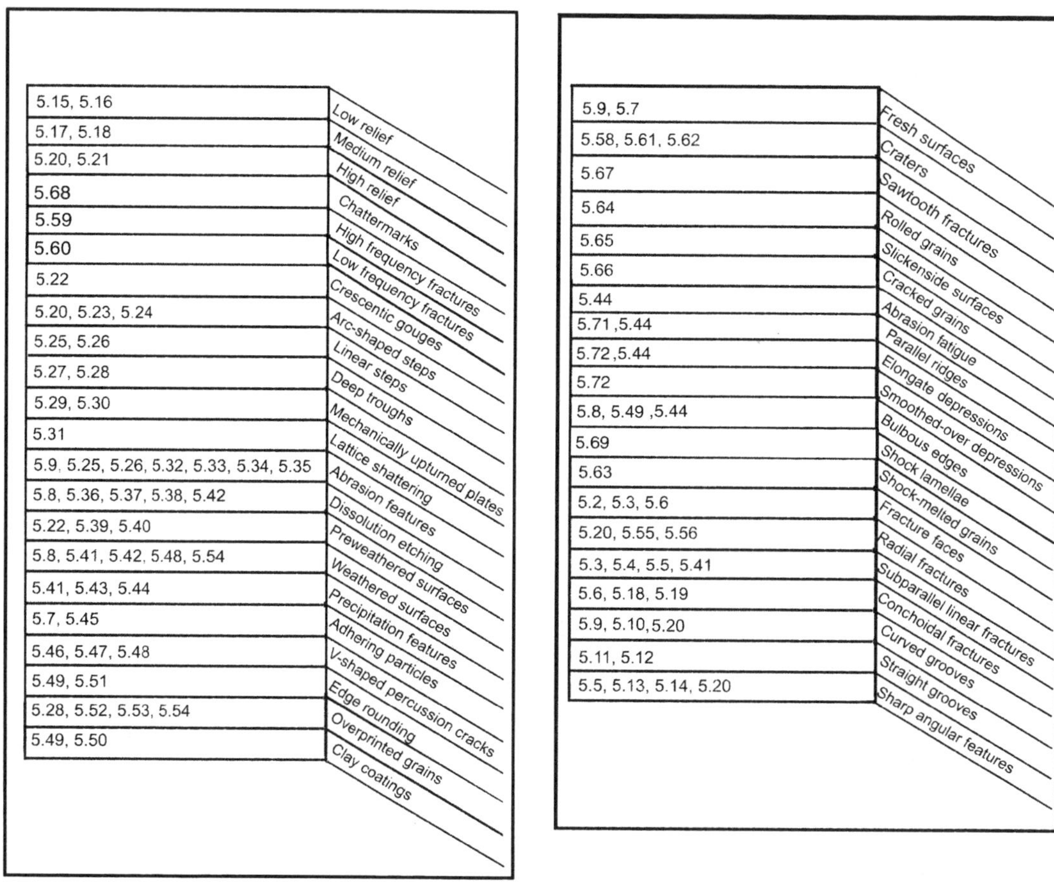

Figure 5.1 Microtexture diagram showing the range of microtextures commonly observed on sands of variable composition. Representative plates are identified for each microtexture.

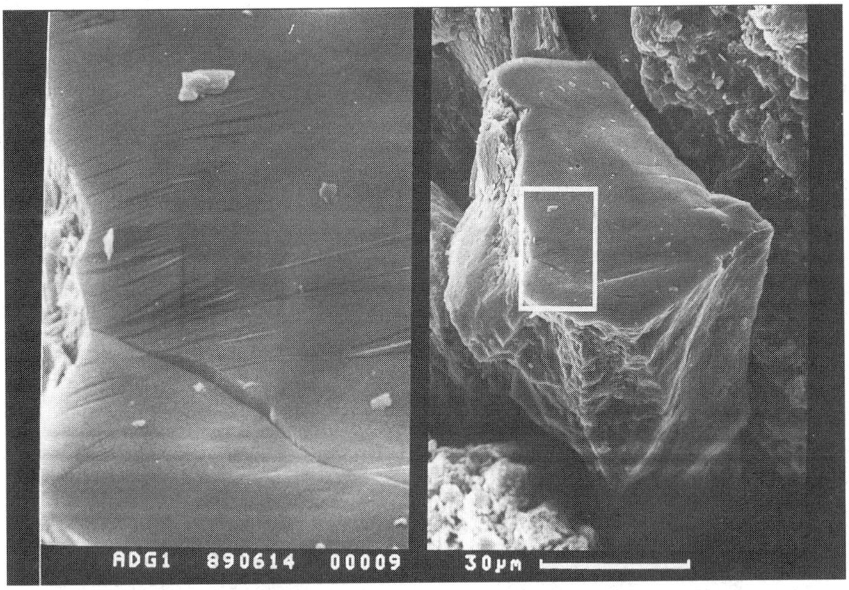

Figure 5.2 Fracture face in quartz (very fine sand). The enlargement is ×5. Note weathered surface below fracture face and shallow subparallel fractures (enlarged). Grain is part of debris mass wasted onto the surface of the Adishy Glacier, Mt. Elbrus, Caucasus Mountains, Russia (see Mahaney et al., 1991).

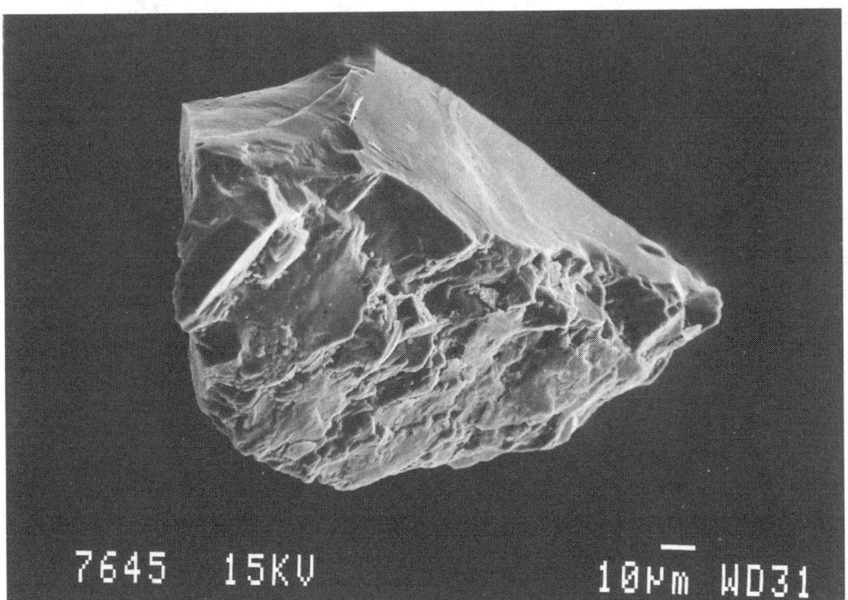

Figure 5.3 Fracture face (upper right) on an angular quartz fragment with sawtooth fractures and minor lattice failure on lower half; fresh whole plate removal (left) and abraded grooves (top left). Source: from the mill of the Boconó Fault, near Apartaderos, Venezuelan Andes, reworked into late Quaternary glaciolacustrine sediment and subjected to minor neotectonic displacement.

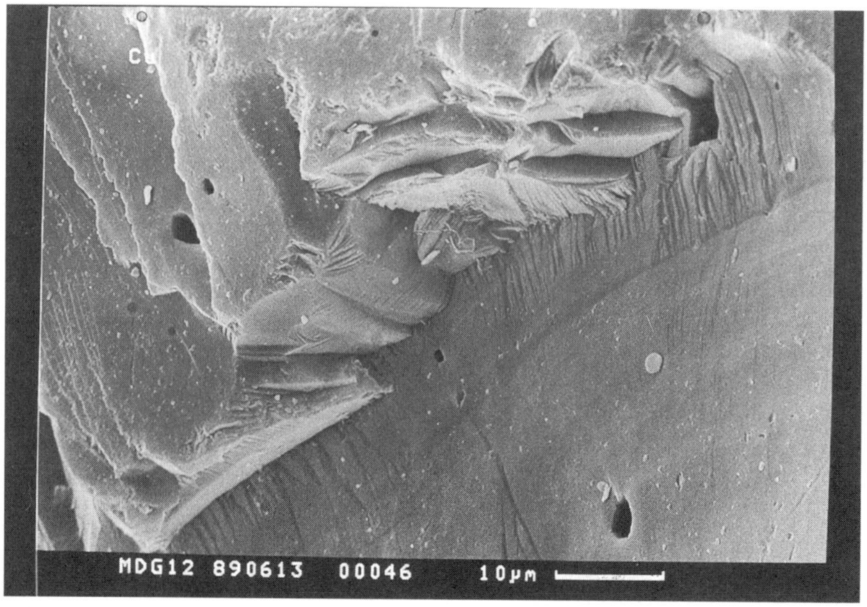

Figure 5.4 Shallow subparallel linear fractures (center) on upper part of a fracture face (lower right) adjacent to deeper linear fractures (upper left) in quartz. Source: the left lateral moraine, Mer de Glace Glacier, France (Mahaney, 1991a).

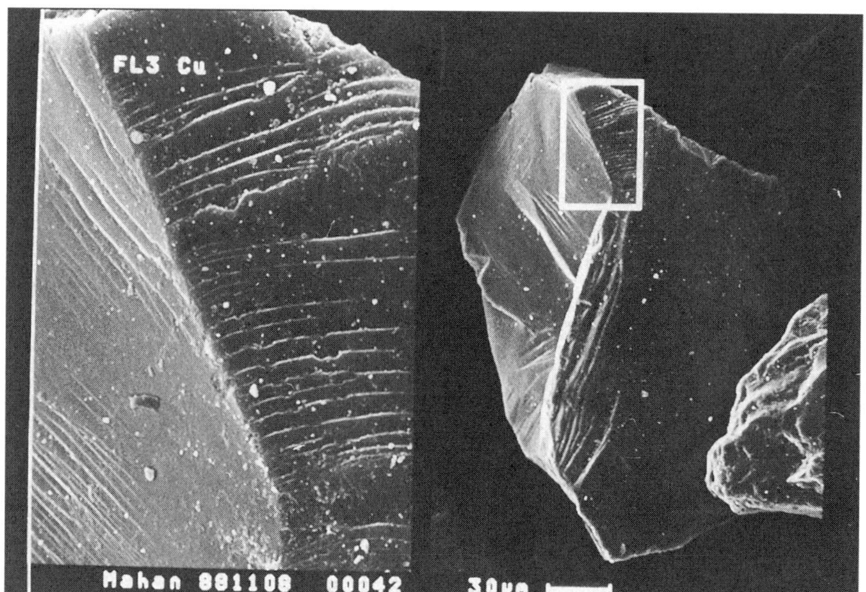

Figure 5.5 Shallow and deeper subparallel linear fractures in quartz from the very fine sand fraction. Note the sharp edges and angularity of the grain. The enlargement (×5) shows the fresh detail of the subparallel fractures. Source: site FL3, till of the Pinedale Glaciation, Fremont Lake, Wind River Mountains, western Wyoming (see Mahaney, 1978). The sample is from the parent material in the inner end moraine complex.

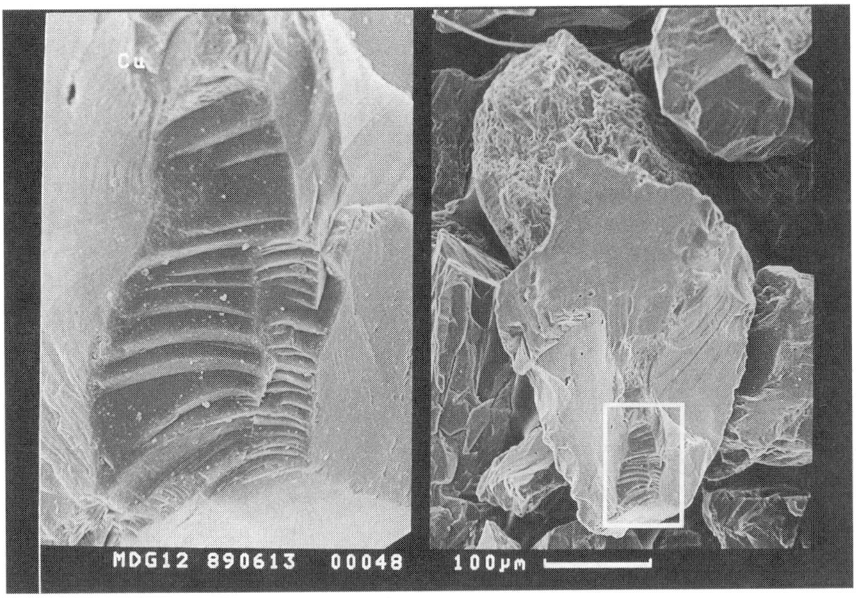

Figure 5.6 Conchoidal fractures (outlined with frame) below prominent fracture face in quartz. Note also the preweathered surface above and the glacially crushed area below. The enlargement (×5) shows the fresh character of the conchoidal fractures. Source: site MDG12, Mer de Glace Glacier, French Alps (Mahaney, 1991a).

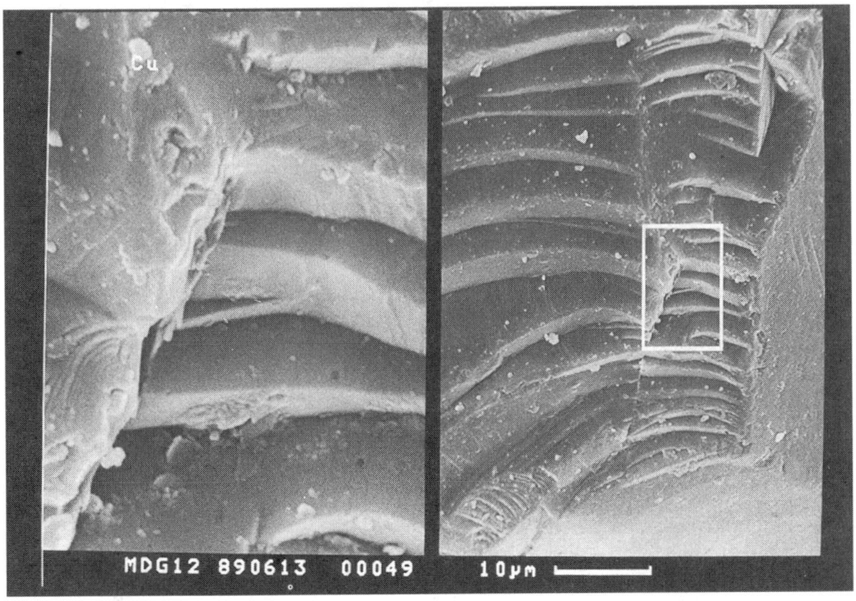

Figure 5.7 Same grain as in figure 5.6. A displacement fracture oriented vertically separates low (left) and high frequency fractures. The enlargement (×5) shows the usual spread of adhering particles on glacial grains and fresh fractures deeply ingrained in the surface, which is typical of glacial grains.

Figure 5.8 Bulbous edges from eolian action on weathered quartz. Dissolution is more pronounced on the flanks of the grain below bulbous microforms. A complex grain that illustrates deep dissolution weathering followed by eolian action, weathering, and abrasion (possibly by glacial action). Abraded areas are flat-bottomed craters or grooves that are probably glacial; the grooves to left are striations. Source: supraglacial load mass wasted from Mt. Elbrus (Mahaney et al., 1991).

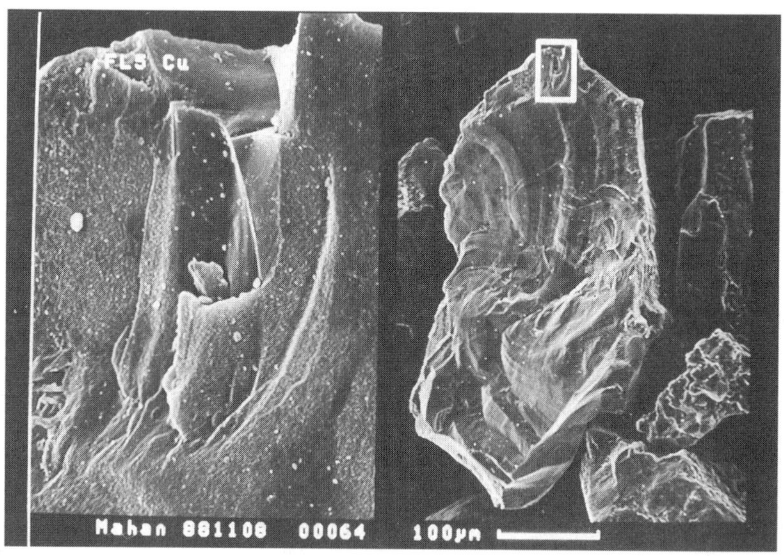

Figure 5.9 Curved deep grooves (top and bottom) and shallow grooves (center) in quartz. Grooves are oriented approximately north–south in the micrograph. The larger grooves are 30 µm wide and up to 15 µm deep; smaller features are around 15 µm wide and 10 µm deep, respectively. Source: fine quartz sand fraction from late Pinedale (= Wisconsinan) till, Fremont Lake, Wyoming. Enlargement (×10) shows weathered remnant with fresh fractures and abrasion. The sample is the parent material for the post-Pinedale soil (Mahaney, 1978).

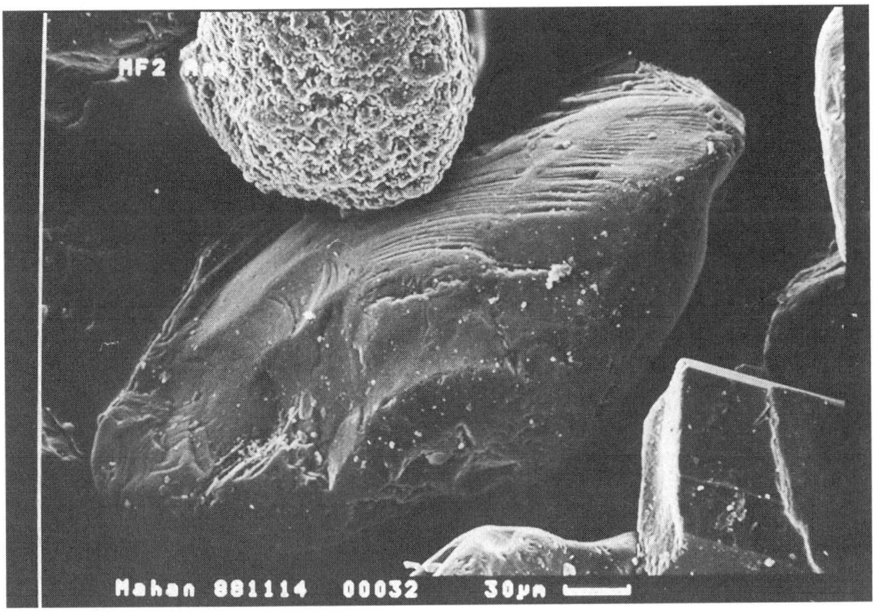

Figure 5.10 Curved grooves (lower half) grading to subparallel fractures in the upper part of quartz grain (center). The highly weathered grain at the top of the figure is plagioclase. Source: La Meseta Formation of early Tertiary age, Seymour Island, Antarctic Peninsula.

Figure 5.11 Shallow straight grooves in quartz with sharp edges and abundant fractures. Enlargement ×10. The sample is from Bull Lake (= Illinoian) till, Fremont Lake, Wyoming. It is the parent material of the post-Bull Lake soil (Mahaney, 1978).

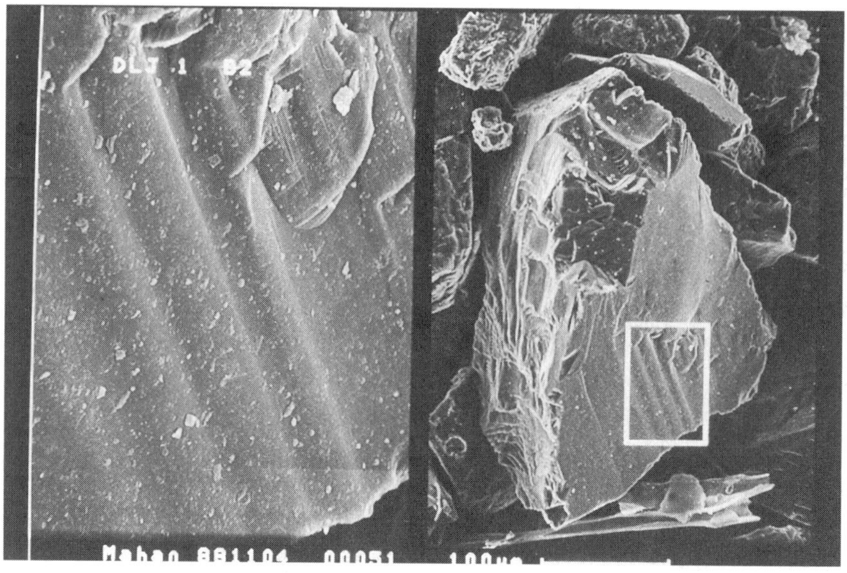

Figure 5.12 Shallow straight grooves in quartz. Surrounding surface is abraded (right) and well fractured (left), with fine-grained clay on surface. Enlargement ×5. The sample is late Wisconsinan (LGM) in age, from parent material of soil formed in LGM till, Dalijia Shan, China (Mahaney & Rutter, 1992).

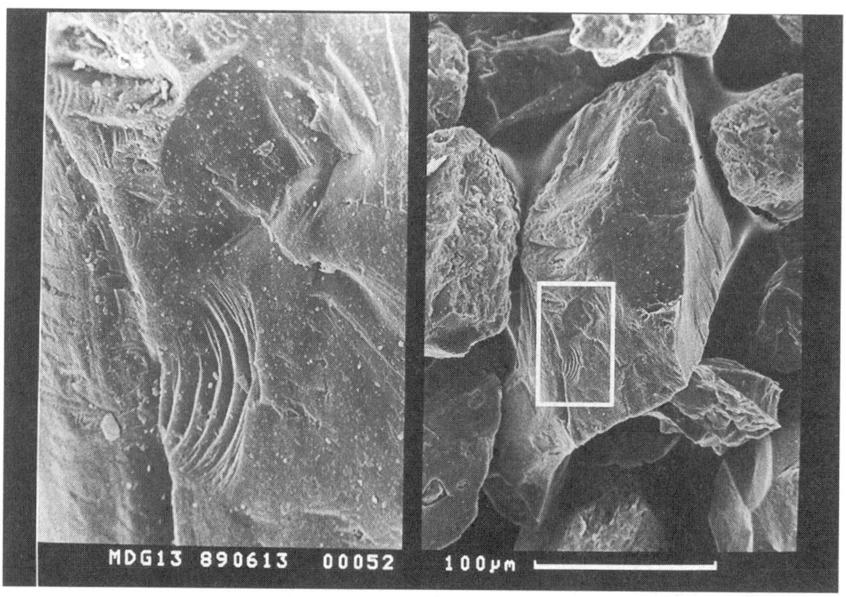

Figure 5.13 Sharp angular features on a triangular-faceted grain from the Mer de Glace, French Alps. A preweathered surface (box and center of grain) survived intense abrasion (right and bottom). Sharp edges, abundant adhering particles, and shallow straight grooves are all representative of glacial grains; conchoidal fractures, while not exclusive to glacial grains, are common. The enlargement is ×5. The sample is from a C horizon of the early Neoglacial soil, Mer de Glace, French Alps (Mahaney, 1991a).

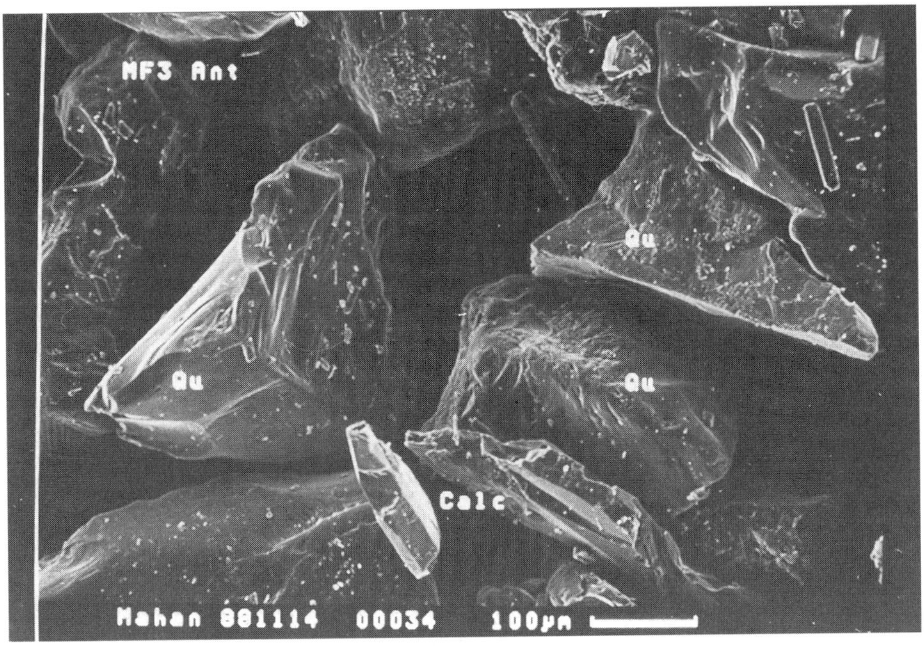

Figure 5.14 Sharp angular microfeatures on quartz (Qu) and calcite (Calc, grain to right). Source: La Meseta Formation of early Tertiary age, Seymour Island, Antarctica.

Figure 5.15 Low relief on quartz in weathered till, TV61A section, Teleki Valley, Mount Kenya, East Africa (Mahaney, 1990a).

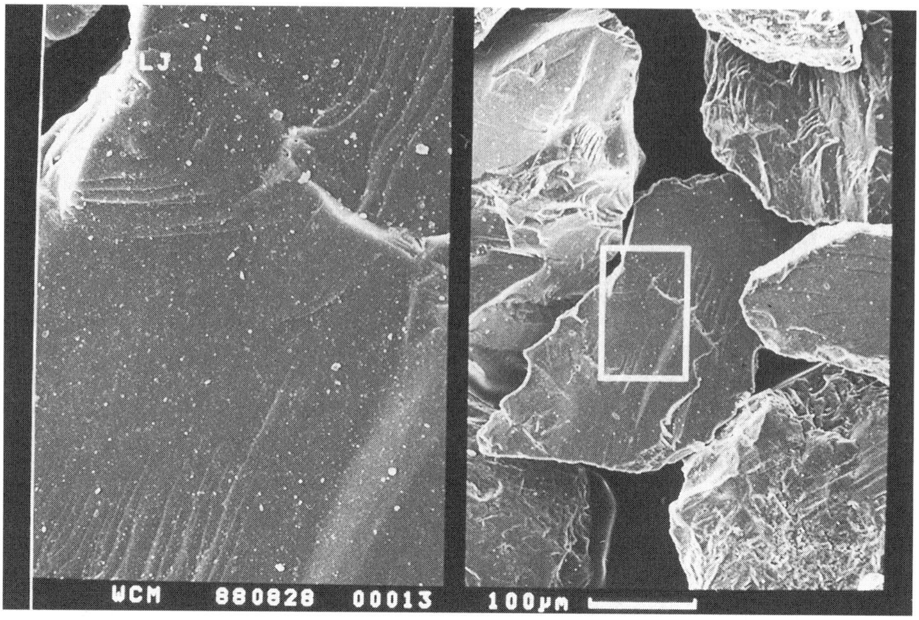

Figure 5.16 Two quartz grains (center and right) with low relief. In this case the low relief may result from uniform abrasion. Enlargement ×5. Source: Wisconsinan age till, Dalijia Shan, north-western China (Mahaney & Rutter, 1992).

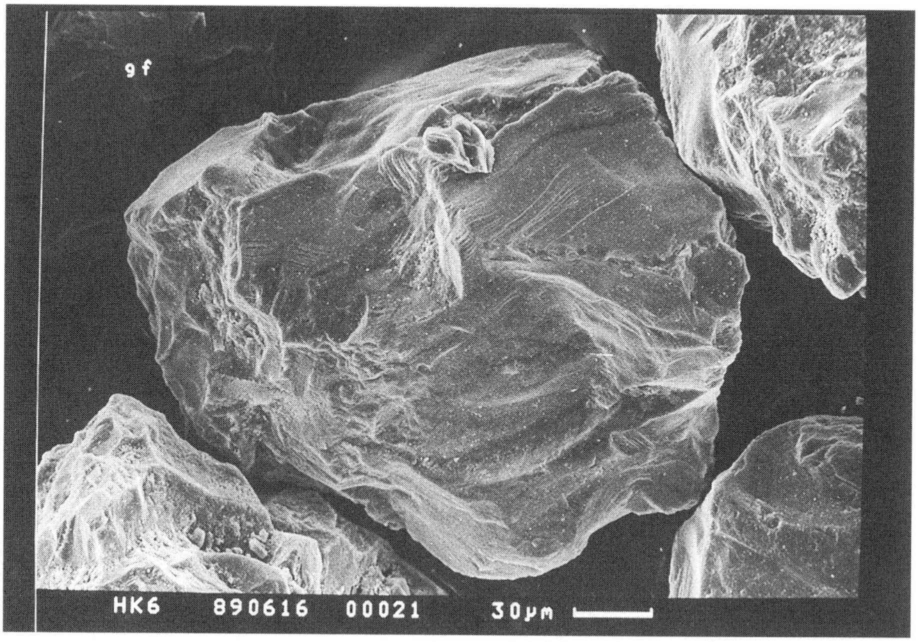

Figure 5.17 Medium relief and slightly rounded corners on quartz from glaciofluvial sediment of late Neoglacial age. Traces of subparallel fractures present in the mid to upper portions of the surface. Source: Zillertal Alps, Austria.

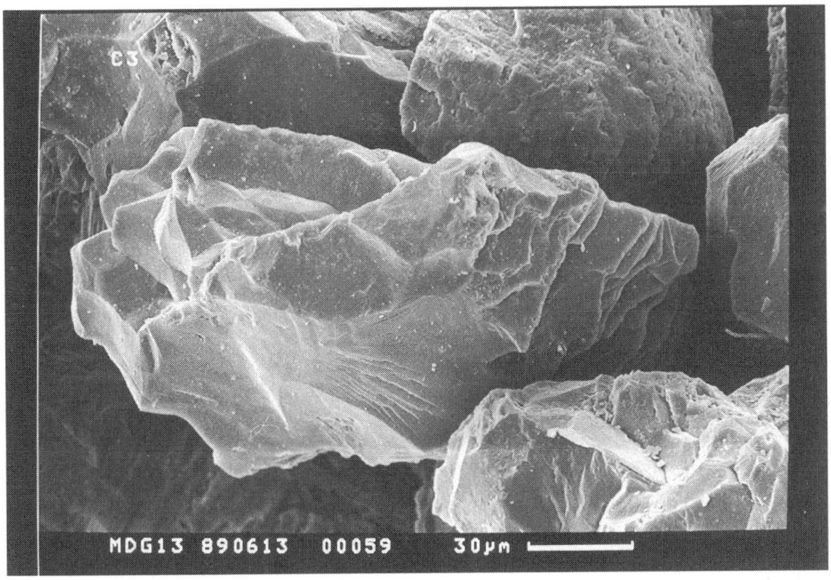

Figure 5.18 Medium relief microfeatures on complex triangular faceted quartz grain. Long and short irregular troughs 5–10 μm deep and 10–20 μm wide, are representative of the glacial environment. The deeply inscribed conchoidal fractures shown in figure 5.19 enhance the surface relief. Two fracture faces on left side; short troughs upper left. Radial fractures lower center. Source: Mer de Glace, France (Mahaney, 1991a). The sample is from the C3 horizon in the early Neoglacial soil.

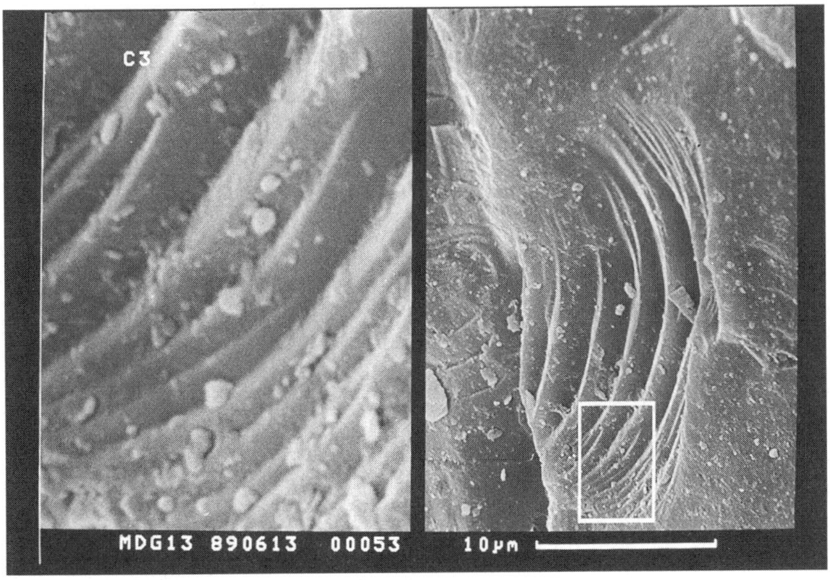

Figure 5.19 Deeply inscribed conchoidal fractures. The enlargement is ×5. The scale bar is equal to 2 μm in the enlarged frame. The sample is from the C3 horizon of the early Neoglacial soil, Mer de Glace, French Alps (Mahaney, 1991a).

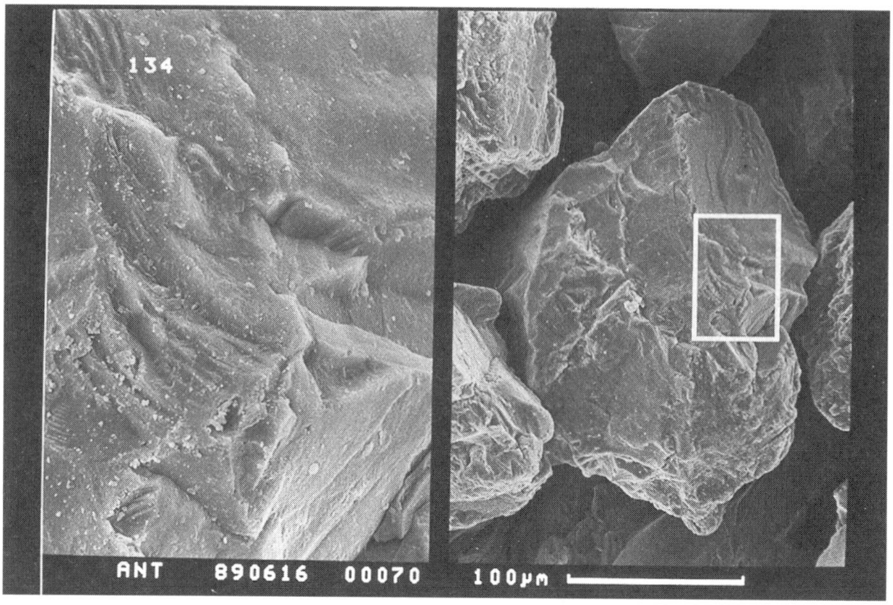

Figure 5.20 Triangular-faceted quartz grain showing high relief features. Microfeatures include sharp edges, multiple craters, radiating fractures, step-like features, and striations, all of which are representative of glaciation. Note the debris on the surface shown in detail in the enlargement (×5). Source: Till 134, Dry Valleys, Antarctica (see Mahaney et al., 1996a).

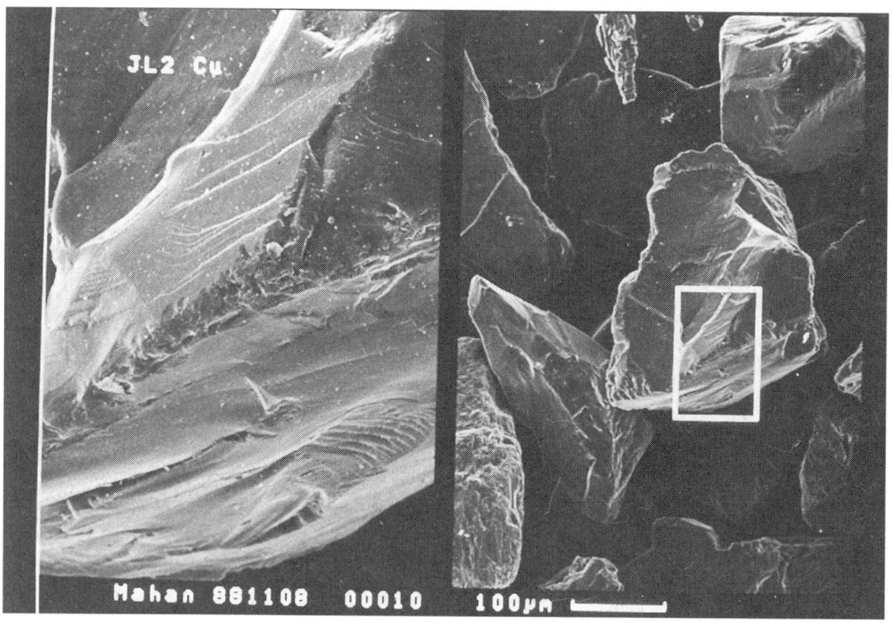

Figure 5.21 High relief on quartz grain, with fractures and abrasion highlighted in the enlarged frame. Partly upturned plate in lower left of the enlargement (×5). Source: Pinedale (= Wisconsinan) till near Jenny Lake, Teton Range, Wyoming. The sample is from the parent material of the post-Pinedale soil.

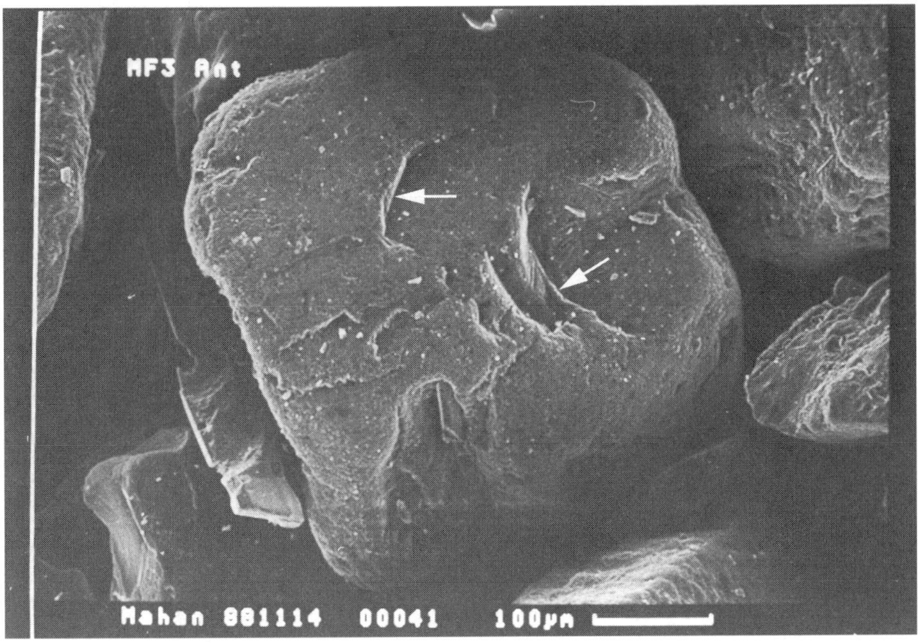

Figure 5.22 Crescentic gouges (arrows) in quartz with low relief and some preweathered surface across the grain. Source: La Meseta Formation, Seymour Island, Antarctica.

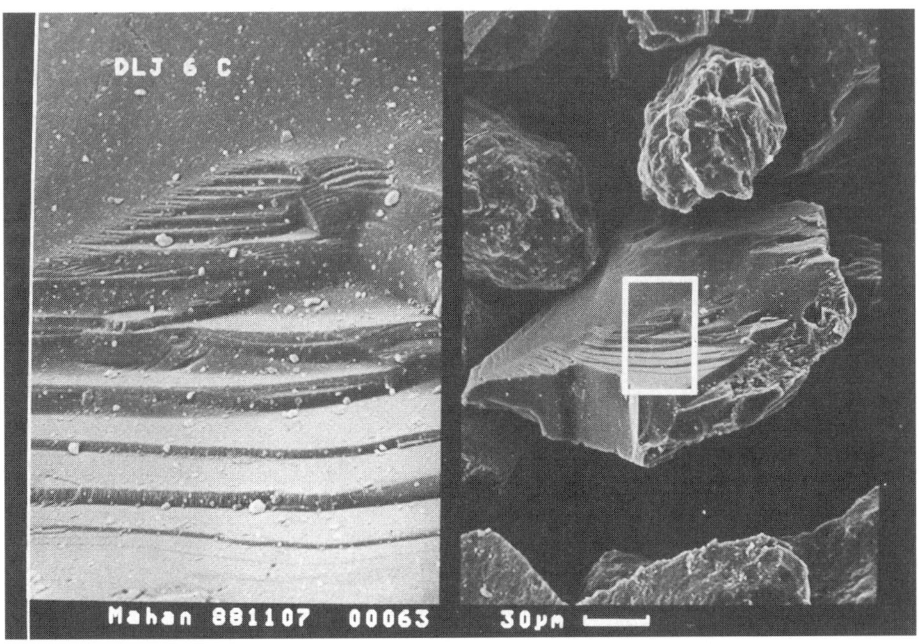

Figure 5.23 Arc-shaped steps in a quartz grain. The enlargement is ×5. Source: Till DLJ6, Dalijia Shan, China (Mahaney & Rutter, 1992). The sample is from the C horizon of the post-LGM soil.

Figure 5.24 Arc-shaped steps in a large trough. Short troughs to left and right. Source: sample 87-106 from Upper Saale age till (equivalent to Upper Ugandi till in Estonian System, 153–206 ka; see Mahaney & Kalm, 1995).

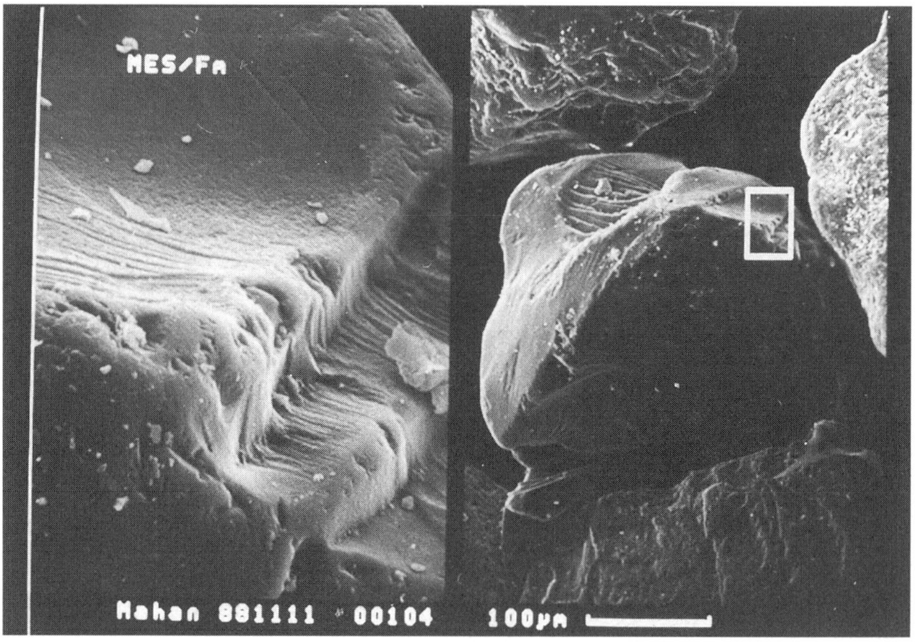

Figure 5.25 Linear steps in otherwise crushed and abraded quartz. The enlargement is ×10 and shows subparallel fractures on the steps. Source: La Meseta Formation, Seymour Island, Antarctica.

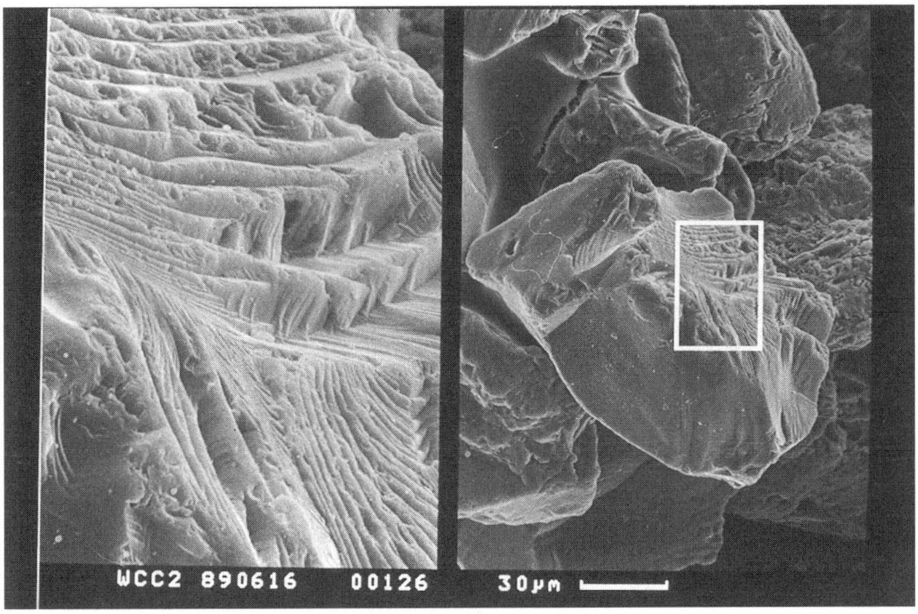

Figure 5.26 Large and small linear steps in quartz. The left side of the grain is heavily abraded; the right side and top are extensively fractured. The enlargement is ×5 and shows the small steps. Source: Illinoian age basal till, Wildcat Creek, Indiana, USA.

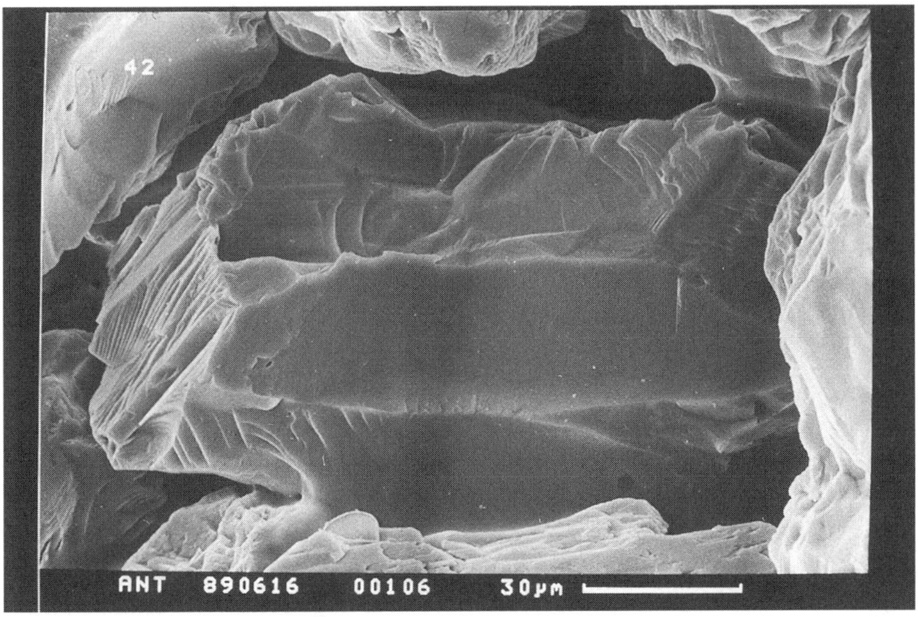

Figure 5.27 Deep troughs 30 µm wide and fresh fractures in quartz. Source: Till 42 (end moraine), Antarctica. See Mahaney et al. (1996a), for details on location of sample.

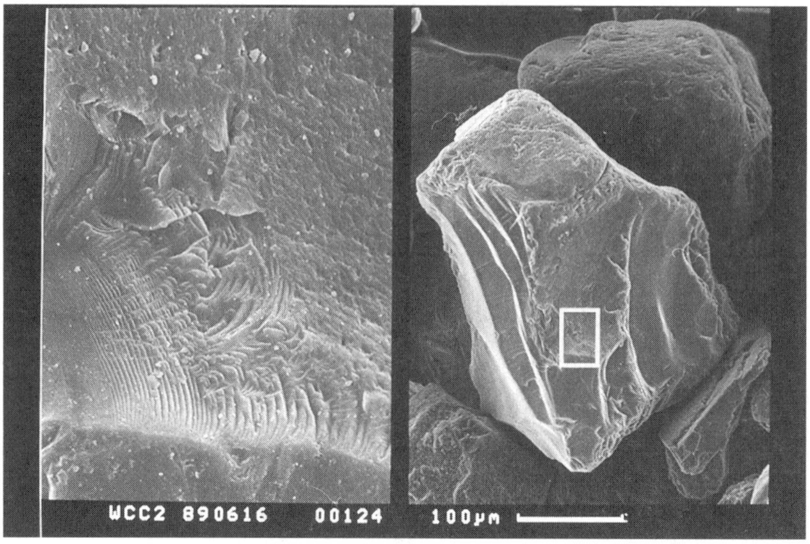

Figure 5.28 Deep troughs in quartz: linear troughs to the left; curved troughs to the right. Pseudo-chattermarks terminating on an upper ridge of the grain (see Folk, 1975, for similar microfeatures on garnet). Quartz (upper grain) is subrounded and carries numerous v-shaped percussion scars. Crossed fractures, much like crossed striae, show high frequency and younger fractures overprinting the low frequency (higher wavelength) features to the right in the enlargement (×10). Source: basal till, Wildcat Creek, Indiana, USA.

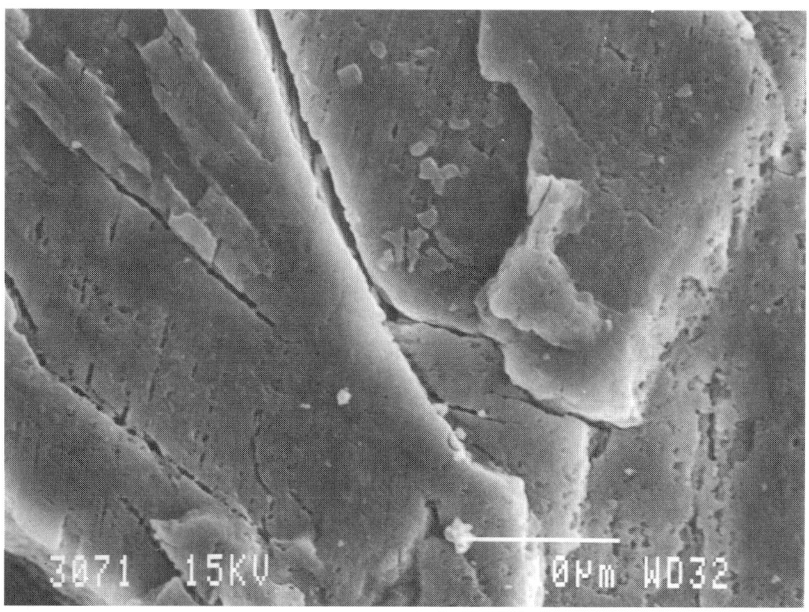

Figure 5.29 Mechanically upturned plate on the upper right and left of a quartz grain. The grain is cracked in several places and the etched surface tends to follow hairline cracks (PDFs) in the mineral fabric. Source: K/T impact zone, Clear Creek, Colorado. (Samples courtesy of Glenn Izett, College of William and Mary, Williamsburg, VA.)

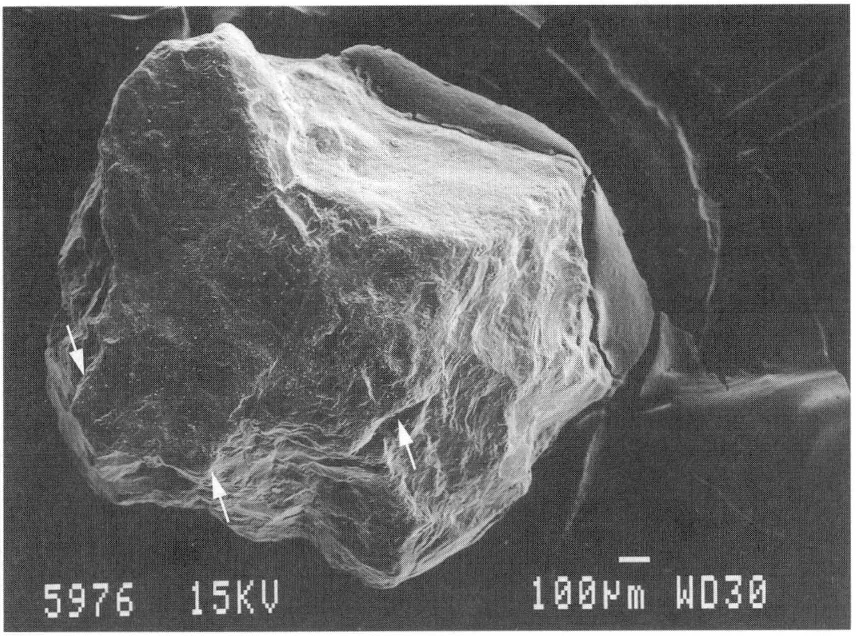

Figure 5.30 Mechanically upturned plates (arrows, lower left; center bottom) on quartz. Plates tend to follow fracture lines. Source: early Wisconsinan outwash in the RF3 Section, western cordillera, northern Venezuelan Andes. See Mahaney et al. (2000a) for details.

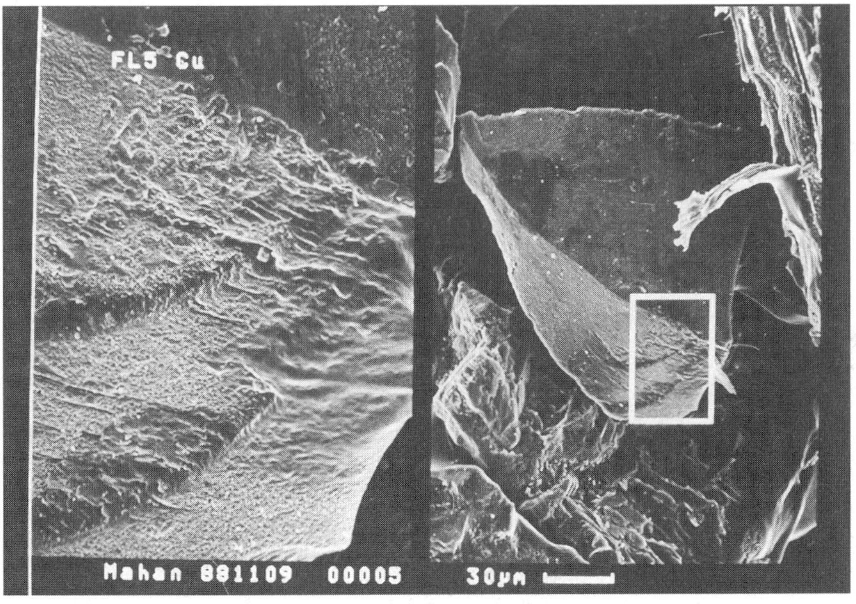

Figure 5.31 Partial lattice shattering of nearly triangular-faceted quartz grain enhanced in enlargement (×5). Lower part coated with Si which gives a fine texture pattern in enlargement. Source: parent material of the post-Bull Lake soil, Fremont Lake, Wyoming, USA (Mahaney, 1978).

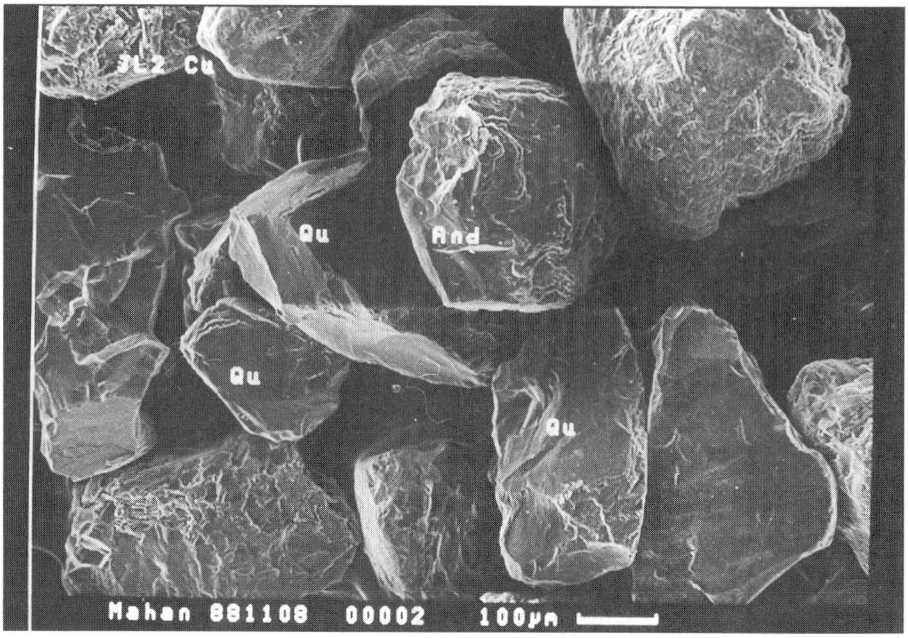

Figure 5.32 Abrasion features on grain faces without edge rounding, as opposed to fluvial abrasion which tends to affect all sides, and hence the entire grain. Mix of quartz (Qu) and plagioclase (And = andesine), the latter partially weathered. Source: Pinedale (= Wisconsinan) till, Jenny Lake, Teton Range, Wyoming. The sample is from the parent material of the post-Pinedale soil.

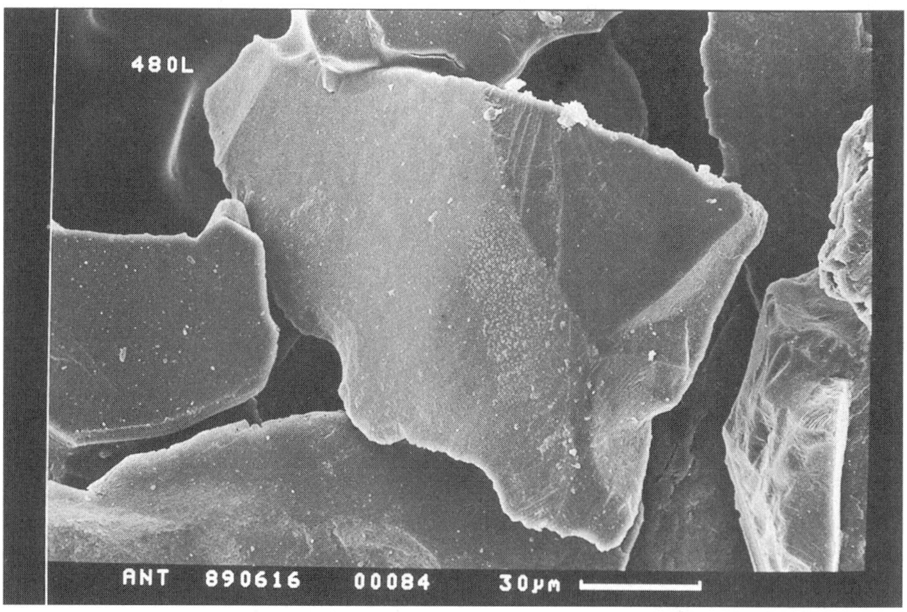

Figure 5.33 Abraded quartz with sharp edges, minor radial fractures (upper center) with a fine Si precipitate below. The radial fractures are older than the sharp edges. Source: end moraine in the Dry Valleys (Mahaney et al., 1996a).

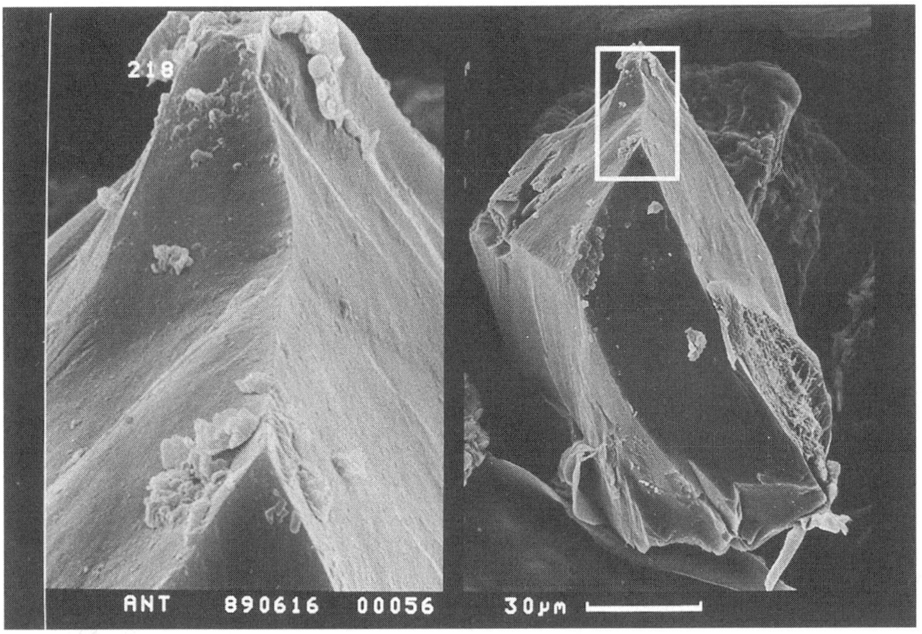

Figure 5.34 Abraded surfaces on angular quartz grain in Till 218, Antarctica. The enlargement (×5) shows detail of an abraded surface and large adhering particles; some etching in lower right. Source: end moraine (see Mahaney et al., 1996a).

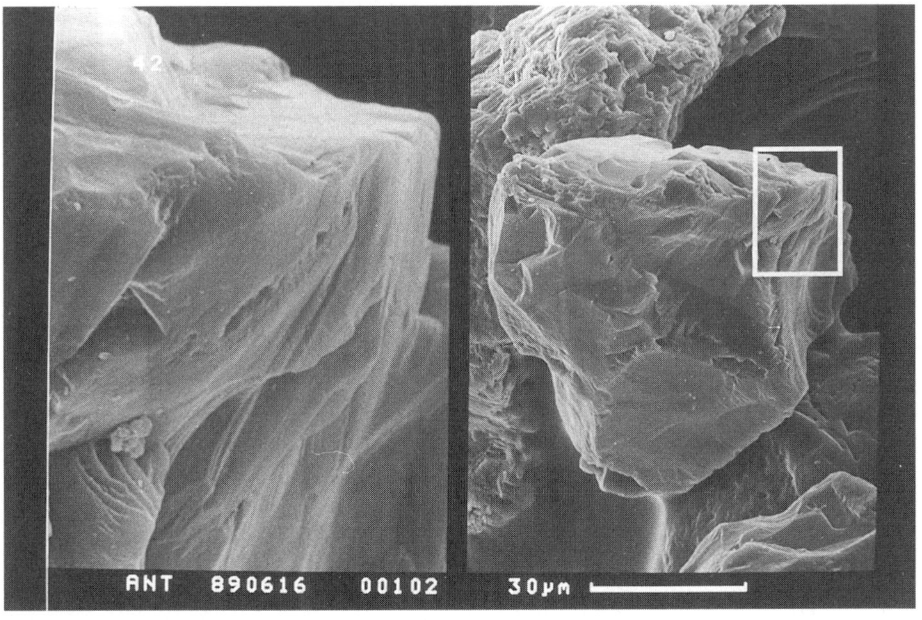

Figure 5.35 Partially abraded quartz grain. Older crushing surfaces still discernible. Note the moderate relief, subparallel linear fractures in the enlargement (×5). Arc-shaped step remnants to the left in the main frame. Source: Till 42, Antarctica. See Mahaney et al. (1996a) for details on till lithology.

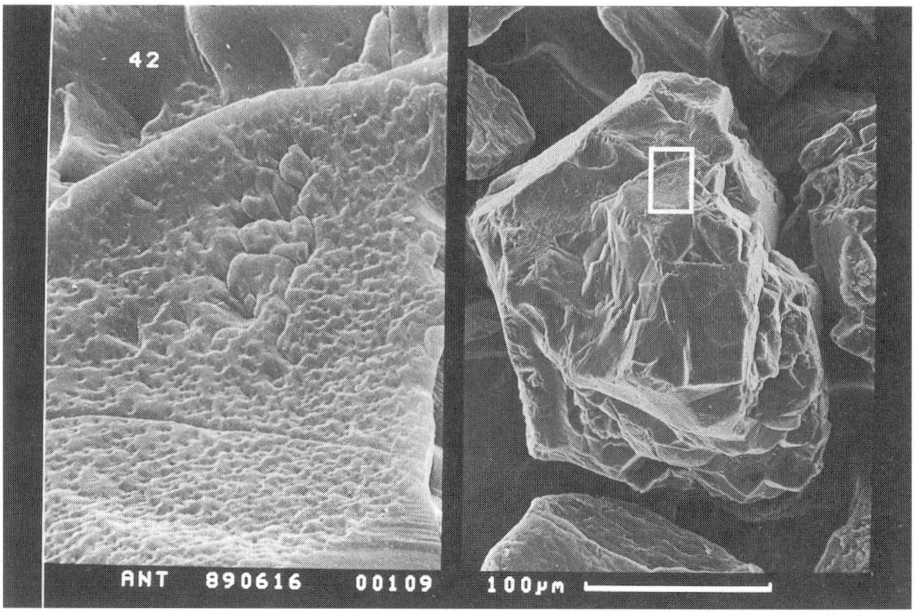

Figure 5.36 Dissolution etching preserved as preweathered microfeatures on glacially crushed quartz. Strong etching shown in the enlargement (×10). Source: Till 42, Antarctica. See Mahaney et al. (1996a) for location and lithology.

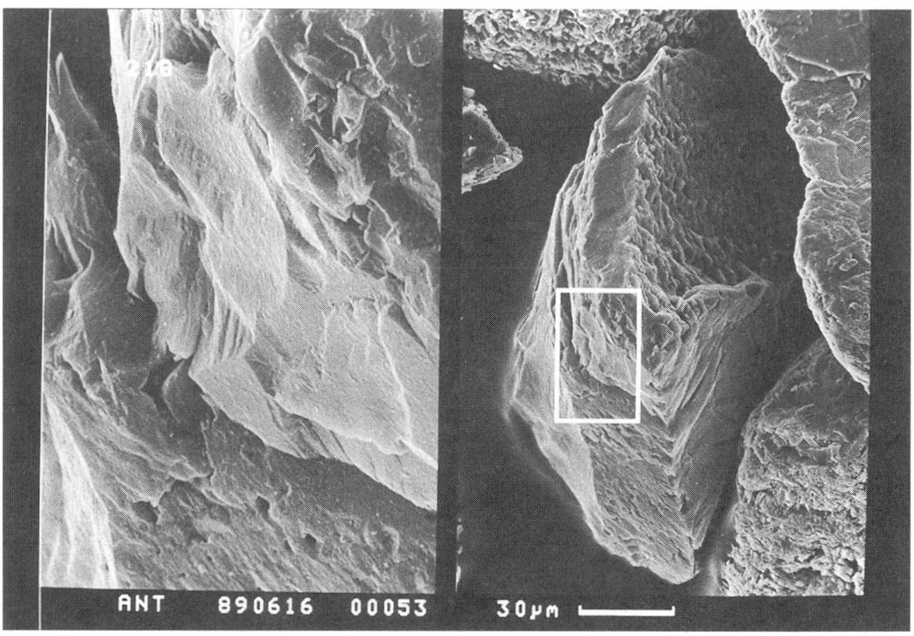

Figure 5.37 Dissolution etching in quartz (top and bottom). The enlargement (×5) shows a fresh (upper) contrasted with a weathered (lower) surface. Exfoliation sheets in the enlargement could produce a fracture face upon dislodgement. Source: Till 218, Antarctica. See Mahaney et al. (1996a), for details on location and lithology.

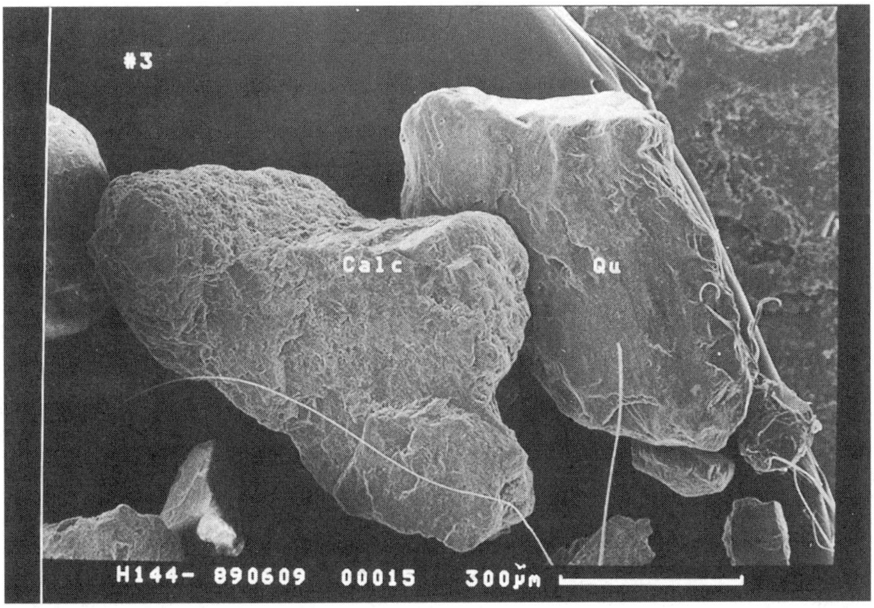

Figure 5.38 Dissolution etching on calcite (Calc) and quartz (Qu); white lines are organic. The quartz grain has minor etching microfeatures; the calcite is in a more advanced weathered state. Source: Sunnybrook Till, Ontario (= early Wisconsinan) (Mahaney and Hancock, 1993). The white lines are artifacts.

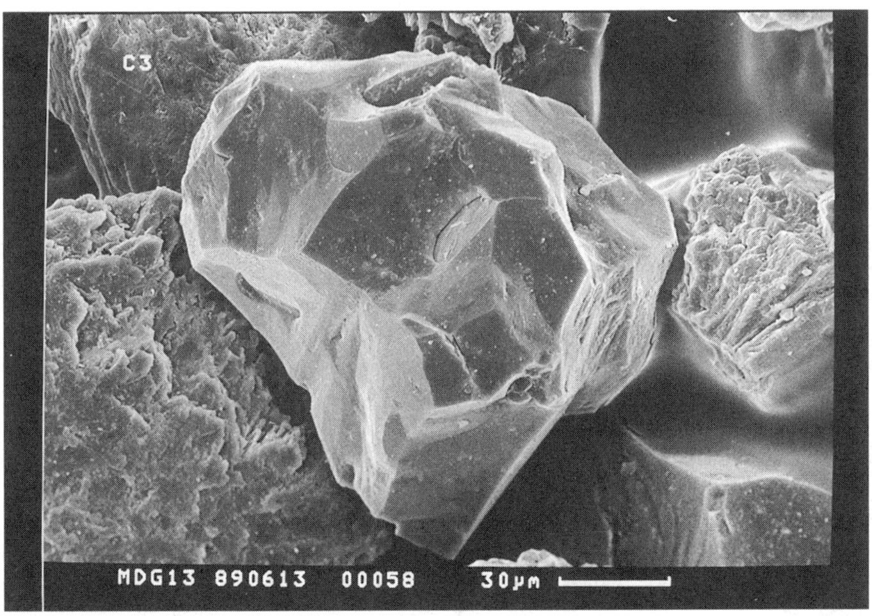

Figure 5.39 High relief, sharp angular features on quartz. A preweathered surface survives to the right. A plate in the center is partially upturned. The large troughs are abrasion features which end abruptly in the lower left indicating the grain is a broken remnant. Source: C horizon in the early Neoglacial soil, Mer de Glace Glacier, France (Mahaney, 1991a).

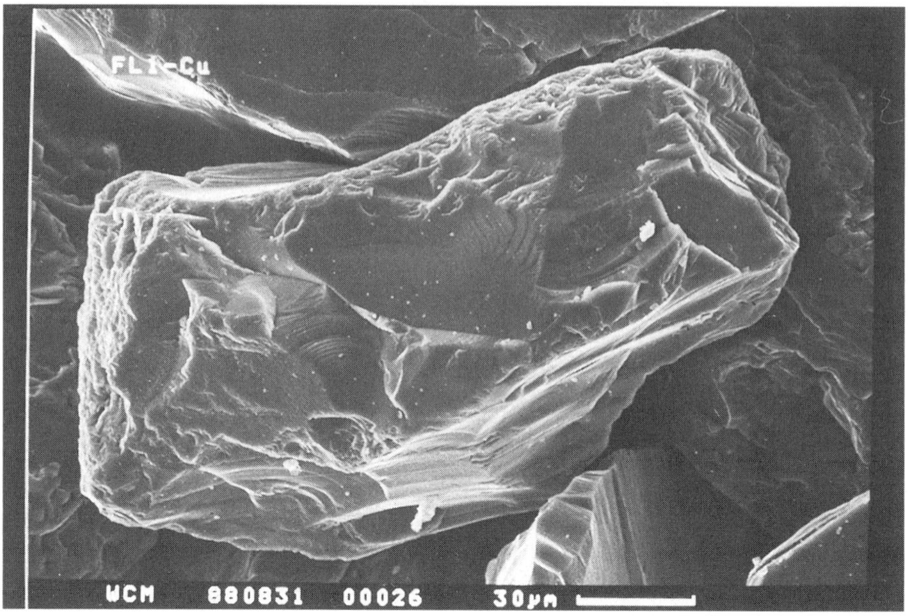

Figure 5.40 Preweathering on left edge and top of quartz. Grain subsequently grooved and fractured. Note incidence of conchoidal fractures in this example. Source: Bull Lake (= Illinoian) till, Fremont Lake, Wind River Range, Wyoming (see Mahaney, 1978).

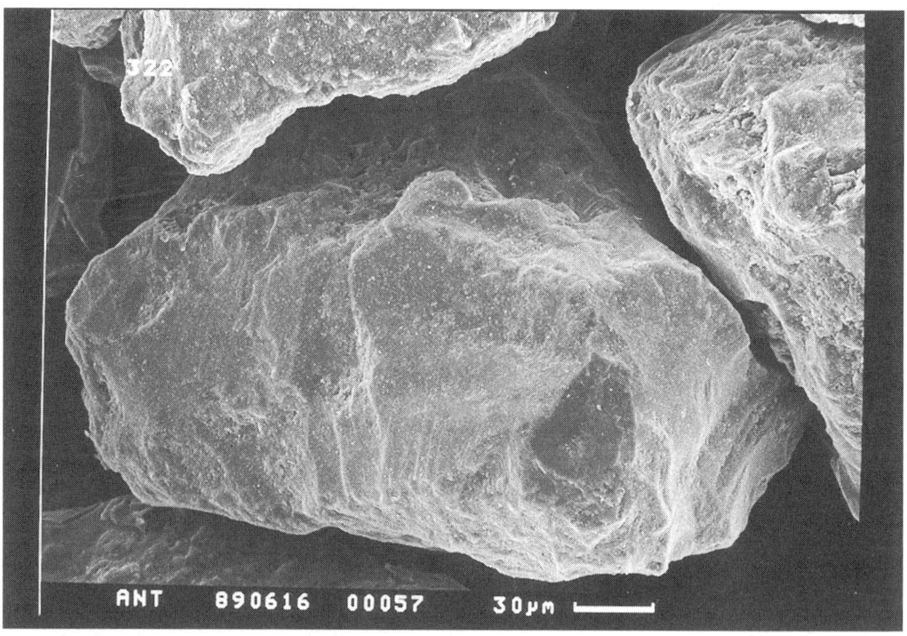

Figure 5.41 Crushing features nearly obliterated by weathering and a thin coating of Si. Note that the crushing features (fracture faces, sharp edges) are now nearly obliterated. Source: Antarctica, Till 322, in an end moraine (Mahaney et al., 1996a).

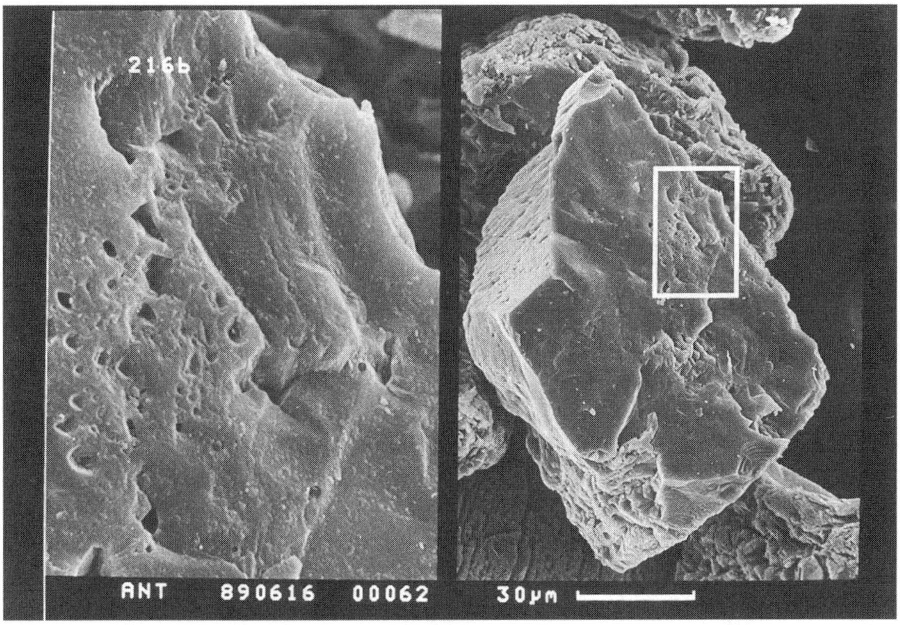

Figure 5.42 Weathering and dissolution etching of quartz. In contrast with figure 5.41, this example lacks abundant crushing features other than general grain shape. The enlargement (×5) shows v-shaped scars enlarged by dissolution effects. Source: Till 216b, Antarctica. See Mahaney et al. (1996a) for details on location and lithology.

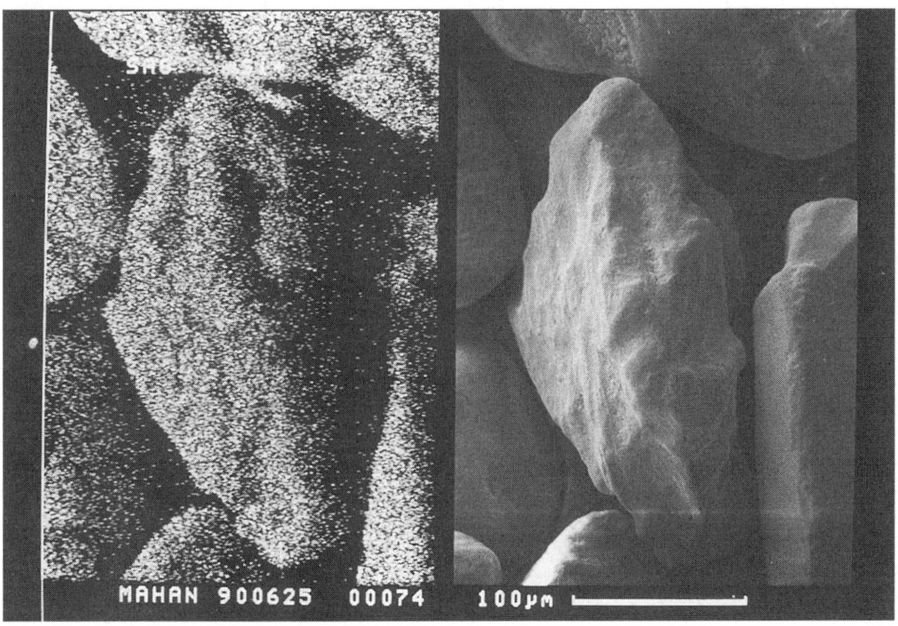

Figure 5.43 Precipitation features on subrounded, almost streamlined quartz; the Si coating is clearly evident on the backscattered image to the left. Source: eolian (erg) deposit. The sample is documented in Mahaney & Andres (1996).

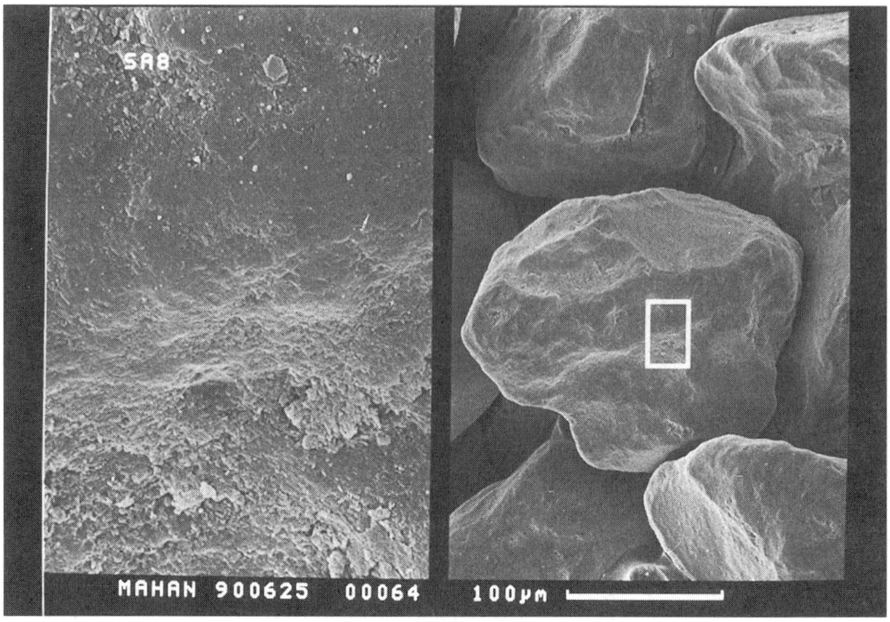

Figure 5.44 Precipitation features: coating on quartz from Saharan erg. Higher areas on grain have coatings Fe, Ca and Al mixed with abrasion surface (Al:Ca = 1.5:1.0; Al:Fe = 3:1); lower areas have only Al coats of fibrous gibbsite lacking Fe and Ca. The enlargement is ×10. Source: eolian (erg) deposit, Sahara (see Mahaney & Andres, 1996).

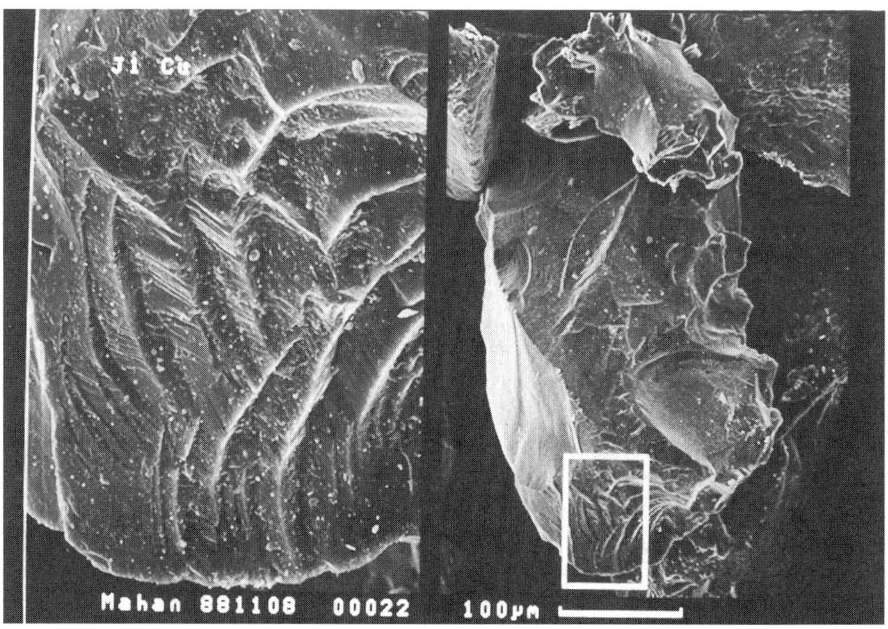

Figure 5.45 Abundant adhering particles on somewhat triangular-faceted quartz with abundant conchoidal fractures in the enlargement (×5). Source: parent material of the post-Bull Lake soil (Mahaney et al., 1984).

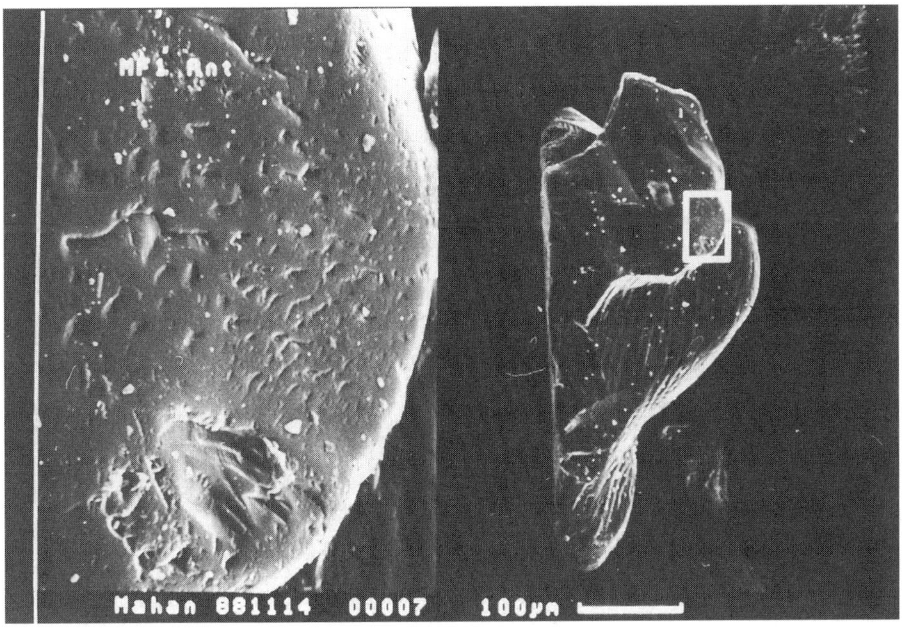

Figure 5.46 Subparallel fractures on quartz. v-shaped percussion cracks modified by weathering (etching) dominate on the enlargement (×10) over glacial-crushing features (troughs, lower right of enlargement). Source: La Meseta Formation, Seymour Island, Antarctica.

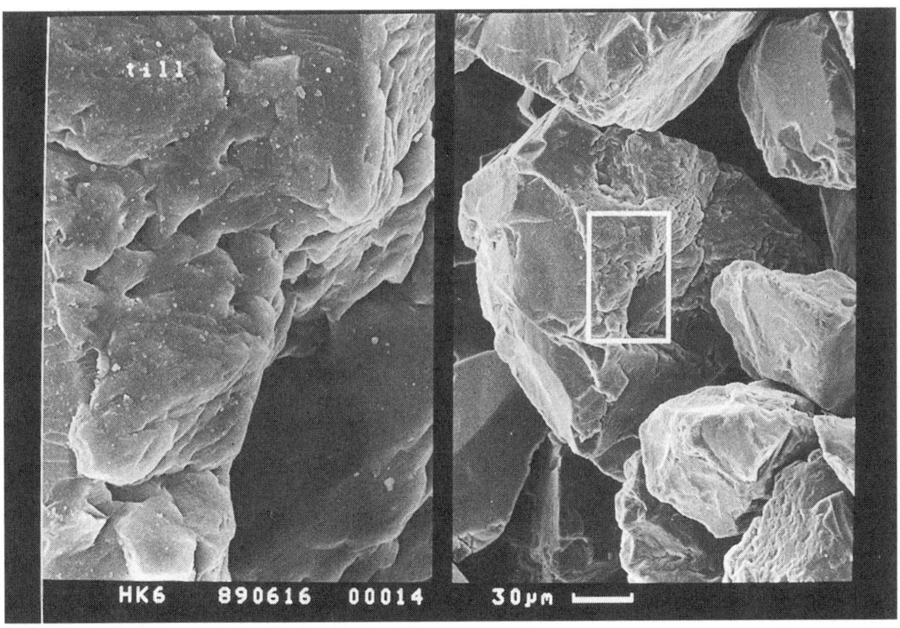

Figure 5.47 V-shaped percussion cracks next to a crescentic gouge in quartz. The enlargement (×5) enhances the v-shaped cracks which could be lattice imperfections. Source: sample HK6 of late Neoglacial age, Austrian Alps (Mahaney et al., 1996b).

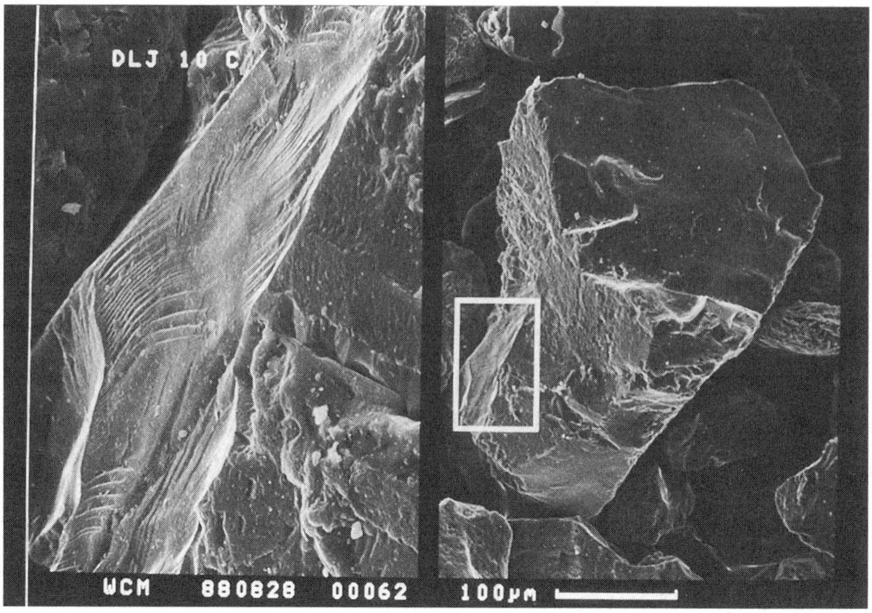

Figure 5.48 Complex grain with v-shaped percussion cracks (right center) in a weathered surface of quartz; fresh crushing to left. Multiple fractures in a groove are shown in the enlargement (×5). The sample is from the C horizon in the post-LGM soil, site DLJ10, Tibetan Marginal Mountains (Mahaney & Rutter, 1992).

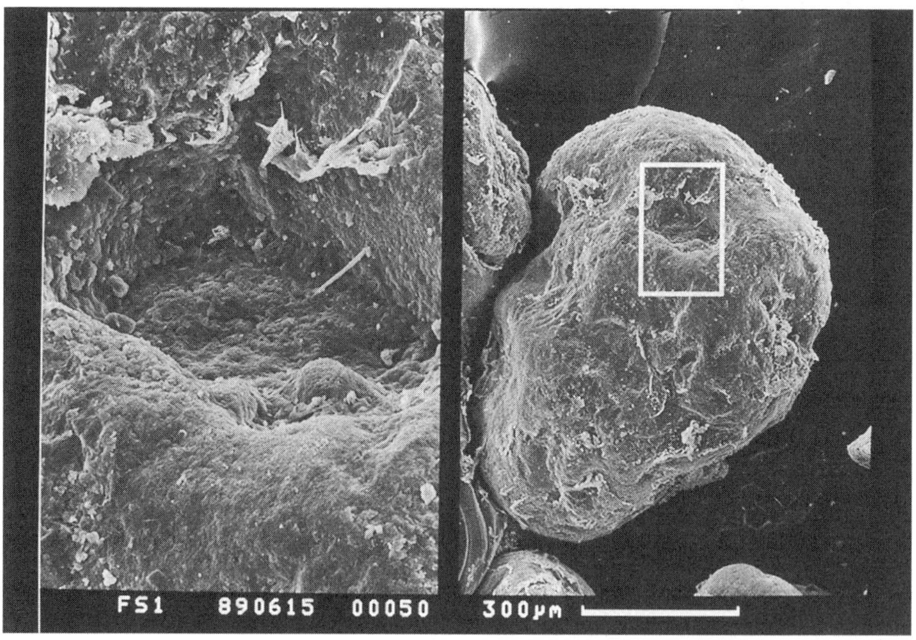

Figure 5.49 Edge rounding (bulbous features) on windblown quartz. The surface roughness is attributed to a clay coating. The enlargement (×5) shows a vug lined with clay. Source: Bunter sandstone of Triassic age, Germany.

Figure 5.50 Enlargement of allochthonous clay particles on the grain in figure 5.49. The clay coating is predominantly dickite (Si:Al = 1:1).

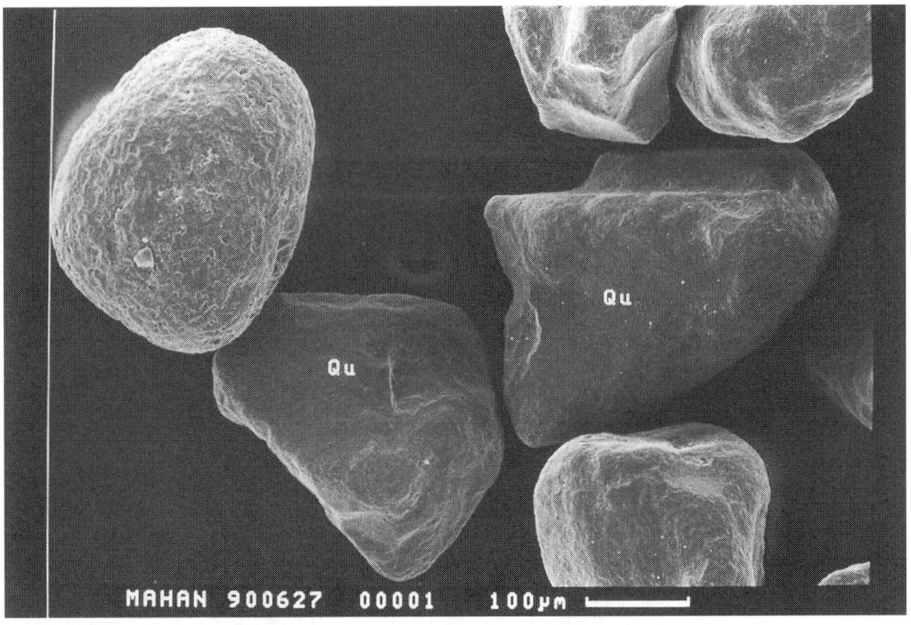

Figure 5.51 Degrees of edge rounding on quartz from Saharan erg. Quartz (Qu) grains in the center have variable coatings (Si:Al = 1:1) of clay. The rounded grain in the upper left exhibits v-shaped percussion scars, upturned plates, and a coating of Si. There is some new growth of quartz on grains in the upper right, with sharp angular edges, probably developed from the reformation of SiO_2. Source: Saharan erg (see Mahaney & Andres, 1996).

Figure 5.52 Overprinted quartz with a pitted, weathered surface and a fresh fractured crater. The grain is complex with multiple histories: weathered → fractured → weathered → fractured (the last episode created the large crater). The fresh microfeatures are from late Pinedale (= Wisconsinan) glacial transport in the Wind River Range, western Wyoming. The enlargement is ×5. Source: parent material of the post-Pinedale soil (Mahaney et al., 1984).

Figure 5.53 Overprinting on quartz with straight and curved grooves and abraded surfaces. Possible new growth (see ×3 enlargement) coated with Si. Source: grain recycled from Ordovician tillite into a Saharan erg and showing little evidence of dune transport (see Mahaney & Andres, 1996).

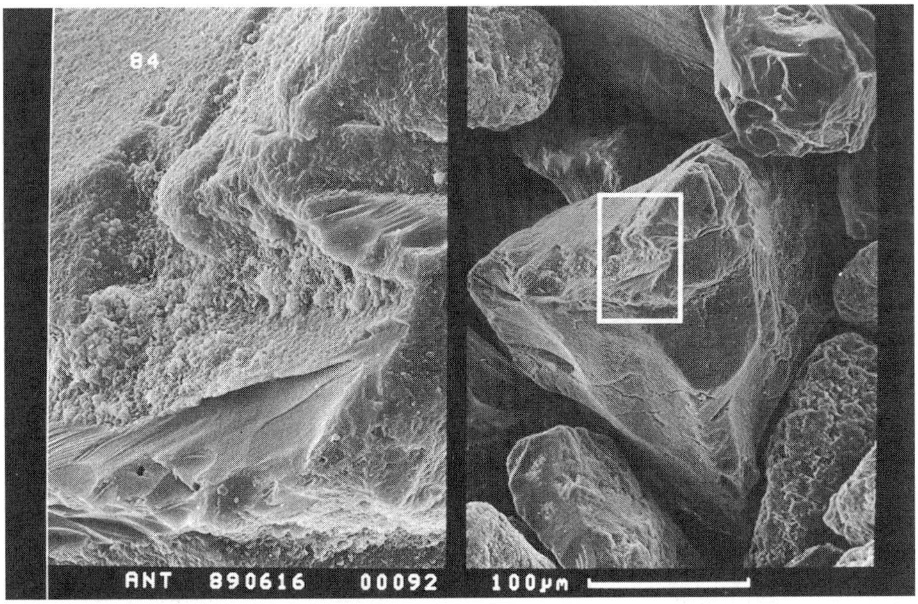

Figure 5.54 Overprinting on quartz. The grain is partly preweathered with a Si coating and fresh fractures, and partly abraded and cracked (center low). An upturned plate is in the bottom center of the enlargement (×5). Source: end moraine debris, Till 84, Antarctica (Mahaney et al., 1996a).

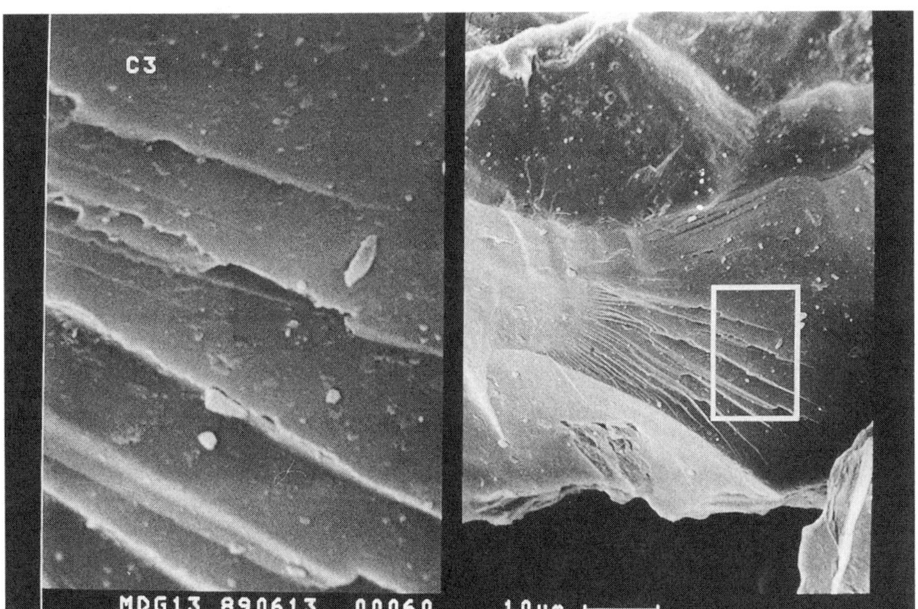

Figure 5.55 Radial fractures on quartz. The fracture pattern radiates from an impact zone. The enlargement (×5) shows slight weathering on fracture planes. Source: C horizon of early Neoglacial soil, Mer de Glace, France (Mahaney, 1991a).

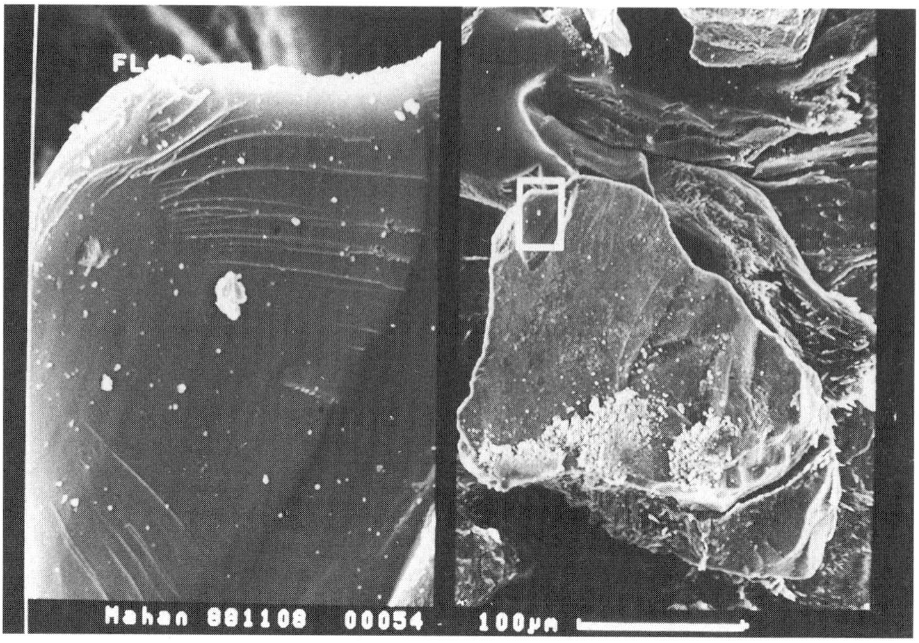

Figure 5.56 Radial fractures on quartz similar to figure 5.55. The light area is charging. There is abrasion on the front of the grain; grooves to the right and sharp edges are evident. The enlargement (×10) shows the point of impact and radiating fractures. Source: Bull Lake till, Fremont Lake, Wyoming (see Mahaney, 1978; Mahaney et al., 1984).

Figure 5.57 Amphibole perfectly preserved in pristine condition following bedrock release and glacial transport. Slight splitting along the cleavage planes. Compare with figure 5.7. Source: Till 480L, Antarctica (Mahaney et al., 1996a).

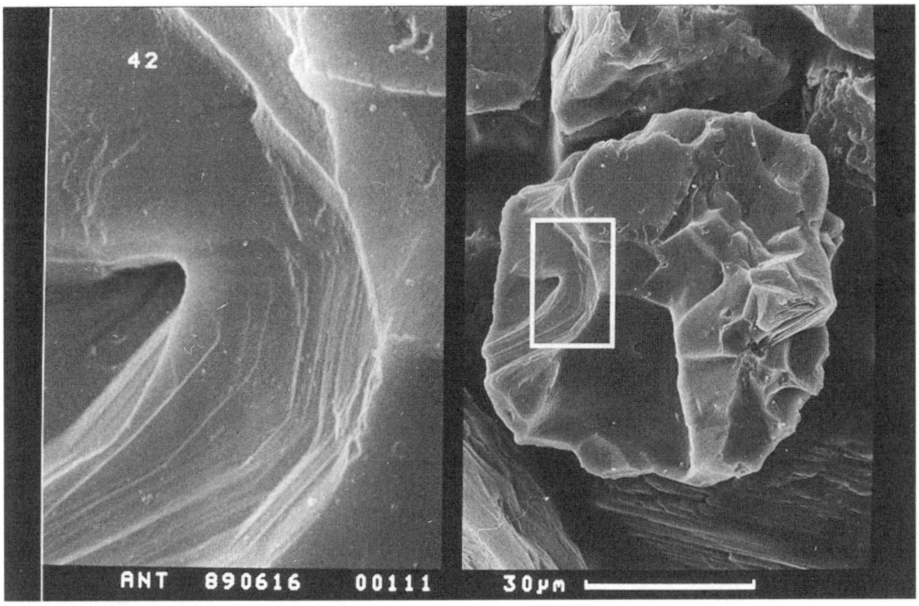

Figure 5.58 Deep cavernous craters in quartz with numerous fractures and distinct fracture faces in the coarse silt fraction. The enlargement (×5) shows fractures within a crater. Source: Till 42, Antarctica, end moraine. See Mahaney et al., (1996a) for details.

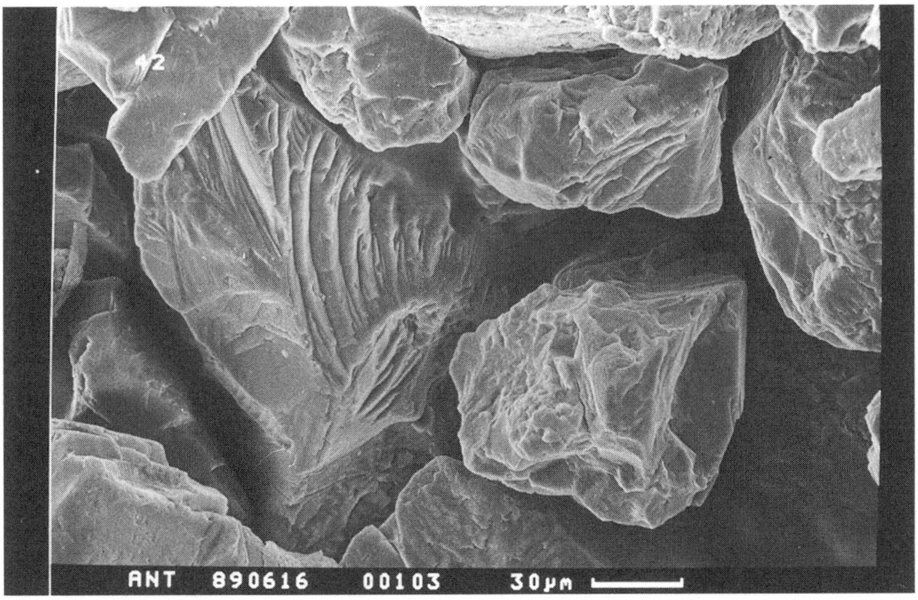

Figure 5.59 High frequency fractures (center) with frequently spaced lines. The sample exhibits a mix of conchoidal or subparallel linear fractures. Source: Till 42, Antarctica (see Mahaney et al., 1996a).

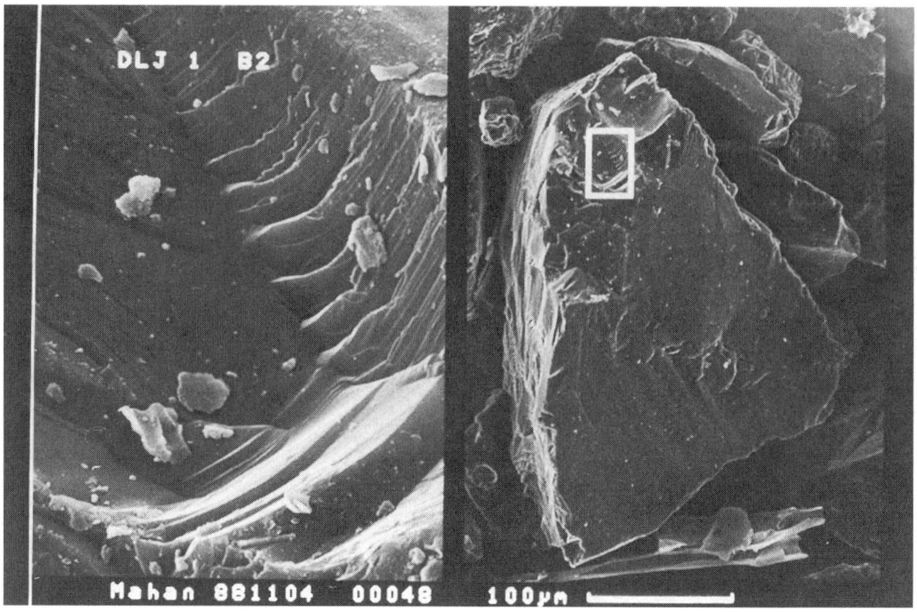

Figure 5.60 Low frequency fractures based on large spacing between fracture planes (as shown in the ×10 enlargement). Source: B horizon of the post-LGM soil, Dalijia Shan, north-western China (Mahaney & Rutter, 1992).

Figure 5.61 Cratered quartz with many smaller quartz and biotite grains to the right. The quartz grain in the center shows large geometrical craters that appear fresh compared with the surrounding grain surface. Source: late Neoglacial till, sample HK6, Zillertal Alps, Austria.

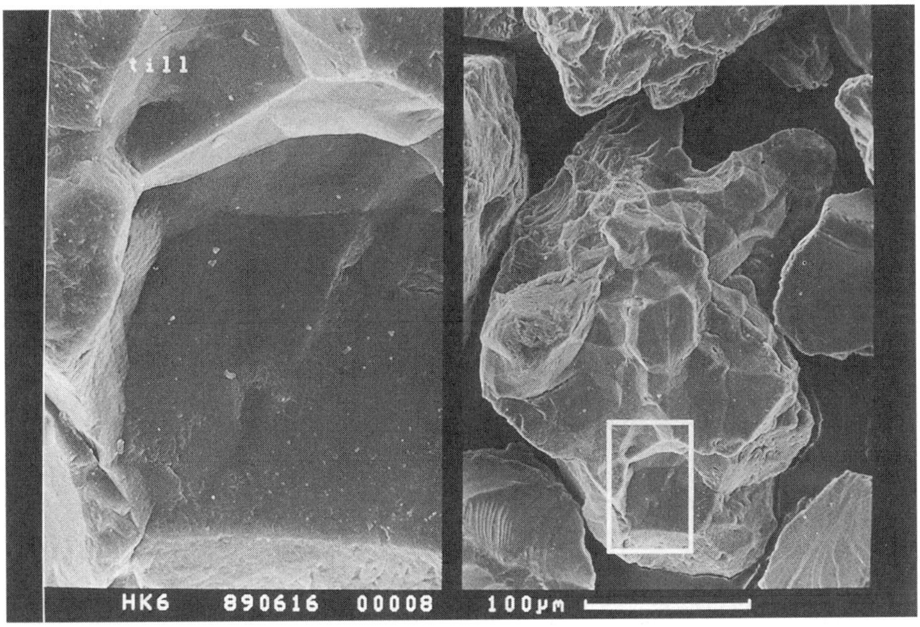

Figure 5.62 Cratered quartz similar to the grain in figure 5.61. Part of the enlargement (×5) could be new growth. Source: late Neoglacial till, Horn Kees, Zillertal Alps, Austria.

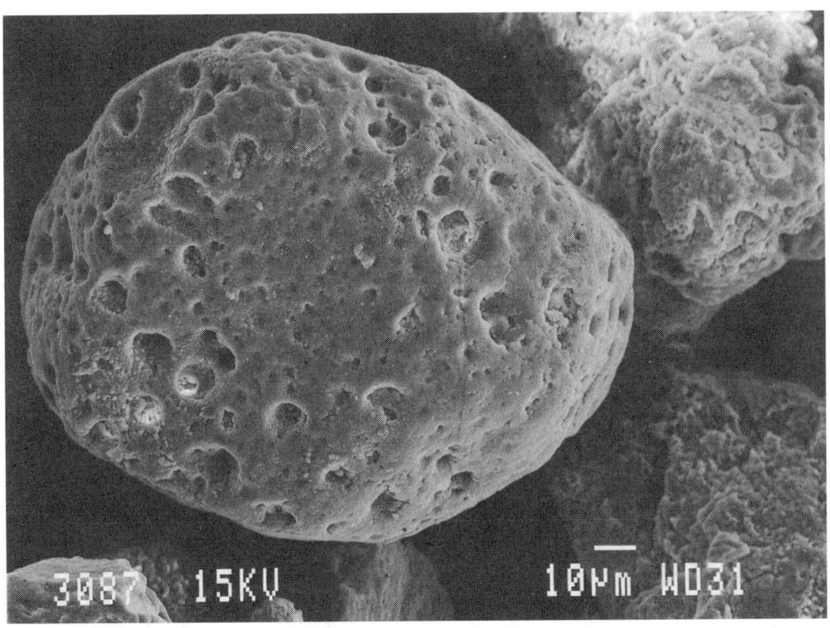

Figure 5.63 Shock-melted quartz with melt spherules. Recrystallized quartz in melt spherules can be seen in the lower left and upper right of the grain. Source: K/T boundary zone, Clear Creek, Colorado. Samples courtesy of Glen Izett, College of William and Mary, Williamsburg, VA (see Izett, 1990).

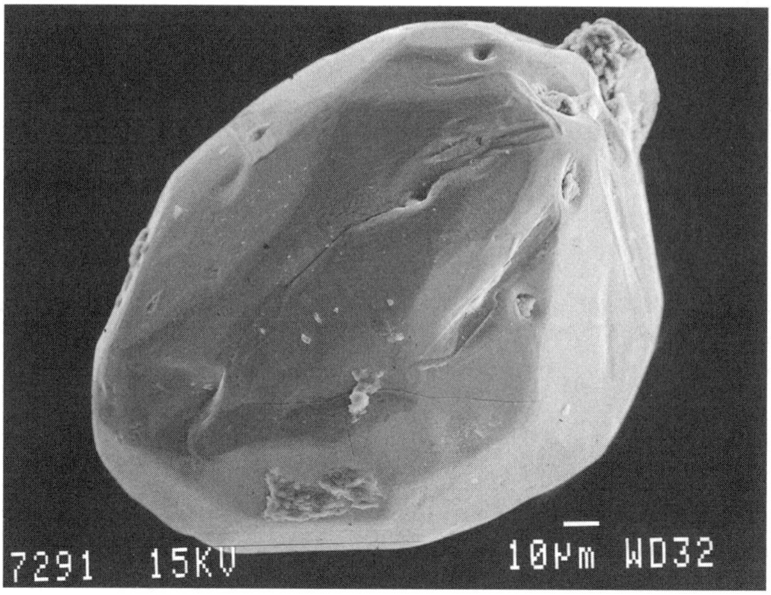

Figure 5.64 "Rolled" quartz with a dislocated lattice structure from the shear zone adjacent to the Boconó-Valera faults. The permanent change in shape of the grain is achieved by a change in the relative positions of atoms or molecules (i.e., movement of lattice defects through the crystal). Also known as intracrystalline deformation (see Poirier, 1985; Passchier & Trouw, 1998). Source: site Vic3, 6 km north-east of Apartaderos, Venezuela, on the water divide between the Rio Chama and Rio Santa Domingo (see Mahaney & Kalm, 1996).

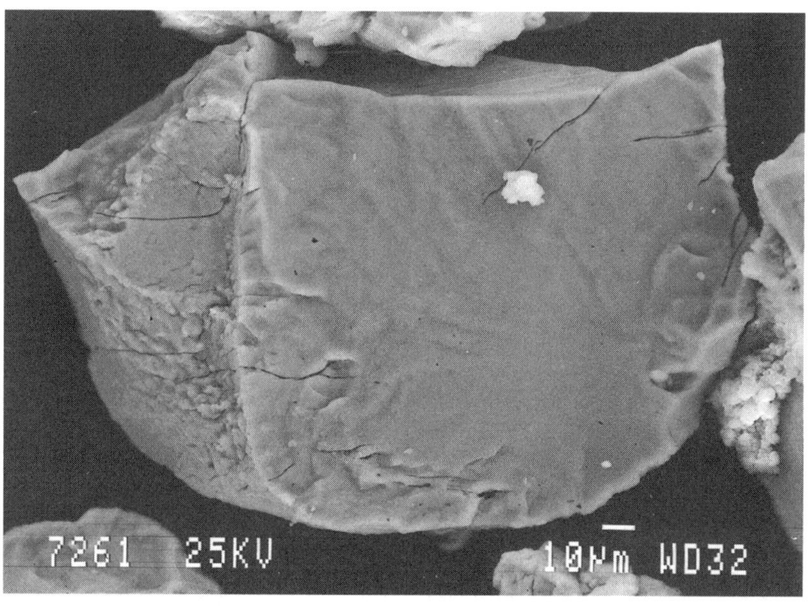

Figure 5.65 Slickenside surface on quartz, partially cracked, with weathered patina on sides. Collected from the Boconó Fault. Source: site Vic3, 6 km north-east of Apartaderos, Venezuela (see Mahaney & Kalm, 1996).

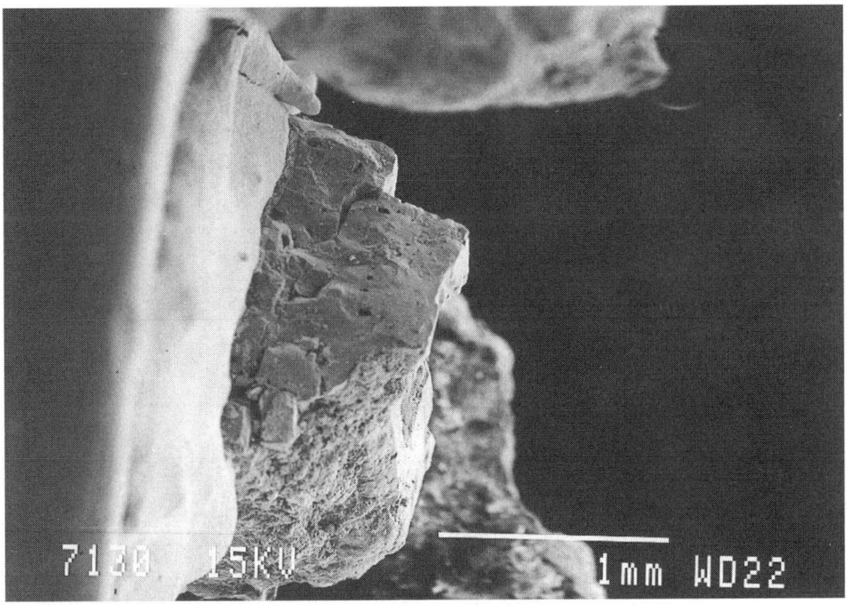

Figure 5.66 Cracked grain with a fresh fracture face above and a weathered face below (on turned stage of the SEM). Partially rough fracture face on the upper portion of the grain; etched lower half. Source: S2 talus, Oregon.

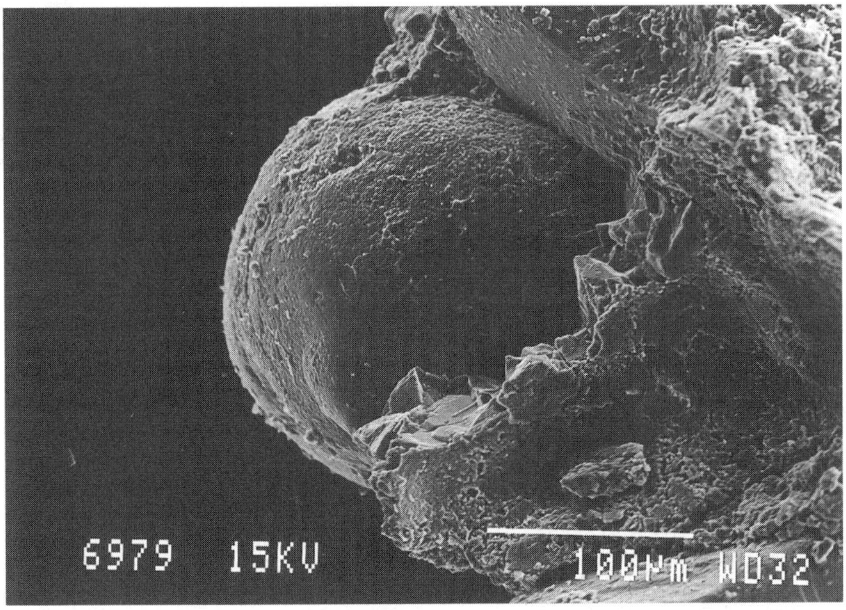

Figure 5.67 Sawtooth fractures in carbonate matrix surrounding spherical quartz ejecta from Neugrund Crater, Estonia (Ordovician). Sample courtesy of Kalle Kirsimäe, Institute of Geology, Tartu University, Tartu, Estonia. Occurrences of this type are rare but are occasionally found in neotectonically released debris in caves (Mahaney & Sjöberg, 1993) and in glacial deposits, especially in basal sediments deposited by thick continental ice (Mahaney, 1995b).

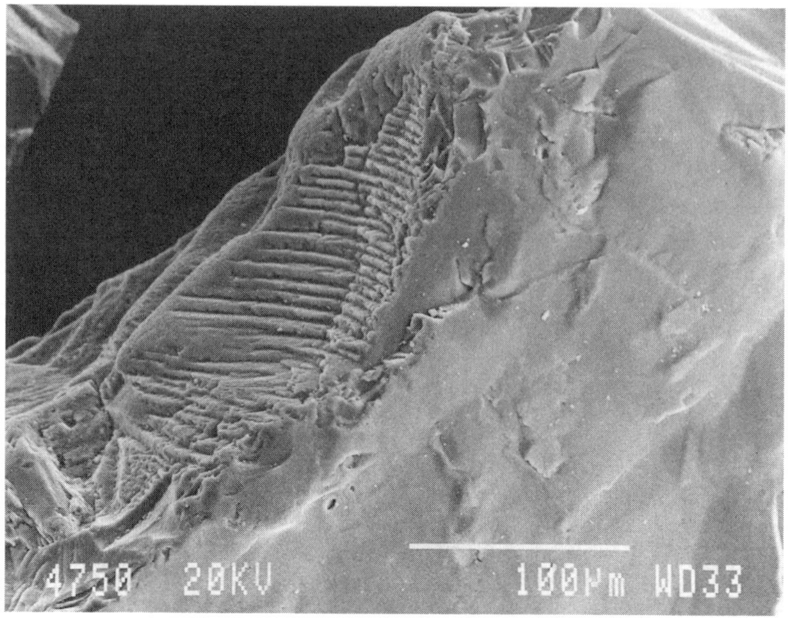

Figure 5.68 Crescentic chattermark trail in quartz from lodgement till in northern Germany. Sample courtesy of Jan Piotrowski, University of Aarhus. The sample comes from the lower till (facies U1), a lodgement till. The most detailed account of the entire cliff section is in Piotrowski (1992, 1996).

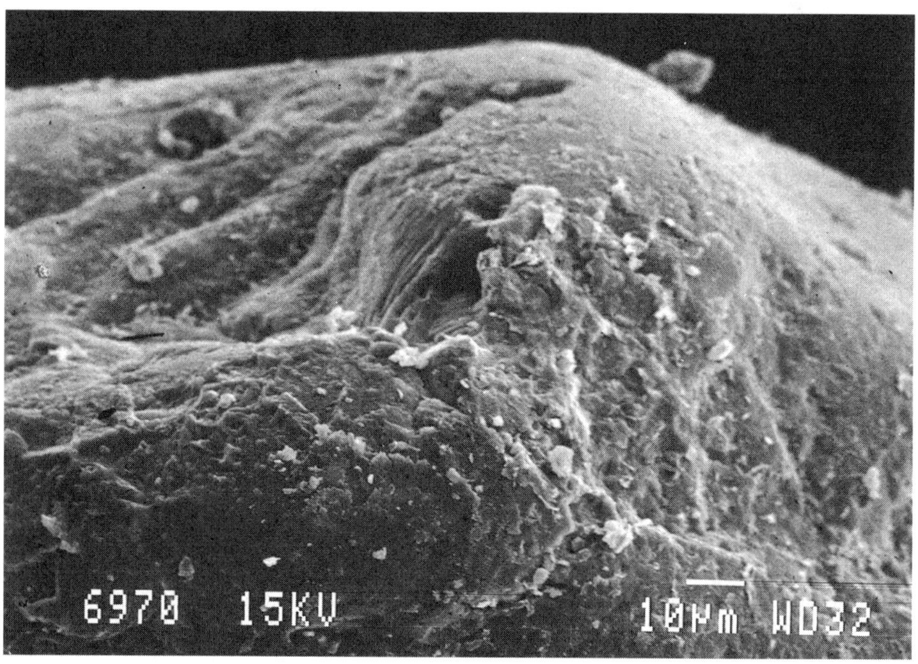

Figure 5.69 Shock lamellae (center) on a quartz grain ejected from the Neugrund Crater, Estonia. See figures 11.2C and D for other examples.

Figure 5.70 Rounded and subrounded quartz grains, Nile Delta, Egypt.

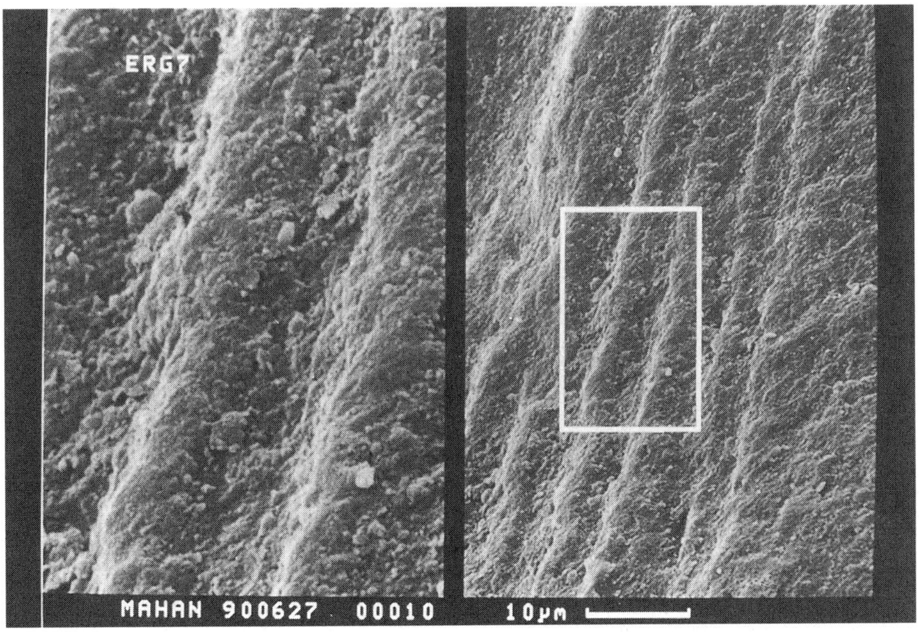

Figure 5.71 Parallel ridges less than 2 μm thick and spaced 3–10 μm apart, Algerian erg (see Mahaney & Andres, 1996). The enlargement (×3) shows the infilling surface with scales that might be derived from torn upturned plates.

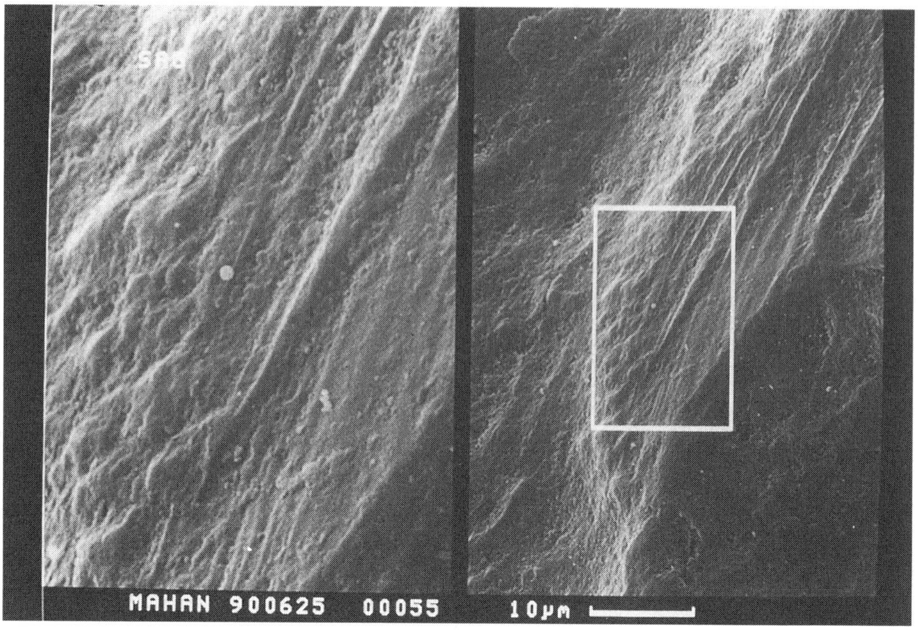

Figure 5.72 Elongate and smoothed-over depressions with parallel ridges. Source: Saharan dunes, Egypt. The enlargement (×3) shows closely spaced elongate depressions and ridges, smoothed to the right and fresh to left.

Summary

The microtextural index indicates the form of each fracture–abrasion–coating surface described later in the text. The index is intended as a guide for investigators describing the microfeatures encountered in the normal routine of SEM investigations. I presume other researchers will add to the list as more microtextural observations are made of sediments in different environmental settings. Many of the microtextural examples included here are from glacial sediments and soils or paleosols formed in till because glaciation and the discrimination of glacial grains from other sedimentary environments have been at the forefront of microtextural analysis. A crossover with other environments is evident from the numerous examples where glacial signatures are mixed with those of water and eolian transport. Recent investigations of grains from mass wasting debris indicate that gravitational processes produce grains which carry a unique signature. The resulting microtextures are related to source rock release and the detachment of large boulder size blocks, which impact and crack smaller pebble to boulder size rock masses, while falling or sliding downslope under the influence of gravity. The multitude of cracked and slickenside grains in talus and rockfall debris seems to be characteristic of mass wasting processes (see chapter 8).

Shock-melted quartz resulting from asteroid impact, and rolled and broken grains milled in the movement of fault planes and in adjacent shear zones have only rarely been analyzed by SEM. In this list some preliminary microtextures are illustrated that are exclusive to the two environments/processes. Heavy minerals, considered to have short recording time spans, and largely avoided by SEM investigators, are intricately linked with paleosols (see chapter 14). Under ideal circumstances, the same suite of microtextures encountered on light minerals such as quartz may also be found on heavy minerals such as zircon and to a lesser degree on ilmenite and magnetite. Indeed, heavy mineral analysis in tills shows quite convincingly that zircon, with its feldspar armor torn away and after reforming at the base of the ice into a "pencil" form, may act like a stylus inflicting severe damage on less resistant grains such as quartz.

Following the illustrations and definitions of basic microfeatures in this chapter, I will examine specific sedimentary environments and the geologic agents that dominate in them, and attempt to identify those microtextures unique to that environment. At the same time I will assess the range of microtextures that cross over amongst two or more environments, but dominate in one. For example, in the past, conchoidal fractures were considered to be the hallmark of glaciation. We now know that whereas they dominate on glacial grains as *deeply embedded microtextures*, they also occur occasionally on eolian and stream transported particles. In the latter group they occur as superficial scars and without the frequency encountered on grains in tills.

6 Eolian Grains

The principal microtextures on sands characteristic of eolian transport will be identified in this chapter along with sedimentary environmental information. These will include both fraction, abrasion, and solution–precipitation/diagenesis microtextures. Possible microtextural origins will be considered and discussed. Mechanical experiments duplicating natural microtextures on Earth and on Mars will be examined in so far as they have been undertaken by other workers (Krinsley et al., 1979; Lindé & Mycielska-Dowgiallo, 1980). Finally, methodological problems will be discussed.

Eolian transport is initiated when wind blowing at greater than threshold strength over an immobile surface meets the leading edge of a deposit of loose material (Bagnold, 1941; Pye & Tsoar, 1990; Barndorff-Nelsen and Willets, 1991; Boggs, 1995). Direct dislodgement from outcrops by wind also may be important in grain transport (Greeley & Iverson, 1984; Anderson et al., 1991; Boggs, 1995). As grains are moved by wind, they collide with other grains in transport as well as on the surface below. If collisions are sufficiently energetic, portions of grains may chip off, or they may shatter. Various types of cracks and plates may form on the parent grains. Much fine material of silt (<63 μm) and clay (<2 μm) may be mechanically produced in wind streams, and transported long distances.

Sand grains usually travel as saltating or creeping bedload (Bagnold, 1941; Glennie, 1970) and may be subjected to a succession of high-energy collisions. During collision, the kinetic energy of each particle is converted, at least partially, into elastic energy in the grain. The energy wave or vibration may bounce back and forth within a given grain a number of times, although it is not known how this affects the grain surface. The result of these collisions, termed "abrasion fatigue" by Pascoe (1961), is a disrupted lattice type of structure on grain surfaces. This makes the surface of the grains physically and chemically reactive and may cause small silt and clay particles to attach to the grains until they are removed by additional abrasion.

"Abrasion fatigue" describes a process that occurs when quartz is ground or abraded in either water or air; a disrupted lattice type of quartz is formed which is considerably thicker after dry grinding than after wet. The product of broken surface structures, the disrupted lattice consists of very small particles less than 3 nm in diameter, which are difficult to distinguish at normal low magnification.

This layer contains small cracks, dislocations, broken Si–O bonds, projecting vertices, and thus abnormal solubility. It also has a lower density than quartz: 2.2 as opposed to 2.6 mg/cm^{-3}. The layer of disrupted lattice material generally increases in crystallinity toward the center of the grain, but if the particle is small and perhaps has spalled off a sand-sized grain, it may consist entirely of disrupted lattice quartz. Krinsley & Smalley (1973) used a high-voltage electron microscope to photograph a grain around 1 nm in diameter, which consisted completely of disrupted lattice quartz (Krinsley & McCoy, 1978).

Some particles subjected to eolian transport have a shiny skin rather than a flaky surface. As postulated by Lindé (personal communication, 2001), if the area of collision is small, then the small volume involved in collision might reach melting temperatures (Lindé, 1978), producing localized melt microfeatures on the grain surface. More experimental work is required along these lines to prove shock melting as a product of eolian transport on Earth; on Mars, which experiences very high wind velocities, one might expect to see shock melted grains (see the last subsection in this chapter).

Diagnostic Microtextures

The microtextures observed on eolian sand grains using SEM in the secondary mode include upturned plates (figure 6.1A), parallel ridges approximately 0.5–10 μm or more thick (figure 5.71), elongate depressions (figure 5.72), smoothed-over depressions and ridges (figure 5.72), and polygonal cracks. The upturned plates produced are probably the resulting cleavage scarps and are frequently modified in the desert environment by solution and precipitation of silica (Margolis & Krinsley, 1971; Krinsley et al., 1976). The relation between grain size and microfeature length is unknown; however, preliminary indications are that the length of the plates increases as individual grain diameter becomes larger. On the other hand, on a given grain, the average plate length at a particular spot may differ from the plate length at a different place on the grain surface. Plates appear to be spaced from about 0.1 μm to about 1 μm apart (Krinsley & McCoy, 1978).

The second feature observed on all grains of sand size larger than about 400 μm consists of equidimensional or elongate depressions (figures 6.6D and 6.7C) approximately 20 μm (or less) to 250 μm in diameter, producing medium relief. They may be the result of direct as opposed to glancing impact between saltating or creeping grains. As grains saltate they also rotate; the effect of this behavior on abrasion is unknown but it may assist in rounding grains and in the genesis of the bulbous edge. Large depressions may also form on grains removed from parent rock, although if a grain is well rounded at source (e.g., from an igneous melt), it is likely to retain this memory. Thus, large-scale depressions in rounded grains are probably of eolian origin.

The third feature observed on most large eolian quartz grains consists of arcuate, circular, or polygonal cracks (about 0.5 μm to several tens of micrometers in length) on grains of coarse silt and fine sand size (figure 6.7E). This microfeature, listed under cracked grains in figure 5.1, is occasionally found on larger sand grains. It is quite brittle and usually removed during eolian abrasion and possibly during solution and precipitation which occur commonly in deserts. For this reason, the microfeature is not expected on fossil eolian sands. It may be the result of physical and/or chemical weathering, probably including the crystallization of salts (Lucci & Casa, 1968, plates XCIII to ICV; Krinsley & Doornkamp, 1973, plate 82; Le Ribault, 1975, figure 91; Krinsley & McCoy, 1978, figure 3).

A fourth microfeature, almost ubiquitous in modern desert grains, is a feature that apparently has smoothed most of the features described above. The tops of upturned plates are no longer sharp, and the valleys between the plates may be filled in. Thus, a rolling topography (figure 6.7F) is created, which generally encompasses the entire grain, often resulting in the production of bulbous edges, as shown in Mahaney & Andres (1996) and Lindé & Mycielska-Dowgiallo (1980).

Environmental Interpretations

Environment interpretations can be explained as follows. In hot desert areas, desert dew can penetrate 2–3 cm into sand and is charged with CO_2 (Engel & Sharp, 1958). Water evaporates during the day and a small amount of moisture may be precipitated on the grains at night. The pH of these water films may rise due to the presence of dissolved evaporites, and small amounts of silica may be removed from individual sand grains. This process is accelerated by the presence of an abnormally soluble layer of disrupted lattice quartz created by abrasion on grain surfaces. Upon complete evaporation, the silica is redeposited as an irregular layer of either opal or silicic acid, which smooths out and subdues the pre-existing mechanical fracture patterns. After the process has proceeded long enough to cover a spalled piece of silt or clay, the disrupted lattice layer plays no further part in chemical action, but the opal or silicic acid may be involved in numerous cycles of solution and/or reprecipitation. This is particularly the case if the particles are too small to be carried in the traction or saltation load (Krinsley & McCoy, 1978).

The fourth feature that is characteristic of desert sand grains (generally larger than 400 μm) is the degree of rounding. Large grains are generally very well rounded, many of them approaching a perfect sphere. The more mature the desert, presumably the rounder the grains. As smaller sand grains are studied, roundness decreases, most probably because the momentum (mass × velocity) involved in grain collisions decreases with decreasing grain size. Finally, silt-sized particles present in deserts are irregular, and are at the low end of the roundness scale. However, the roundness variability of even large desert grains is considerable, and frequently very spheroidal grains may be in the minority. This may result from differences in original shape, or possibly because many desert grains may spend much of their time buried at the bottom of active dune forms.

An interesting study using Fourier shape analysis and SEM textural analysis of the St Peter Sandstone (Middle Ordovician) was done by Mazullo (1987) at Texas A & M. There have been suggestions that the textural analysis technique could not be used with fossil sands (Pettijohn, 1975), as diagenesis was presumed to eliminate mechanical and chemical textures formed during previous cycles.

However, Mazullo was able to separate the sands in the St Peter into two groups: well-rounded grains that exhibited surface textures diagnostic of intensive abrasion during a phase of eolian transport; and irregularly shaped grains that had undergone little abrasion while being transported in an aqueous media. An examination of the morphology of these unabraded grains indicated that most of the St Peter was derived from shallowly buried, partially to completely indurated sandstones, with lesser amounts of sand being derived from igneous and metamorphic rock sources. Thus, Mazzullo was able to show that SEM textural features represented valuable evidence regarding the origin of the St Peter Sandstone.

Mountain Loess

Fine sands and coarse silt, sourced from outcrops and outwash trains in high mountain valleys, are transported by gravity winds and convection currents of thunderstorms and deposited in alpine soils. First described by Nelson (1954) in the Rocky Mountains of Colorado, many workers have used high silt and fine sand distributions in soils and paleosols to argue for an eolian origin (Birkeland, 1973, on Mount Sopris; Mahaney, 1978, in the Wind River Mountains, Wyoming). Few have used the SEM to study the shape and microtexture of eolian grains in mountain loess (Mahaney and Rutter, 1992), or to combine micromorphological investigations with rare earth geochemistry to determine source areas for loess (Mahaney et al., 1996b).

Select samples of mountain loess from the surface horizons of soils and paleosols in the Teton Mountains (figure 6.1A, B), the Dalijia Shan of northwestern China (figure 6.1C, D) the Rhine Valley (figure 6.1E, F), and the Zillertal Alps of Austria (Figure 6.2A–D), reveal variations in shape that result from distance of transport. These samples show a mix of grains with a primary origin under glaciation followed by slight modification during eolian transport. Most grains lack adhering particles despite a primary glacial origin and short transport distances, the latter confirmed in some instances by rare earth geochemistry (Mahaney et al., 1996b). While the shape of many grains is distinctly glacial (figures 6.1A, C, D and 6.3A–C), the bulbous shoulders of some grains (figure 6.1A, D–F) are distinctly eolian. Upturned plates on some grains (figure 6.1C, E, F) could have either a glacial or eolian origin. However, the most distinctive eolian microfeatures are found on grains in the Rhine loess (figures 6.1E, F), which are particles subjected to the longest airborne transport. Grains from the Zillertal Alps, Austria (figure 6.2) exhibit sharp edges, little edge rounding, and a high degree of weathering, which indicates reworking of previously weathered material along with the incorporation of fresh material.

Continental Loess

The surface epipedons of paleosol–soil profiles, situated in topographically high portions of ground moraine in continental North America, often have a fine texture of silt and fine sand that is considered to have an eolian origin

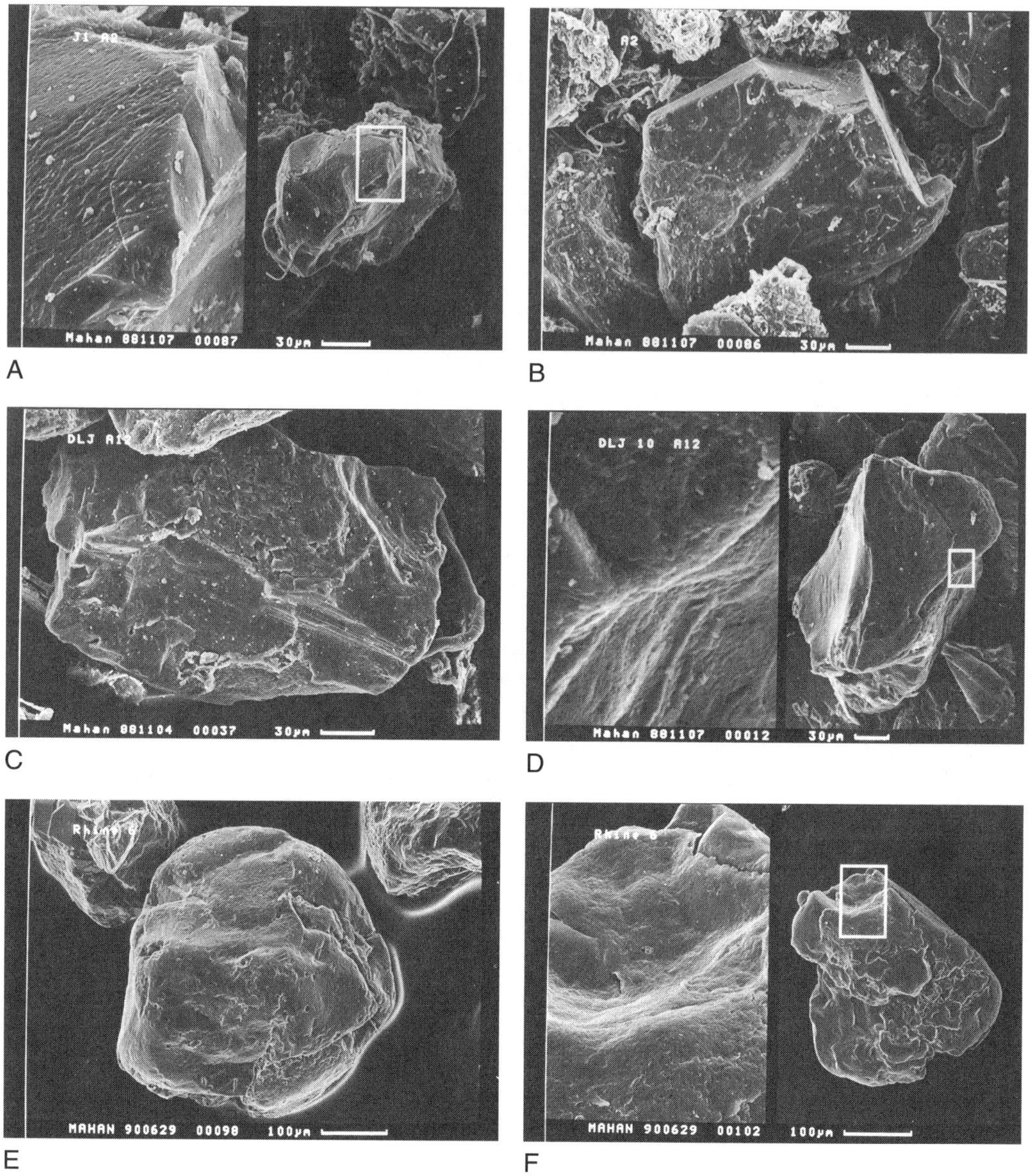

Figure 6.1 (A) Quartz from the E horizon of a Spodosol in Jaw Cirque, Teton Range, western Wyoming, with sharp edges, fractures and parallel ridges. (B) Nearly equant grain from the same horizon with fractures, broad shoulders and minor upturned plates. (C) Quartz grain with sharp edges from the A horizon of an Inceptisol in the Dalijia Shan, north-western China, with some upturned plates (center). (D) Nearly triangular-shaped quartz with one bulbous edge (upper right). The enlargement (×10) shows ridge and trough topography that could be abrasion fatigue. (E) Quartz with bulbous edges and upturned plates from the Rhine loess near Koblenz. (F) Quartz from the Rhine loess with bulbous edges and numerous upturned plates. The enlargement (×5) shows a crack (upper right), upturned plates and abrasion fatigue.

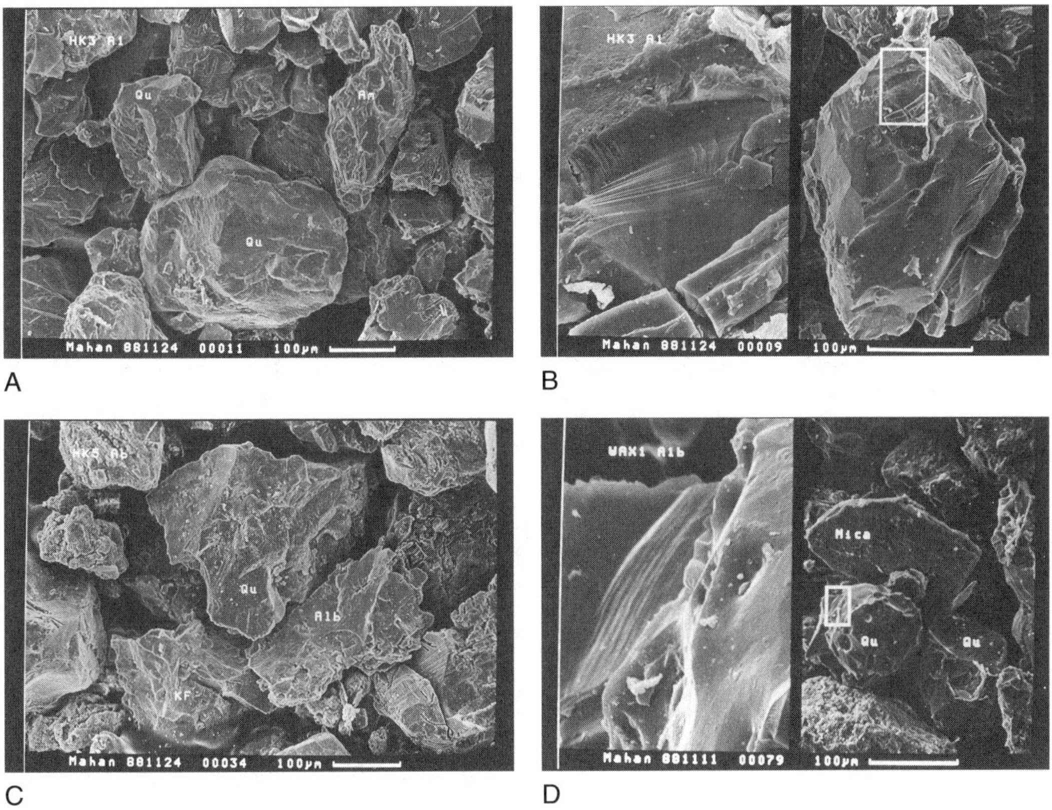

Figure 6.2 Quartz grains from the Zillertal Alps, Austria. (A) Quartz (Qu) and amphibole (Am) in the A horizon of an Inceptisol with sharp edges and multiple fractures. (B) Quartz with distinctive glacial features including multiple fractures partly abraded and irregular grooves in the same horizon as in (A) above. (C) Quartz (Qu), albite (Alb), and orthoclase (KF) grains with sharp edges and considerably more weathered features than in (A) in the Ab horizon of a pedostratigraphic complex in the Horn Kees, Zillertal Alps, Austria. (D) Mica and quartz grains with no distinctive eolian features in the Ab horizon of a buried Spodosol, Waxeckkees, Zillertal Alps, Austria.

(Mahaney & Hancock, 1993). Usually, the distribution of REEs down profile, from the eolian "cap" into the till parent material, indicates the sediment source comprises local materials, the only evidence for an eolian origin being the texture and the partial modification of grain shape. As shown in figure 6.3, some grains retain a glacial shape (figure 6.3E), whereas others are only slightly modified with parallel ridge and trough microtopography that is only equivocally eolian. As with mountain loess, the fine sand fraction contains grains that are remarkably free of adhering particles, the frequency of which is generally high on the surfaces of glacial grains (see chapter 9). Theory would predict an increase in eolian microfeatures with increasing distance of transport, as suggested from the Rhine loess samples in figure 6.1E, F, but this has not been tested as yet. However,

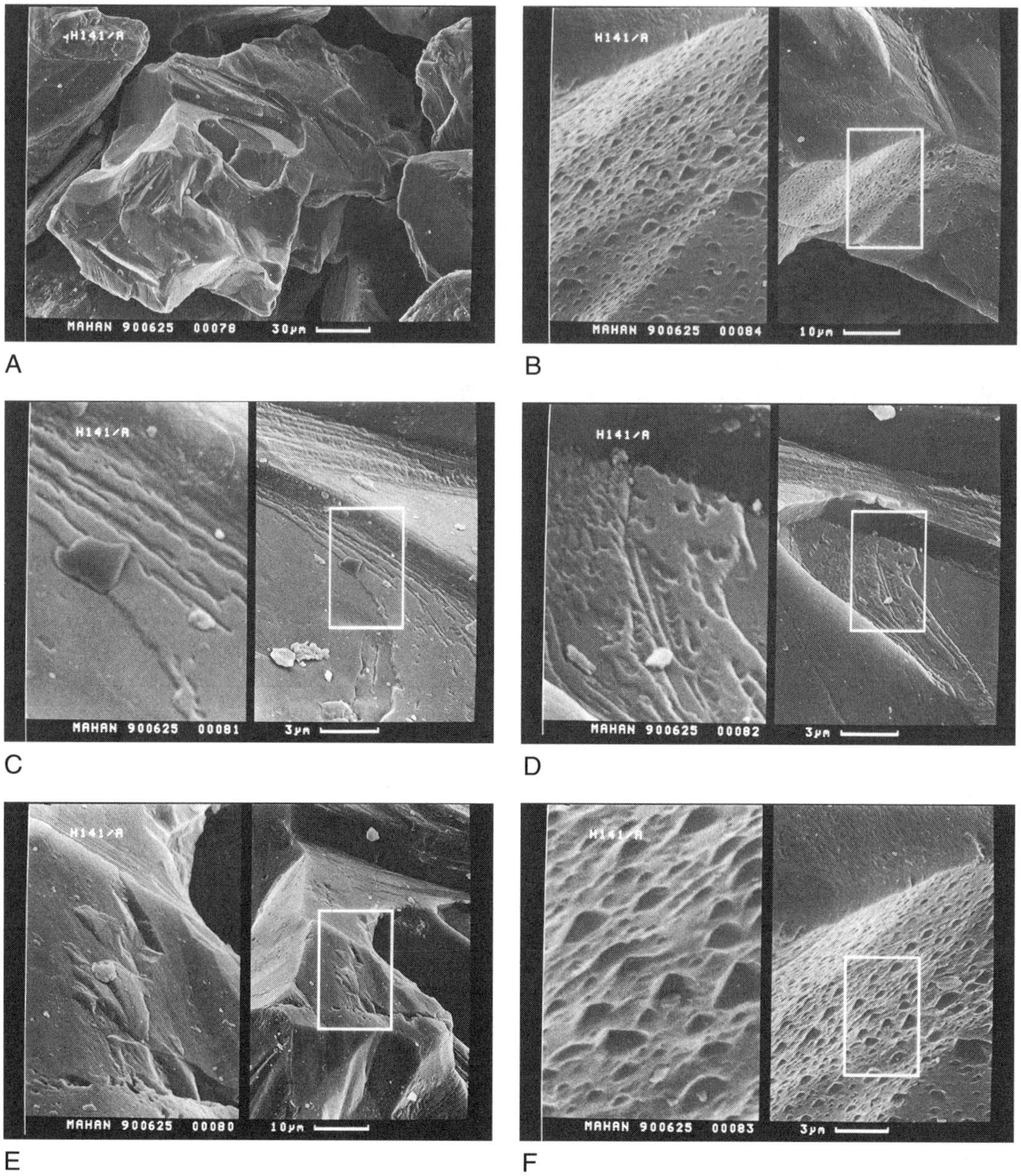

Fig. 6.3. (A) Quartz grains with sharp edges, large craters and irregular grooves in the A horizon of the R48 profile, Rouge River, Ontario. (B) New growth of quartz on a bulbous protrusion of a quartz grain. The enlargement (×5) shows ridge and trough topography that could have an eolian origin (expanded in F below). (C) Exfoliated skin of quartz with ridge and trough microfeatures in the enlargement (×5). (D) Exfoliated surface in depression on quartz. (E) Sharp-edged, abraded limbs of a quartz grain inherited from glacial action. (F) New growth of quartz on quartz contrasting with the expanded view of ridge and trough topography (upper wall).

increased eolian transport tends to produce grains with bulbous edges, parallel ridge and trough microtopography that often requires high magnification (figure 6.3F) for analysis, and clean grains with few adhering particles.

Saharan Ergs

A north–south transect of ergs in Algeria provided numerous sampling stations in the north-central Sahara Desert (Mahaney & Andres, 1996). As part of an extensive and deep sand sea, only medium to very coarse sands were studied. The fine fractions, possibly originating in the Atlas Mountains, or farther afield in other areas of the Sahara, were not studied as they might introduce records of other sedimentary environments. For the complete record of the giant erg field between the Hoggar and the Atlas Mountains, see Mahaney & Andres (1996). What follows is a brief summary of the microtexture record of quartz grains in samples from Ergs 1 and 3.

The parallel ridge and trough microtopography commonly reported on desert grains is common on less than 20 percent of grains studied from Algerian ergs. Examples are seen in figure 6.4A–C. As pointed out previously, it may be that parallel ridges are more pronounced and more frequently encountered on larger sands in the medium and coarse fractions. Mechanically upturned plates, common on less than 5 percent of the grains studied, are quite possibly removed by saltation and they may even be modified by solution and reprecipitation of Si (Margolis & Krinsley, 1971). Upturned plates are common along the parallel ridges seen in figure 6.4A.

For modern deserts, sand seas such as the Great Western Erg of Algeria, where the average wind velocities are known (Landsberg, 1972), it is possible to correlate plate spacings with wind speed. The average wind velocity for Algerian sand is 13.5 m s^{-1}, which is well within the range of wind speeds of 10–15.7 m s^{-1} for actively saltating sand grains. While it is not possible with the limited data available to accurately determine wind speed from plate spacings, future research may show that modern and ancient wind velocities can be quantified from plate spacings on quartz grains.

As pointed out previously, abrasion fatigue results from dry grinding of quartz which produces a disrupted lattice due to broken surface bonds with particles smaller than clay minerals that might collect together to form individual clusters of particles packed in troughs between ridges on grain surfaces. Such clusters are frequently encountered on quartz in Saharan ergs and presumably occur on quartz in other deserts of the world. As shown in figure 6.5, such particle clusters are reactive and are coated with Ca, Al, and Fe.

California Coastal Dunes

Eolian transport of sand along the California coast has been studied by Douglas Sherman (University of Southern California) using vertically integrated sand traps located across the beach. Traps located at different heights segregate the vertical components of saltation from 0 to 30 cm above the bed. Analysis of saltated grains

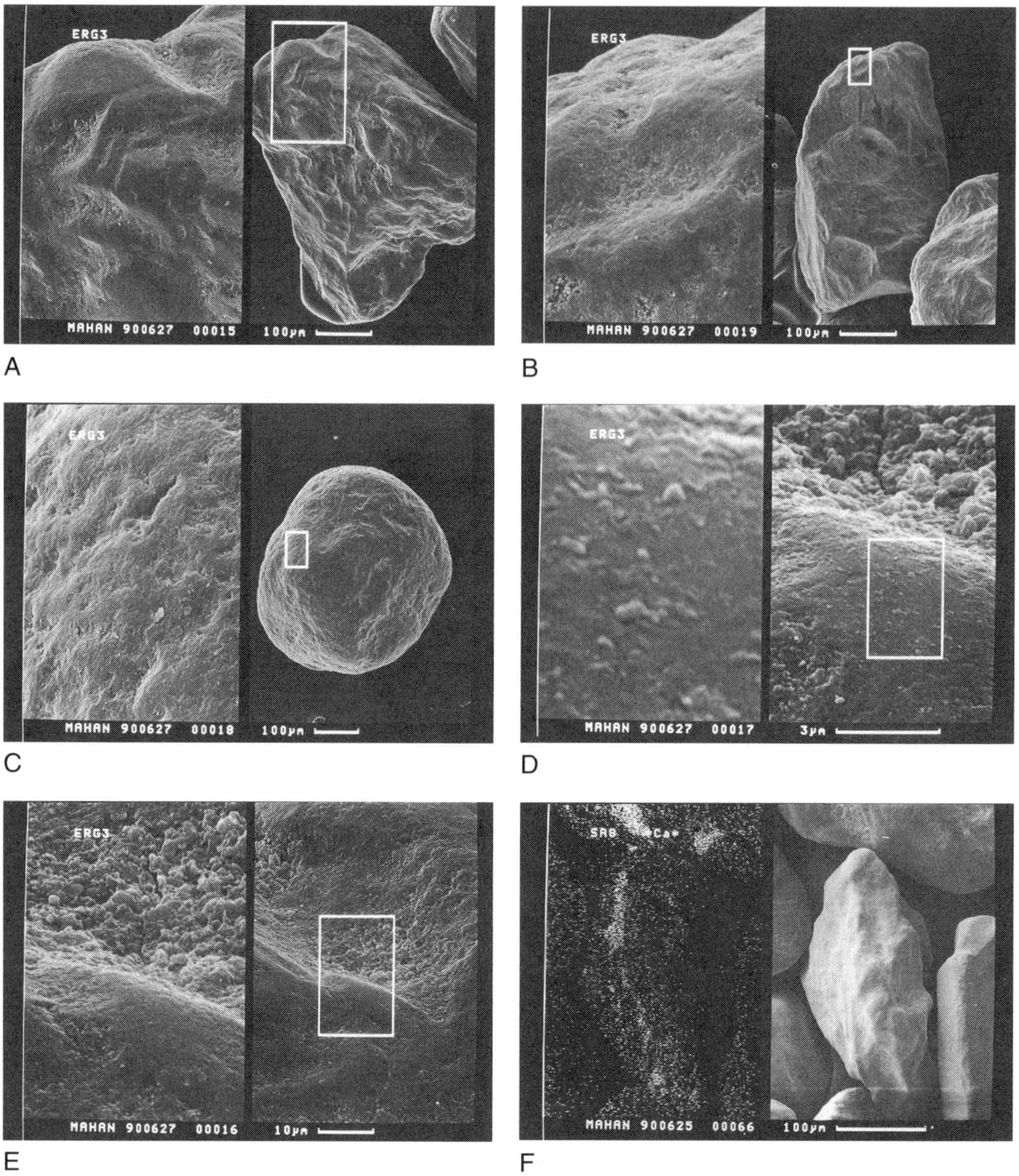

Figure 6.4 (A) Rounded but sharp-edged quartz grain showing ridge and valley microtopography with clusters of fine adhering particles shown in the enlargement (×5). (B) Atypical subangular elongated quartz grain with minor dissolution pockets and accumulations of Si flakes (abrasion fatigue) in troughs as shown in the enlargement (×10). (C) Subrounded quartz grain of low relief with bulbous edges and ridge and trough microtopography shown in the enlargement (×10). (D) Partly weathered skin of quartz with fresh area and adhering flakes of Si in enlargement (×3). (E) Lower resolution micrograph of the area shown in (D). (F) Back scattered image (left) showing distribution of a Ca coating on quartz grains.

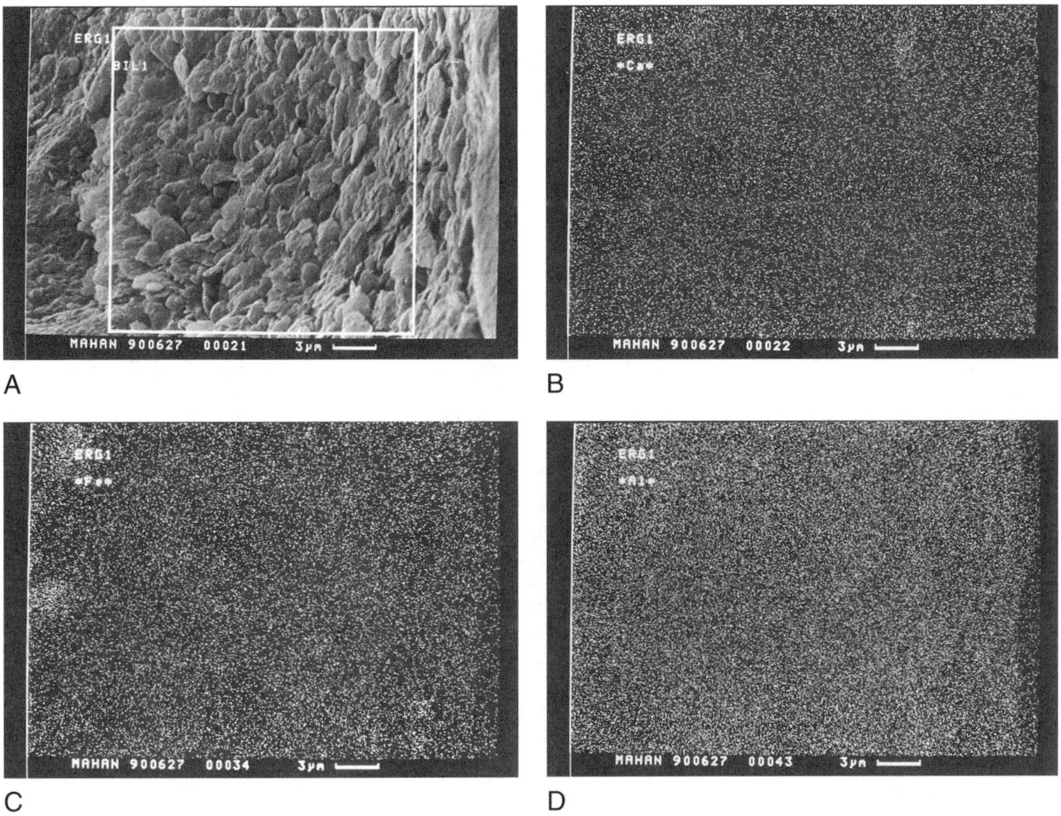

Figure 6.5 (A) Coated grains in trough on quartz. The frame depicts the area where the majority of coated grains are located. (B) Back scattered image shows Ca coating is nearly uniform across the surface. (C) An Fe coating is uniform over the area, with packed particles and it is slightly stronger to the left where fewer packed particles adjoin the grain surface. (D) Al is nearly uniform across the surface either from clay minerals or from adhering aluminosilicates.

from various traps provides a data bank of information on local eolian processes and the degree to which the sedimentary record is formed and maintained.

The frequency of occurrence data in figure 6.8 summarizes the microtextures that are common on the California coastal plain sands. Low relief dominates the suite of samples studied, along with a low percentage of fractures and a high percentage of fracture faces, the latter presumably inherited from release of source rock. Mechanically upturned plates and lattice-shattered grains are relatively common along with abrasion, etched, preweathered, and weathered microfeatures. Adhering particles are scarce as most grains are cleaned off in transport. Edge rounding, including bulbous edges, are relatively high along with overprinted grains. A complete absence of troughs, steps, and crescentic gouges (commonly found on glacial grains) serves to differentiate the eolian from glacial sedimentary environments.

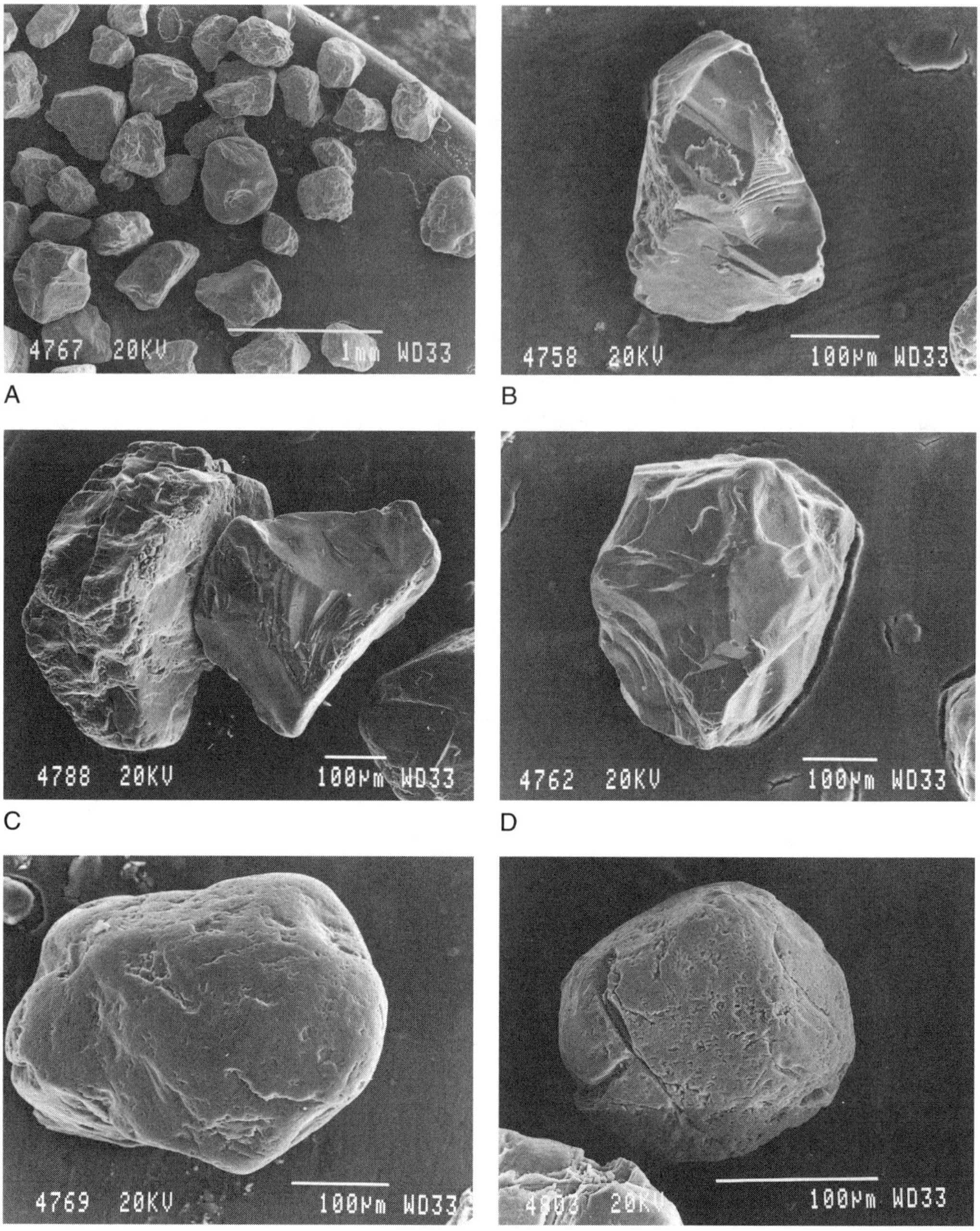

Figure 6.6 (A) Medium to fine subangular sand grains remarkably free of adhering particles. (B) Triangular-faceted grain with sharp edges and multiple fractures. (C) Subangular medium sand grains: bytownite to the left with dissolution features, and quartz to the right. (D) Subangular quartz with upturned plates to left of and below center. (E) Rounded K-feldspar with bulbous edges to the right and upper left and v-shaped percussion scars. (F) Nearly round grain of ilmenite in pristine condition with cracks and v-shaped percussion scars.

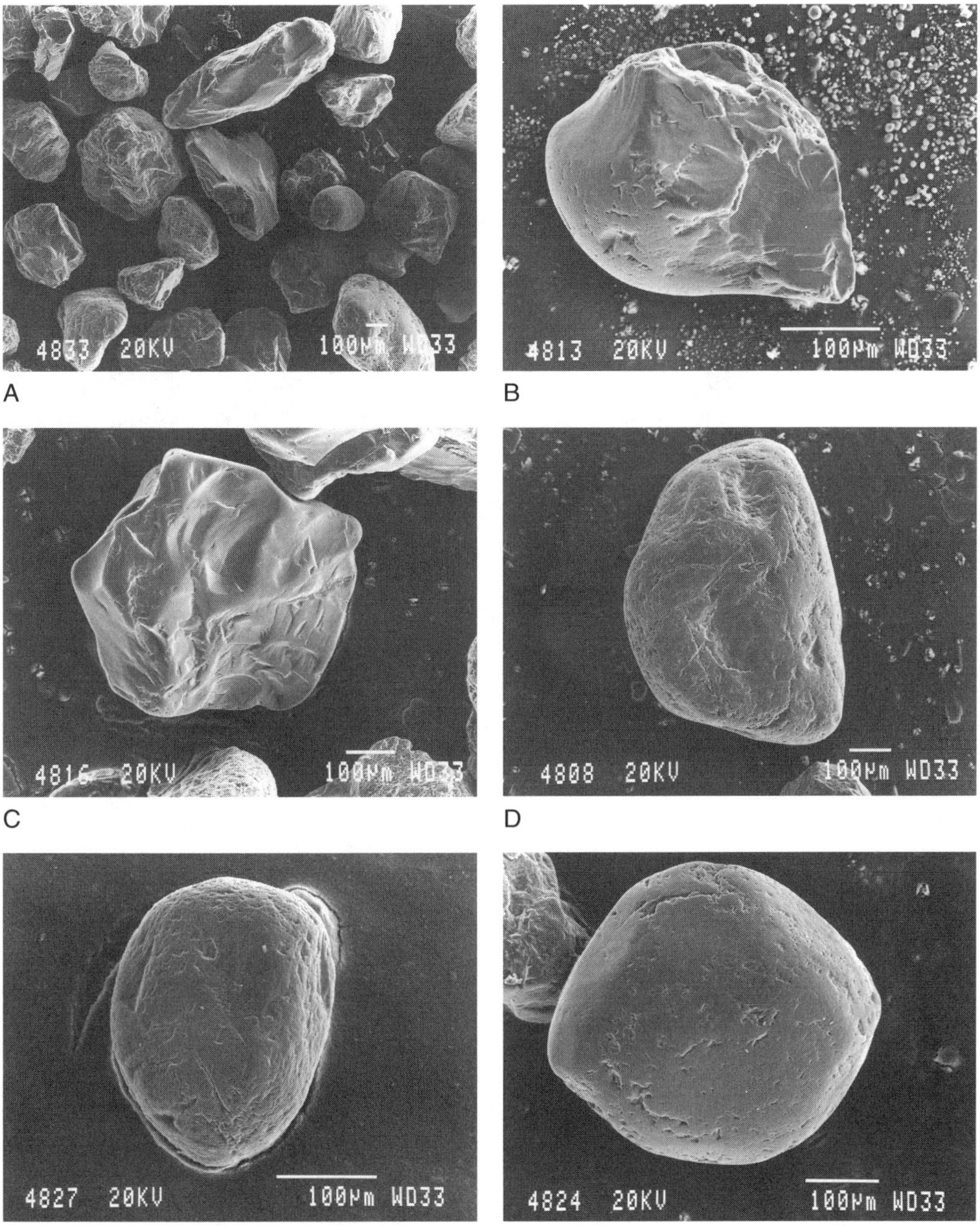

Figure 6.7 (A) General frame of mainly quartz grains in the 5 cm trap with subangular to nearly round forms. (B) Subangular quartz grain (from 30 cm trap) with minor dissolution microfeatures, extensive abrasion, v-shaped percussion scars and bulbous edge on left. (C) Subangular quartz (30 cm trap) with rounded edges and several upturned plates. (D) Subrounded bytownite grain (30 cm trap) with bulbous edge on left. (E) Round amphibole (5 cm trap) with many upturned plates, dissolution microfeatures and bulbous edges. (F) Nearly round quartz (5 cm trap) with many upturned plates and v-shaped percussion scars.

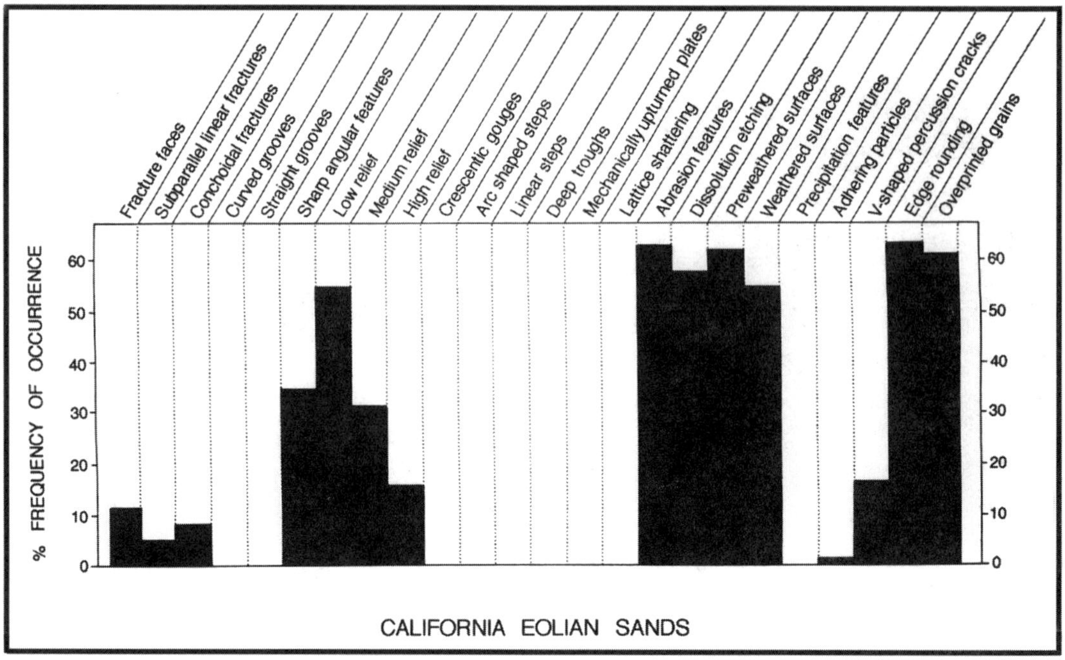

Figure 6.8 Summary record of frequency of occurrence of different microtextures on samples from the California coastal zone.

Mars Analogue

Global wind regimes are related to the pressure differentials associated with global-wide circulation patterns. In contrast, local wind anomalies are generated by physical characteristics within a specific area or region. Local wind anomalies, for example, may reflect a difference in the thermal properties of surface materials: the greatest wind activity is often observed in those environments where large temperature variations exist in the air layer immediately above the ground. Such conditions are common in deserts, along seacoasts, or in areas with pronounced diversity in elevation, such as the break in slope between a mountain range and plains-forming terrain. Wind-scoured landscapes and wind-deposited landforms are often a result of the direction, velocity, and degree of turbulence of wind among other variables. Thus, wind-related features, including sand seas, ripples, dunes, and streaks, which are visible remotely from orbital platforms (including radar) on the surfaces of Earth and Mars, are not only markers of present-day circulation patterns and local conditions, but may also archive past conditions, including environmental and topographic change.

Similar in many respects to the Earth, wind-related processes have also modified the surface of Mars. Viking images, for example, have revealed the existence of an enormous circumpolar sand sea covering an area of about 7×10^5 to 8×10^5 km^2 around the residual ice cap of the north polar region of Mars (Tsoar et al., 1979). Linear, transverse, and barchan dunes have been mapped in this sand sea (Breed et al., 1979; Tsoar et al., 1979; Mainguet & Moreau, 1982; Dial, 1984). The dune field is caused by the extreme temperature variations in the north polar region between the cooler ice cap and the warmer plains of the Vastitas Borealis Formation. During the martian summer, these temperature variations produce strong winds that exhume materials from many sources, including weathered lava, volcanic ash, and silicic dust from the Vastitas Borealis Formation and the layered polar unit (Dial, 1984; Tanaka & Scott, 1987).

Varying wind regimes are observed in the north polar region of Mars. Dominant south-west-trending winds, for example, travel through a broad (~50 km wide) curvilinear canyon, Chasma Boreale, and intersect minor counterclockwise winds from the residual north polar cap at the interface between the juncture of the mouth of the canyon and plains-forming terrain. At this juncture, a transition is seen from isolated barchans to collective barchan chains forming barchanoid ridges (Dial, 1984; Dial & Dohm, 1994), similar to transitions between dense types seen on Earth (Bagnold, 1941). In addition, at the mouth of Chasma Boreale, part of the dune field lies within a shallow depression where both linear and barchan dunes are present. The dune field extends southward from the depression onto level terrain. Here, isolated barchan dunes form barchanoid ridges on both the east and west edges of the linear dunes. Transverse dunes south of the linear dunes have low perpendicular slipfaces, indicating the collision of two wind regimes of equal strength (Tsoar et al., 1979). Bagnold (1941) suggested that barchan ridges and linear (seif) dunes are the result of a positive maximum sand supply and indicate bi-directional winds. Although Breed et al. (1979) suggested that eolian activity on Mars may have taken place early in its history, dune fields on Mars are likely archives of both ancient and recent wind regimes and potential markers of past environmental changes (Baker et al., 2000).

Martian Simulations

Photographs of the martian surface taken by the Viking Landers show distinct eolian features (Cutts, 1973; Cutts & Smith, 1973; Sagan, 1973; Krinsley & Greeley, 1986; Greeley & Williams, 1994), although no surface samples have yet been returned to Earth. Carr (1996) and Moore et al. (1977) suggest that dunes cover much of the martian surface. Light dunes are likely composed of soft material, possibly sulfates, which appear to be common. It is also possible that electrostatic aggregates (Greeley, 1979; Krinsley & Leach, 1979), originally present as sand-size particles, were later cemented within dune forms by sulfate. Greeley et al. (1999) also note the widespread presence of dune forms and suggest their light color may indicate the presence of felsic material. It is believed that the bright material on stoss slopes consists of grains small enough to be ejected into suspension via saltation impact, or of sand-sized silt–clay aggregates that are incapable of surviving saltation (Edgett & Malin, 1998). Particles smaller than sand might again aggregate and form sand, assuming that low moisture levels allow electrostatic adhesion.

In an attempt to produce eolian textures on sand grains representative of the high wind velocities and low atmospheric pressures characteristic of the martian surface, experiments were performed with the so-called Mars Erosion Device (MED). This is a paddle wheel driven by a variable speed motor that slings particles at a given rate against each other and the sides of the vessel, and self-abrades the grains (Krinsley et al., 1979). The inner surface of the MED was coated with latex to minimize particle-wall collisions. The device was operated at normal pressures and at a pressure of a few millibars (the global average is estimated to be 5.6 mbar; Kieffer et al., 1992), simulating martian pressures (Krinsley & Leach, 1981).

Samples of fresh, crushed Brazilian quartz (175–850 μm) were examined and photographed with SEM in the secondary mode. Subsamples of quartz were abraded at 8 m s^{-1} under Earth (1 bar) and martian (4 and 5.3 mbar) atmospheric pressures for periods of 20 min, 1 h, 4 h, and 8 h. Fresh, angular grains contained a series of irregular fractures. During eolian abrasion at both high and low atmospheric pressures, mechanical action started to reform grain edges and with time worked its way over entire surfaces, thus rounding the grains, as reviewed by Greeley et al. (1984). The textures produced are similar to those on quartz grains from hot deserts (Krinsley et al., 1976). However, under the low atmospheric surface pressure of Mars, the process appears to operate more rapidly. Thus, grains of the same size run at low pressures are more abraded than those at high pressures, all other factors including velocity being equal.

Other observations include the following:

1. After extended abrasion, the larger grains (at both pressures) were similar to modern eolian dune sands in terms of roundness and surface texture. When Earth and martian grains are compared, the only major discernible difference is a lack of solution and precipitation on the latter. However, it should be noted that Krinsley & Leach (1981) indicated some solution and reprecipitation in their experimental eolian device.
2. The larger grains were more rounded after 3 h than the smaller ones after 8 h at both high and low atmospheric pressures, probably because of the increased momentum associated with the larger sizes.
3. Spacing of "upturned plates" on small grains abraded at low pressures appears to be closer than on those abraded at high pressures. This may be due to the greater number of collisions and greater momentum generated at low pressures as a consequence of the greater speed. Just as Krinsley & Wellendorf (1980) showed that plate spacing is related to wind velocity in Earth deserts, it would be interesting to determine whether pressure and wind velocity on Mars could be related to plate spacing when martian samples are returned to Earth.

Mars Return Mission

Later in this decade, surface material on Mars is likely to be collected and returned to Earth by a robotic lander or rover. Because of its chemical stability, quartz is expected to dominate on the martian surface, with sand-sized material likely enriched in quartz. Thorough mixing of surface sediments by eolian processes should insure a representative sample of the martian crust (Schieber et al., 2002), providing information on source rocks and transport history.

With the continual refinement and miniaturization of SEM–EDS equipment, it may well prove possible to conduct a remote analysis of samples collected on the martian surface. As pointed out by Schieber et al. (2002), remote analysis of samples with SEM–CL and/or SEM–EDS is possible with a light-weight, compact column mounted on a rover. Producing a vacuum in the low-pressure environment of the martian surface would present less of a problem than on Earth. Until the first sample return to Earth is possible, remote equipment will likely provide the first of martian sediments.

7 Fluvial Grains

Fluvial systems vary from slow, sluggish streams with a lower flow regime, and fast-moving upper flow regime rivers with high discharge, to flood-release jökulhlaup-type discharges capable of moving boulder-size clasts. Streams may have light to heavy sediment loads which, broadly speaking, depend on the lithology of the source rocks (weathering limitations) and discharge factors governed by channel cross-section and volume of water (transport limitations). When streams become choked with sediment, they merge with debris and mudflows, and eventually, with greater viscosity and reduced turbulence, it becomes less likely that transported grains will be mechanically rounded or suffer fractures.

As with glaciers, streams cause continuous particle abrasion in free turbulent suspension, especially when grains are ground together, or when, in low-velocity collisions, kinetic energy is released into the water instead of the grains. With high-speed direct collisions, vibrational energy, released through the grains, results in the familiar v-shaped percussion crack or v-shaped wedge-like microfeature. This is the most diagnostic characteristic found on quartz and other mineral grains that have been transported by water.

In theory, the degree of abrasion and edge rounding, and the frequency of v-shaped percussion cracks, increases with stream discharge due to increased stream competence, sediment concentration, and particle velocities. The microtextural analysis is frequently complicated when it is impossible to determine the source area and/or the distance of transport for the grains under investigation. Most of the author's experience deals with previously glaciated environments, where sediments carry a glacial signature as well as a fluvial one. However, as will be seen later on in this chapter, there are instances where discharge through breached moraines onto outwash fans allows, for example, the investigation of grains with known transport distances from glacial margins. Although such grains may have been affected by water during their transport through a glacier, it is possible to assess this by reference to samples recovered from the breached moraine. Such a study was carried out by Jackson (in Mahaney & Kalm, 1996) in the Pedregal Fan in the northern Venezuelan Andes. Reference will be made to this study and other allied investigations later in the chapter.

In many tropical mountain areas (Mahaney, 1990a; Mahaney et al., 2000a,b), warm-based glaciers were melting for most of the year, even when they were at their maximum Pleistocene limits. In some areas where meltwater release was accompanied by a heavy load of coarse clastic debris, rock flutings or Nye channels were produced, sometimes with the most exquisite deep flutings carved in crystalline rocks. The catastrophic release of meltwater in several catchments provides an opportunity to investigate the most extreme effects of fluvial action and upper flow regime turbulence on the production of microtextures. The Teleki Valley and Burguret catchments on Mount Kenya (Mahaney, 1990a), La Canoa Catchment of the western Venezuelan cordillera (Mahaney et al., 2000a), breached moraines of the Mucuchache, in the eastern cordillera of the Venezuelan Andes (Mahaney & Kalm, 1996), and the massive flutings beneath the ancestral Humboldt Glacier in the eastern Venezuelan cordillera (Mahaney et al., 2000b), in particular, have yielded material suitable for testing by SEM. Before discussing the evidence for catastrophic flow and resultant microtextures at these sites, it is necessary to look at lower flow regime systems using material from a catchment in southern Ontario where glacial and glaciolacustrine grains are being reworked after having been emplaced by the Laurentide Glacier during different stages of the Wisconsinan Glaciation.

Lower Flow Regime Grains

Rouge River

The Rouge River drains an area of 520 km^2 extending from the Oak Ridges interlobate moraine southward to Lake Ontario (Mahaney & Sanmugadas, 1986). Like many small rivers in southern Ontario, the Rouge has incised the till plain north of Lake Ontario and exposed considerable thicknesses of glacial, glaciofluvial and glaciolacustrine deposits of Wisconsinan age. Elevations in the catchment range from 75 m a.s.l. at Lake Ontario to ~300 m a.s.l. in the headwaters. The stream that initially carved the valley was most likely at the low end of an upper flow regime, perhaps capable of moving large pebble (32–64 mm) size clasts at maximum velocity. However, the present stream is underfit and best classified as a low flow regime subject to peak flows and occasional flooding during spring melt. Most floods are associated with ice dams and the diversion and release of impounded water with discharge rates sufficient to move sand. Normally, however, the suspended load is fine silt and clay.

Around 120 samples of fluvial and glacial sediment from throughout the Rouge Catchment were assessed to determine whether differences in flow regime and the effect of higher and lower discharge could be detected on sand grain morphometry. Because glacial grains in the catchment originated below the névé zone in the Laurentide Ice Sheet, and in some cases may have been subject to turbulent subglacial and englacial discharge regimes, v-shaped percussion scars and edge rounding features found on sands in the floodplain, and in flights of terraces linked to meltwater channels carved into the till plain

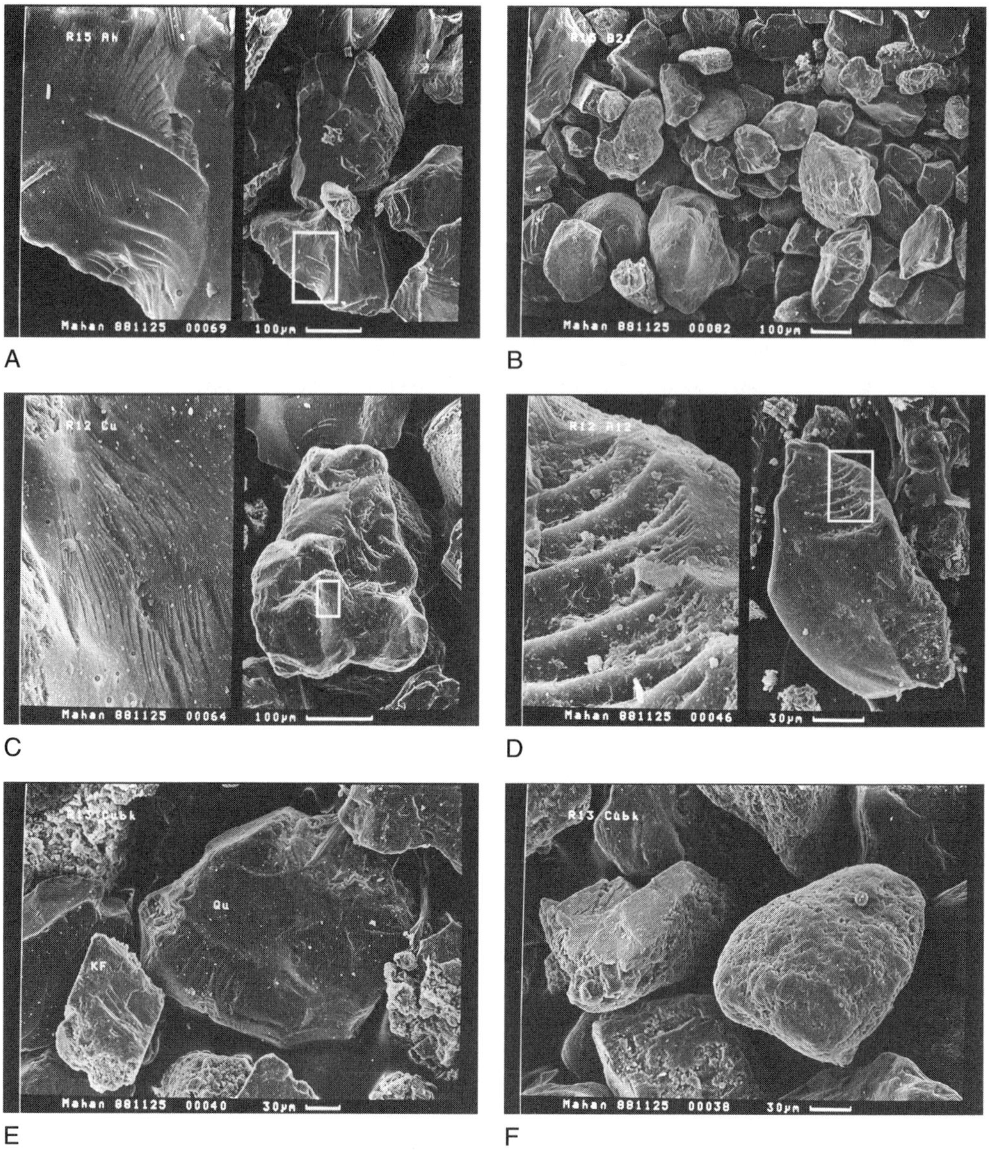

Figure 7.1 (A) Representative angular grains of quartz and orthoclase, showing minor dissolution, in the A horizon of an early Holocene/late Pleistocene paleosol in the Rouge Catchment, southern Ontario. The parent material is eolian mixed with stream sediment. The enlargement (×5) shows deeply embedded relict glacial microfeatures. (B) In contrast with (A), this is representative mix of angular and rounded quartz grains in the B horizon of an early Holocene/late Pleistocene paleosol in the Rouge Catchment. The parent material is lag gravel. (C) Quartz with bulbous edges in a middle Holocene lag gravel. The enlargement (×10) shows subparallel elongated ridges and depressions somewhat abraded with nearly circular dissolution features younger than the fracture pattern. (D) Long angular quartz grain with extensive abrasion, sharp edges, some adhering particles and fresh fractures at top (enlargement ×5). (E) Subangular quartz (Qu) and K-feldspar (KF) with sharp edges in the parent material of a floodplain soil. The weathered grains are a mix of calcite and orthoclase. (F) Nearly elongated grain of quartz (center) with minor dissolution surrounded by calcite and plagioclase, all subrounded to subangular with minor to severe dissolution features.

north of Lake Ontario were compared with sands from till outcrops (figure 7.4).

There is considerable abrasion on 30–50 percent of the grains present in figure 7.1A–F; however, most grains are subangular to angular with microfeatures that may have originated from glacial grinding and/or meltwater transport within the ice. While there is some dissolution due to pedogenic processes (figure 7.1A), about 30 percent of the grains show edge rounding, as a result of water transport. Grains are fresh with little modification, and, with the exception of R13 (figure 7.1E, F), contain no carbonate coatings. Despite greater weathering in the B horizon than in the A, about 25 percent of grains show v-shaped percussion cracks from water transport (figure 7.1B). As shown in figure 7.1C, many elongated grains are present, possibly from percussion and fragmentation in the lag gravel, but substantially similar to sands in till. These carry subparallel fracture planes shown to be disrupted in the enlargement (×10), possibly as a result of turbulent water transport. Grains above and below the quartz grain carry thin carbonate coatings. Both edges of the quartz are abraded, possibly by stream action. As with the older grains in R15, this grain is little affected by stream transport and retains a fracture history from transport in the Laurentide ice.

The bulbous edges on the quartz shown in figure 7.1C indicate it could have an eolian history. The enlargement shown here contains subparallel fractures somewhat abraded with nearly circular dissolution features younger than the fracture pattern. Curved deep troughs (right of center) are directional features considered to be glacial in origin. This is a complex grain with a multicyclic history: eolian → glacial → weathering (diagenesis?) → fluvial. The angular quartz grain in figure 7.1D, with extensive abrasion, sharp edges and few fresh fractures, contains no fluvial signature despite its location in the A horizon of a soil formed in a stream lag gravel.

In the younger floodplain deposit (figure 7.1E), matching quartz and K-feldspar grains show lighter areas produced by charging on carbonate coatings. Both equant grains show little effect of fluvial transport and emplacement as overbank sediment. The quartz grain in figure 7.1F (center) contains rounded edges and dissolution features, whereas the grain to the left (orthoclase) is partly abraded and coated with a thin carbonate precipitate. Because the age of this younger horizon is 360 ± 60 yr BP (see Mahaney et al., 1989), all grains in it are considered to be reworked from nearby till and/or lacustrine deposits.

Studying the effect of fluvial transport in a glaciated catchment is subject to problems and pitfalls, but it is still possible to compare grain shape characteristics as a rough guide to the overall effect of lower flow regime transport in alluvial systems. To fully assess the impact of variable flow regimes it is necessary to find a way to analyze the effect of glacial transport, or to move outside the margin of glaciation and analyze grains derived directly from source rock with no glacial influence. With regard to the former it is possible to compare SEM imagery of the fluvial grains with grains recovered from till (see figure 7.5B–E). In this instance (figure 7.1F) it is apparent that lower flow fluvial regimes move grains over short distances at low velocity with little or no change in grain morphometry.

Upper Flow Regime Grains

Laurentide Ice Sheet

Within tills deposited by the Laurentide Ice Sheet, sand beds of variable thickness contain quartz and other mineral grains that have proven to be recorders of turbulent water transport. Water-transported grains found in tills within Wellsch Valley, Saskatchewan (as shown in figure 7.2B, C), exhibit rounded edges and v-shaped percussion cracks. Moraines constructed by the last ice sheet in North Dakota contain fluvial facies with rounded grains (figure 7.2E) and till facies (figure 7.2D) combined. Correlative grains become slightly rounded with water transport as shown by esker material 1.5 km from the distal flank of the Soo Moraine (figure 7.2F).

As shown in figure 7.3, quartz grains within the esker system retain much of their angularity, but contain a number of large v-shaped percussion features thought to be the product of upper regime flow, or possibly critical flow within a jökulhlaup regime. Transport of grains within many eskerine systems probably does little to round sands, but produces v-shaped percussion cracks as evidence of grain-to-grain contact. There is an overall similarity between grains in the Rouge River Catchment inherited from Laurentide tills and the eskerine grains shown in figure 7.3.

Systematic sampling of deposits in the R48 section (figure 7.4) of the Rouge River, a section with lag gravel at the surface overlying Halton Till (Upper Wisconsinan), sands of the Thorncliffe Formation (Middle Wisconsinan), and Sunnybrook Till (Lower Wisconsinan), produced the imagery in figure 7.5. Grains recovered from the lag gravel at the top of the section (figure 7.5A) are rather similar in shape, relief, and fracture/abrasion record to those from Halton Till (figure 7.5B–E) and Sunnybrook Till (figure 7.6B–D), and different from the glaciolacustrine grains shown in figures 7.5F and 7.6A. The greater degree of edge rounding shown in the lower Halton Till (figure 7.5E), adjacent to the Upper Thorncliffe beds (figure 7.5F), is probably due to reworking of the glaciolacustrine material. As with the Rouge River Catchment samples, stream action seems to have contributed little to reshaping grains inherited from till, whereas glaciolacustrine action has modified many grains by edge rounding, and in some cases complete reformation into rounded shapes has occurred (see figure 7.6A). Close examination of grains from Sunnybrook Till show the triangular-shaped forms that mostly result from grain-to-grain contact within the ice, but the v-shaped percussion scars in figure 7.6C bear witness to water transport. A considerable number of glacial grains undergo water transport before leaving the glacial system.

Siberian Ice Sheet

Outwash from the Siberian Ice Sheet at Ice Mountain on the Lena River was analyzed to determine the effect of high-energy water transport on quartz originally transported by ice. As expected, some grains are little affected by water transport, whereas others are well rounded after being transported approximately 15 km from the glacier terminus. About 60 percent of grains

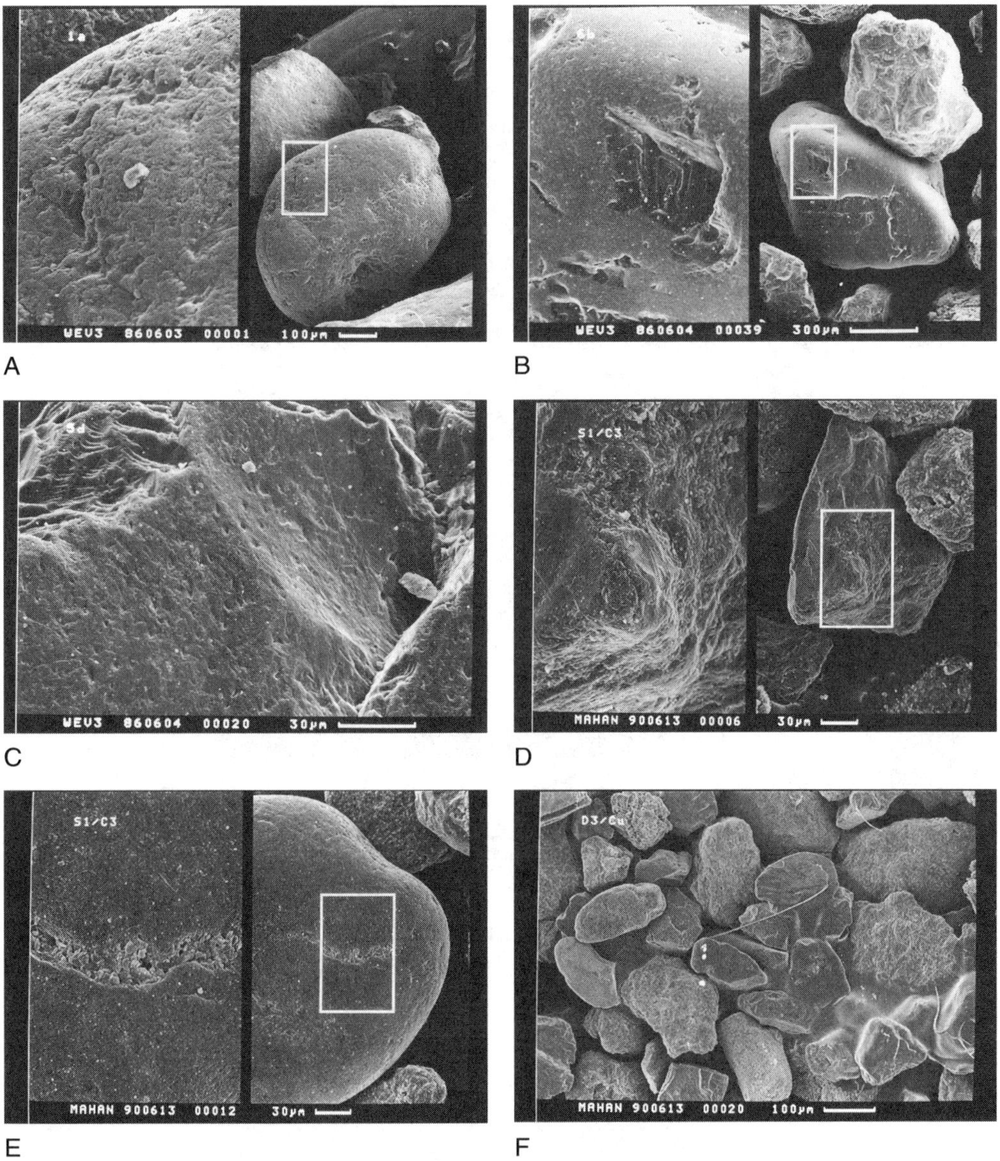

Figure 7.2 (A) Two subrounded quartz grains with numerous v-shaped percussion cracks in Till 1, Wellsch Valley, WEV3 Section. (B) Quartz with rounded edges and v-shaped percussion cracks in Till 6, Wellsch Valley, WEV3 Section (see Mahaney & Stalker, 1988, for stratigraphic details). (C) Enlargement of quartz grain showing a multitude of v-shaped percussion scars in Till 5, Wellsch Valley, WEV3 Section (see Mahaney & Stalker, 1988, for stratigraphic details). (D) Angular quartz in the Soo Moraine, North Dakota, showing fractures and adhering particles. (E) Rounded quartz, with percussion cracks and silt particles packed into an irregular-shaped trough, Soo Moraine, North Dakota. Details are shown in the enlargement (×3). (F) Mixed angular and subangular grains of quartz and orthoclase from the Dahlen Esker, 1.5 km from the distal flank of the Soo Moraine, North Dakota.

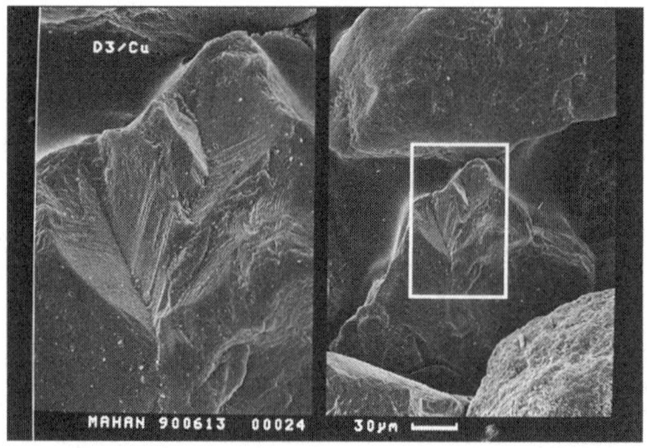

A

B

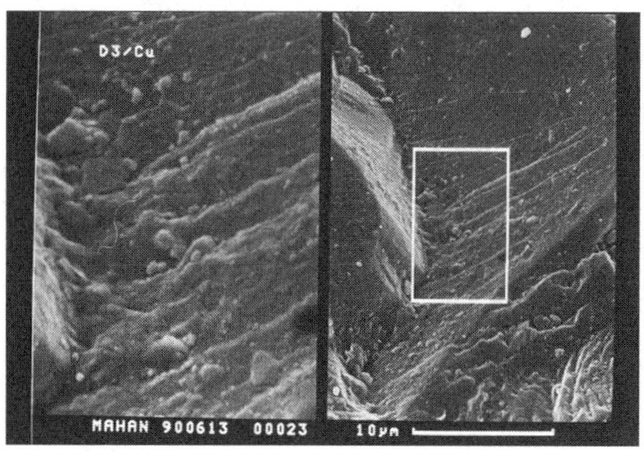

C

Figure 7.3 Upper flow regime quartz grains from the Dahlen Esker, North Dakota. (A) Angular to subrounded grains of quartz (center and top) with large v-shaped fractures. (B) Multiple v-shaped craters on quartz. (C) Enlargement of v-shaped fractures in quartz.

maintain an angular shape as shown in figure 7.7A, whereas 40 percent develop subrounded forms as in figure 7.7B, C. The radial fracture pattern shown in figure 7.7A is similar to microtextures seen on grains from upper flow regime outwash and alluvium in other areas. The impact producing the

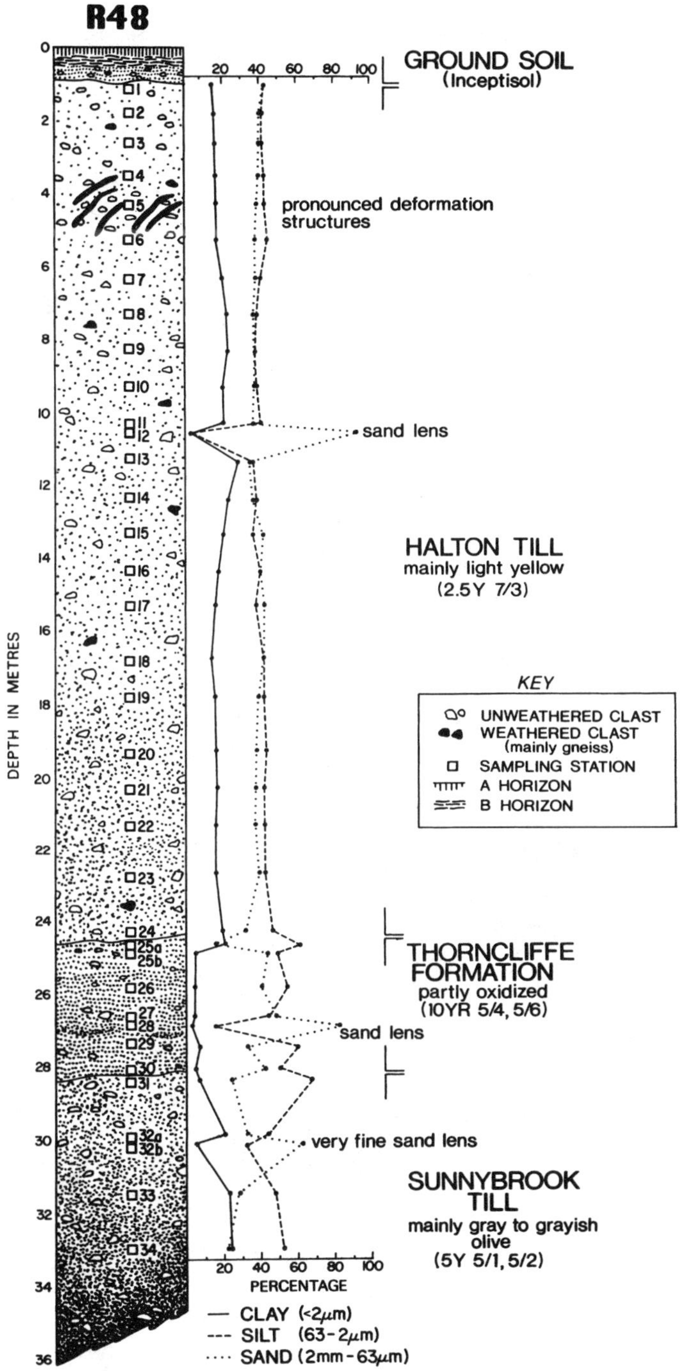

Figure 7.4 Stratigraphy at the R48 Section, southern Ontario. The soil at the top of the section is formed in lag gravel deposited in the ancestral Rouge River following retreat of the Ontario Lobe of the Laurentide Glacier. Samples 1 to 24 are in Halton Till; 25a to 30 in lacustrine sediment (Thorncliffe Formation); and 31 to 34 in Sunnybrook Till (early Wisconsinan age).

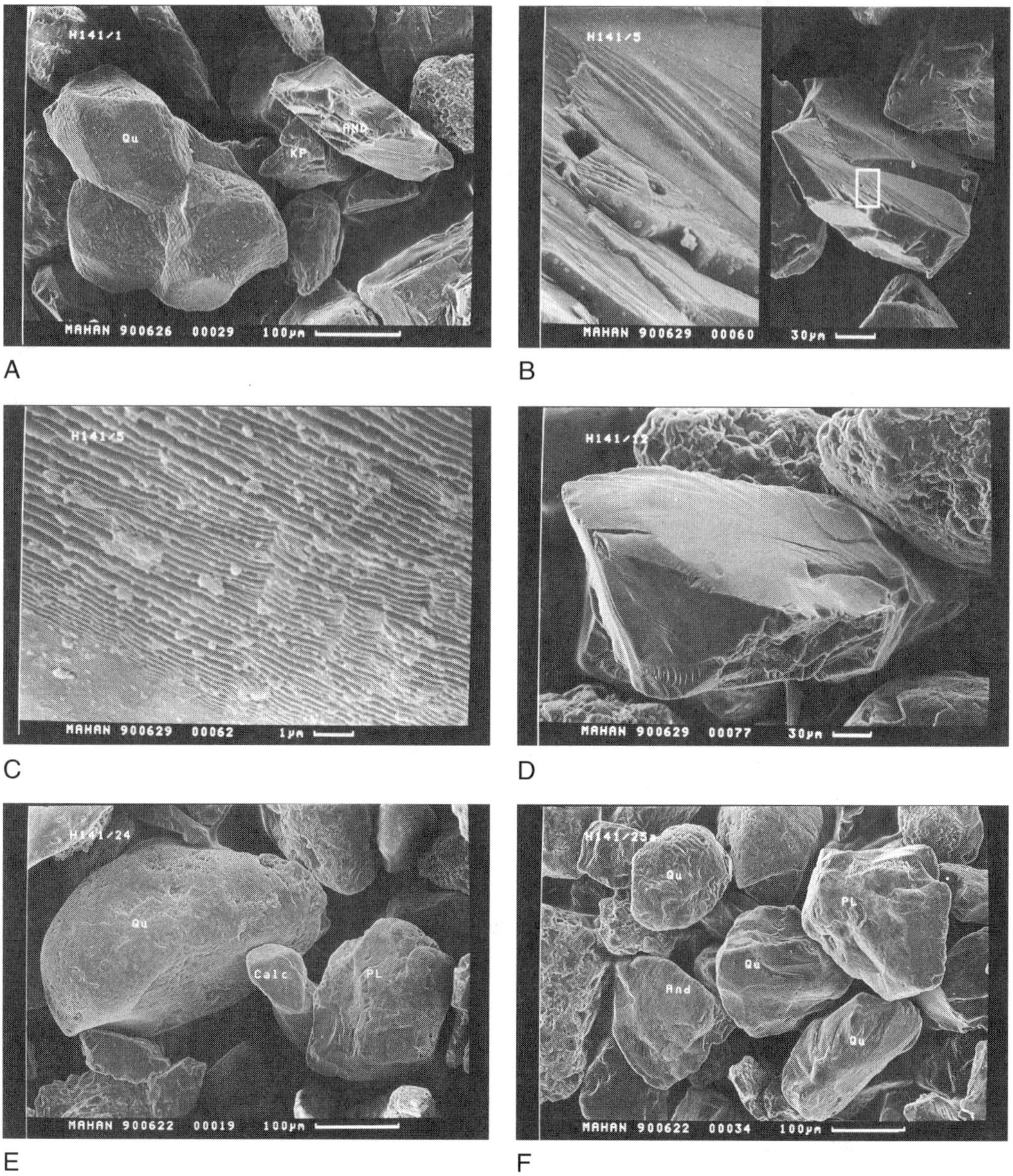

Figure 7.5 (A) Mixed grains at the top (sample 1) of the Halton Till Sheet showing subangular quartz (Qu) with bulbous edges, and subangular–subrounded grains of orthoclase (KF) and andesine (AND). The andesine grain is well crushed. (B) Quartz with upturned plates, sharp edges and multiple fractures in sample 5 (see figure 7.4). (C) Enlargement of the subparallel linear fractures in B. (D) Well-abraded and fractured quartz in sample 12 (see figure 7.4). (E) Rounded mixed grains at the base of the Halton Till Sheet. The quartz (Qu) and plagioclase (PL) grains carry multiple v-shaped percussion fractures indicating water transport. (F) Edge-rounded grains of mixed mineralogy including quartz (Qu), andesine (And), and plagioclase (PL) in sample 25a (see figure 7.4).

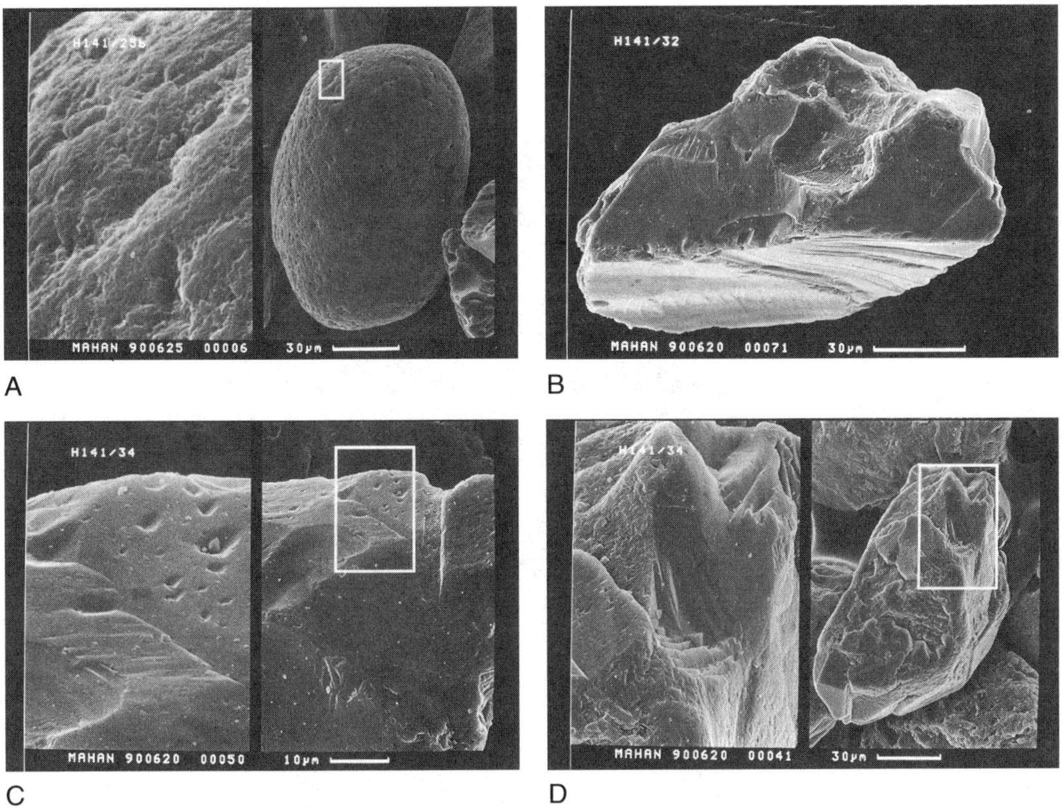

Figure 7.6 (A) Rounded quartz with multiple v-shaped percussion fractures in sample 25b (lacustrine) (see figure 7.4) (enlargement ×10). (B) Fractured and severely abraded quartz in sample station 32a (till) (see figure 7.4). (C) Extensively abraded quartz in the Sunnybrook Till of early Wisconsinan age showing multiple v-shaped percussion fractures (sample 34, figure 7.4, (enlargement ×3). (D) Quartz with multiple upturned plates and a deep cavernous trough with multiple deeply embedded striations and v-shaped percussion scars (sample 34, figure 7.4) (enlargement ×3).

radial fracture pattern could result from stick–slip movement in a glacier or from water transport in a high-energy stream. Numerous v-shaped percussion cracks are present on the bulbous edges of the subrounded grains, suggesting an eolian origin followed by water transport. The roundness results from water transport.

Venezuelan Andes

Middle Pleistocene sands and gravels at the base of the RF3 Section (Mahaney et al., 2000a) in the western cordillera of the Mérida Andes attest to high-velocity discharge from a pre-Wisconsinan glacier prior to the last interglaciation (see Mahaney & Kalm, 1996, for stratigraphic details). Found with inclusions of

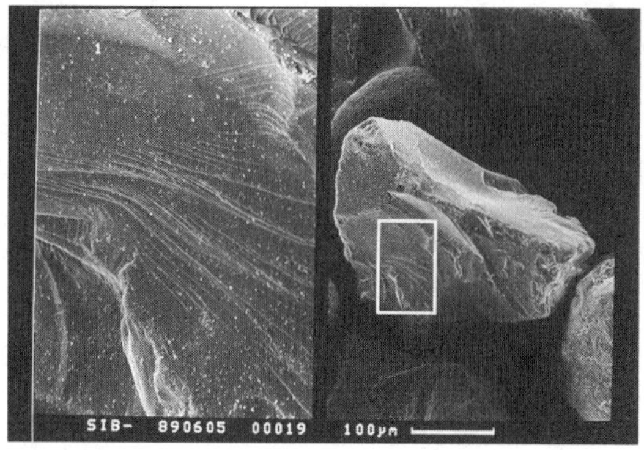

A

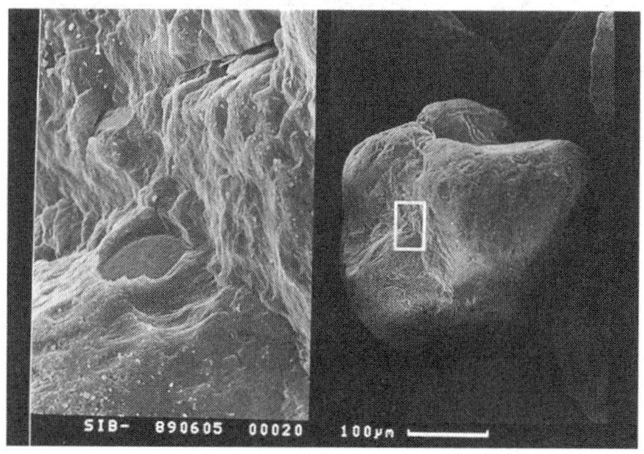

B

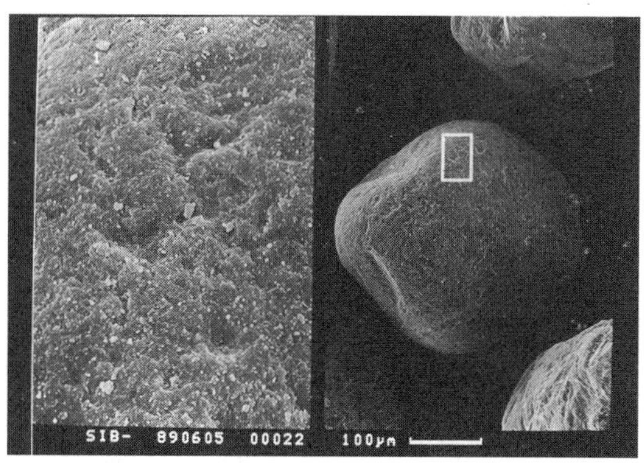

C

Figure 7.7 (A) Angular, well-crushed quartz from outwash showing radial fractures emanating from an impact point (enlargement ×10). (B) Quartz with bulbous edges, percussion cracks on edges and a small cavern filled with imbedded fine-grained material (enlargement ×10). (C) Subrounded quartz grain with Si precipitates (enlargement ×10), one bulbous edge and multiple v-shaped percussion cracks.

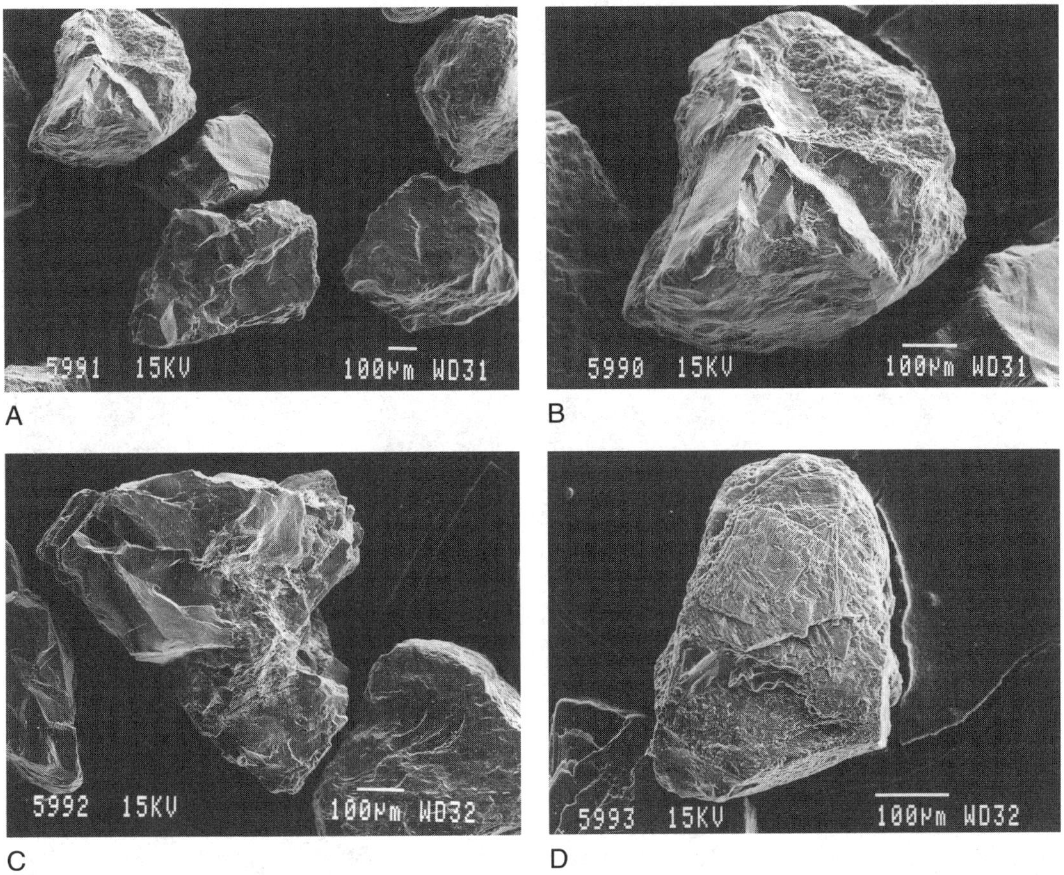

Figure 7.8 Middle Pleistocene outwash sands, RF3 Section (Mahaney & Kalm, 1996; Mahaney et al., 2000a). (A) Angular to subangular sands showing abraded fractures with deep craters and v-shapes ranging from 100 to 25 µm in depth. (B) Enlargement of abraded fractures on grain shown in A with deep v-shaped cracks. Left facet of grain shows smooth fracture face probably indicating a split grain. (C) Deep v-shaped craters that could be percussion cracks indicating high-velocity discharge on several grains. The grain at the top has undergone lattice failure from grain-to-grain collision. (D) Sheet rupturing and exfoliation, with upturned plates possibly from high-velocity collisions.

middle Pleistocene till, these well-sorted glacitectonized sand to fine pebbly fluvial beds are indicative of high-velocity discharge. Froude numbers are estimated to have been >0.6 in some beds, and grains taken from these beds show severe tensional stress culminating in the tearing of the surface skin and loss of chunk-like masses leading to the production of unusually large v-shaped percussion cracks. As shown in figure 7.8, these microfeatures occur throughout the suite of samples studied and they are dramatically different from the glacial grains discussed in chapter 9.

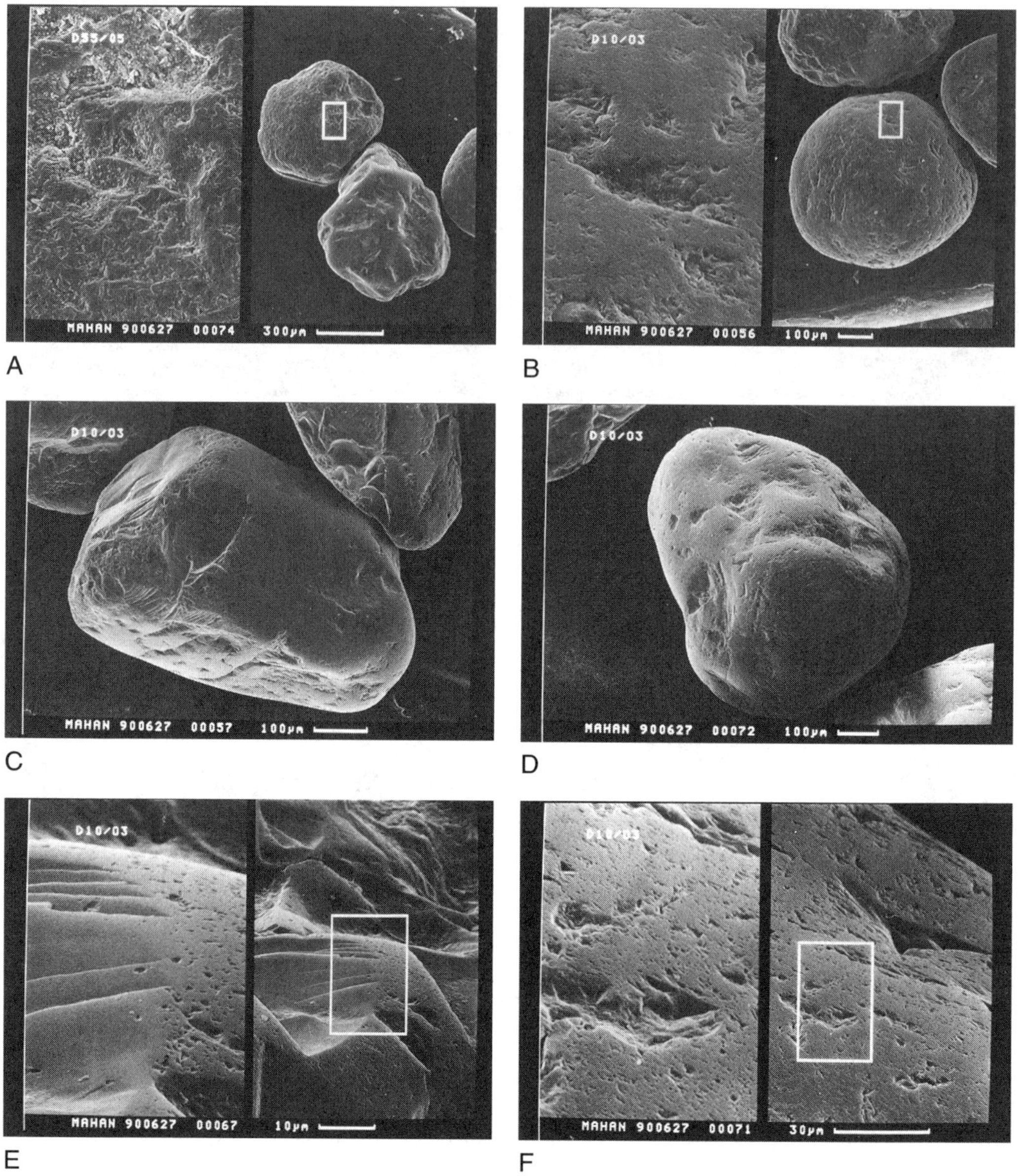

Figure 7.9 (A) Subrounded and rounded grains from the Nile Delta with multiple v-shaped percussion cracks and fine particles in depressions with details in the enlargement (×10). (B) Rounded grains in the Nile Delta sediment with multiple v-shaped percussion cracks. The enlargement (×10) shows details of the percussion cracks. (C) Subrounded anhedral quartz grain from the Nile Delta with v-shaped percussion cracks and some fractures partly abraded. (D) Subrounded grain with bulbous edges at the bottom and top and multiple small and large percussion cracks. (E) Quartz with surface showing radial fractures and multiple v-shaped percussion cracks on the leading edge of the fracture zone. The enlargement is ×3. (F) Enlargement (×3) of small and large v-shaped percussion cracks. The larger cracks have developed into small fracture zones.

Nile Delta

Quartz sand grains in the Nile Delta range from subrounded to rounded, and contain minor linear and conchoidal fracture patterns, as well as a multitude of v-shaped percussion cracks (figure 7.9). All grains are abraded to some degree and many show abraded fracture patterns suggesting they might have blown in from surrounding desert environments before being abraded during water transport. The high degree of rounding and great frequency of v-shaped percussion cracks on quartz illustrates the effect of turbulent water transport over long distances.

Conclusions

The most diagnostic microfeature of the fluvial process is the v-shaped percussion fracture. Well preserved in quartz, percussion scars are also found on other grains, including heavy minerals. The higher the flow regime, the greater the frequency with which this microtexture appears on a random sample of grains. It is also present on glacial grains, especially grains emplaced by warm-based glaciers, where water is available within and at the base of the ice, and moves material at high velocity (Mahaney & Kalm, 1996).

In upper flow regime systems, including flood bursts from breached moraines, coarse-grained outwash where velocities are high, and eskers, v-shaped percussion cracks become excessively large and frequent. The presence of these microfeatures is taken as evidence of near-catastrophic flow; however, much more work needs to be done with respect to correlating microtextures with increasing stream velocity.

Abrasion, seen primarily as a rounding factor with quartz, is also important in wearing fractures down to a plane surface, and in some cases obliterating them. The available evidence indicates that upper flow regimes produce greater degrees of abrasion and rounding than middle or lower flow regimes and tend to obscure any previous microtexture signature, especially over long transport distances. From the data available it appears that low flow regimes do not appreciably change the shape of grains inherited from till outcrops.

8 Mass Wasted Grains

All forms of mass wasted deposits are among the most poorly studied materials with regard to surface textures. From colluvium (Bull et al., 1987) and rockfall debris (Mahaney et al., 1991), which have received some attention, to landslides, avalanche debris, talus, and solifluction, which have received none, little information is available concerning microtextures on gravitationally transported materials. Mass wasted landforms either have a bedrock or constructional landform source; the transport mechanism may be either fast or slow. Velocity ranges from millimeters per year to several hundred meters per second, and one might therefore expect either minor abrasion (slow process) or impact microstructures and intense abrasion (fast process). As shown by both Bull et al. (1987) in South Africa, and Mahaney et al. (1991) in the Caucasus Mountains, the source rock may be highly chemically altered prior to movement.

From the limited data available it seems that bedrock weathering may produce grains with minor surficial fractures (Brown, 1973) and clean fracture faces (Mahaney et al., 1991; Molén, 1992). On a theoretical basis there should be no abrasion microfeatures present on grains released from source rock unless they were inherited from some previous transport episode in the rock cycle. With this scant baseline data bank in mind, I set out to investigate grains from supraglacial debris originating as rockfall in the Caucasus Mountains, landslide debris in catchments in southern Ontario, and talus from the northern Andes of Venezuela, the coast mountains of Oregon, and coastal cliffs along the St Lawrence River in Cape Breton, Nova Scotia. The objective was to tally the sum total of microtextures observed on selected representative grains in these deposits, separate source rock from active landform matrix material, and compare the range of microtextures with other geologic transport agents. And, of course, there was always the prospect of finding new microtextures that could be genetically linked with gravitational transport.

Rouge River Catchment

The Rouge River Catchment, one of the last nearly pristine drainage basins in southern Ontario, is little affected by encroaching urban sprawl and agricultural activity (Mahaney & Ermuth, 1975). It has been the subject of Quaternary investigations (Mahaney & Sanmugadas, 1986; Mahaney & Hancock, 1993) designed to investigate the composition and source of glacial sediments in the region. Like most neighboring catchments, it is a basin that was formed initially by meltwater draining across the Lake Iroquois plain from the Oak Ridges Interlobate Moraine (20 km north of the present northern boundary of Lake Ontario), ultimately incising its bed during at least three major stages in the Holocene (Mahaney & Sanmugadas, 1986).

During the initial stage of stream deposition, lag gravel was deposited on the Lake Iroquois lakeplain by a meandering river flowing south into the lake. With increased melt and discharge, this stream quickly cut a channel 25 m deep, constructing the oldest fluvial terrace complete with a lag gravel that includes boulder-size clasts. While still undated, this older terrace level was probably present either in very late Pleistocene or early Holocene time. It is this terrace that provided the source material for the landslide mass at site R43, situated on a lower terrace, considered to range in age from 5000 to 6500 ka. The total subsidence of the landmass is approximately 15 m along the proximal side of the high terrace, and along a slide plane of approximately 18 m, with a lobate-shaped mass of debris 10 m in radius curvature and 2 m high.

Details of the soil profile formed in this landslide are seen in figure 8.1. Sands recovered from the Bm and Ck horizons of the profile show that a number of microtextures are inherited from glacial and fluvial transport, with some surfaces overprinted with new microfeatures. Weathering is slight, given the young age of the soil, and as a result quartz dissolution is minimal. What dominates is a coating of smectite (mainly montmorillonite) and vermiculite in the Bm and Ck horizons, which is attributed to imperfect drainage in the paleosol. Details on microtextures in the R43 paleosol are given in chapter 12.

Microtextures found here that might be attributed to mass wasting include fracture faces, micro "tectonic" features, *en echelon* fracture faces, closely spaced deep grooves, and radiating grooves. None of these microfeatures have been found on grains from any other sedimentary environment, although radiating fractures are known on impact structures (bolide grains) and infrequently on glacial grains. When movement of a landmass is initiated, grain collisions are probably capable of dislocating and chipping large (50–200 μm) pieces off the grain surfaces (figure 8.2A). As shown in figure 8.2B and C, this is likely a widespread process, even for small-magnitude mass wasting such as occurred at Site R43 in the Rouge River. This process also leaves vestiges of the original grain surface with older dissolution microfeatures. The original grain surface shown in figure 8.2D, complete with numerous fractures and minor dissolution microfeatures, probably results from glacial transport followed by minor weathering in the first half of the Holocene (a weathering event in the higher terrace source area). The lower part of this grain carries a fresh fracture face.

Deep grooves, both straight and curved, are more common in glacial grains than in grains from other sedimentary environments. In the limited sample run presented here, deep narrow and wide grooves are common both as longitudinal and radiating microfeatures. I have not observed either variety on glacial grains, although radiating fractures are known on both glacial and asteroid-impact grains. These features are probably formed by abrasion with rough-edged grains

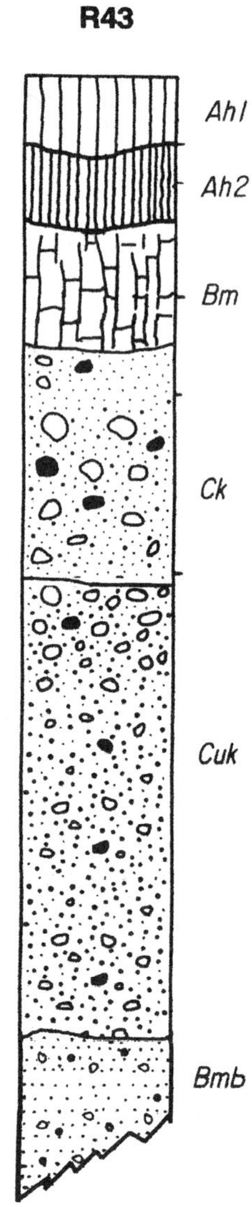

Figure 8.1 Pedostratigraphic profile in landslide debris (surface unit) and mid-Holocene weathered stream terrace gravel, Rouge River Catchment, southern Ontario. Soil description follows Canada Soil Survey Committee (1977).

during transport (figure 8.2F) and impact, followed by splitting and abrasion (figure 8.2E).

Caucasus Mountains

The Adishy Glacier, located in the central Caucasus Mountains, Russia, southeast of Sochi and south of Mount El'brus (5633 m a.s.l.), is situated between 2450

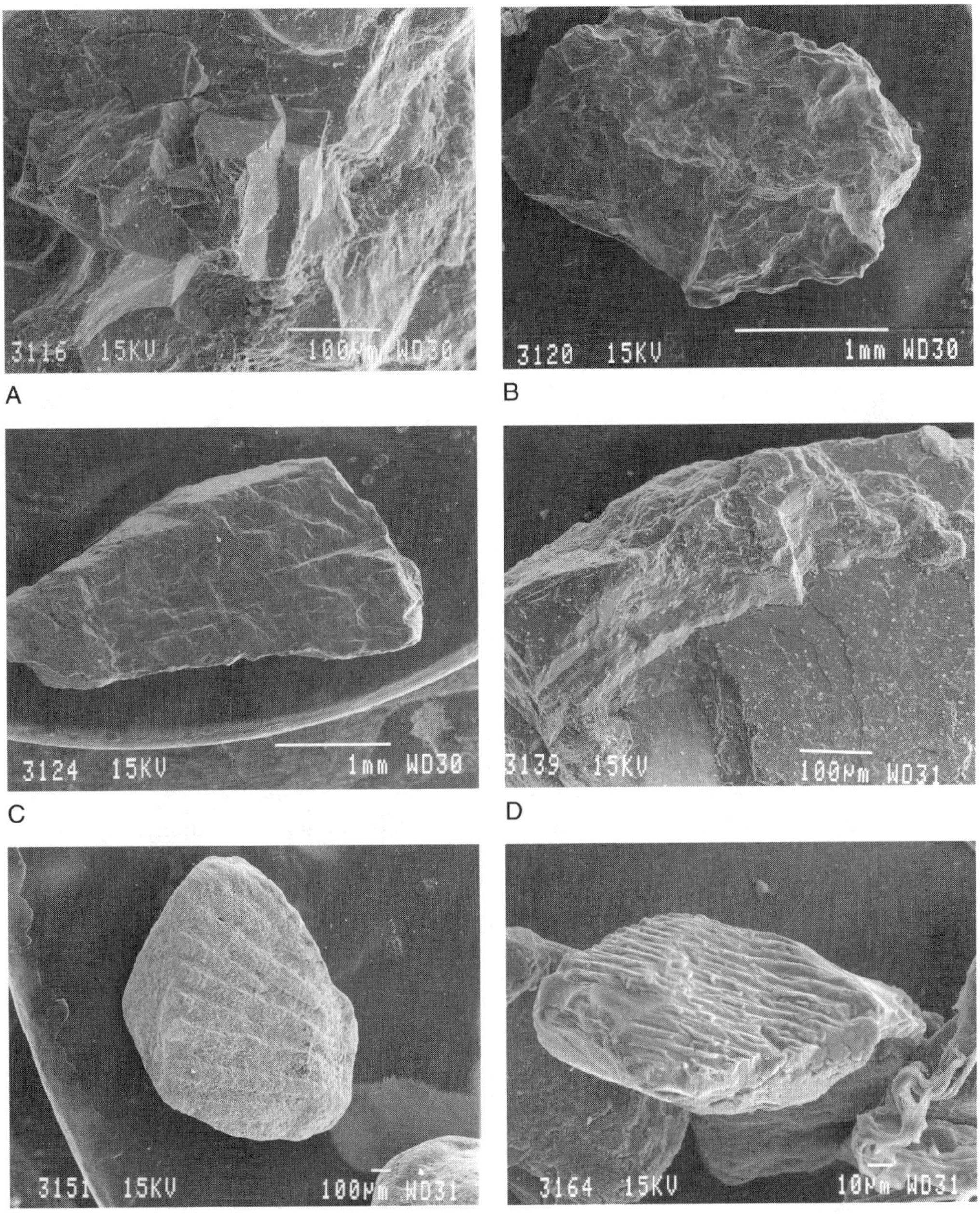

Figure 8.2 (A) Magnified quartz grain with fractured surfaces, including many fracture faces, ripped off the surface (center left) which may expose cleavage, a "downfaulted" surface (center right), and an upturned and rotated plate (top left). (B) Quartz with many small fracture faces covered with precipitated silica. (C) Quartz with fracture faces *en echelon*. (D) Quartz with fresh fracture faces covered with adhering particles. (E) Calcite with radiating grooves; weathering is more advanced in the carbonates relative to the quartz grains shown in B and C. (F) Quartz with deep parallel grooves.

and 2850 m a.s.l.. With an ice volume estimated at 400,000 m^3, it is preserved under a thick rock mantle mass wasted from Mount Tetnuld at the crest of the range. Because mechanical weathering of source rocks prior to glacial entrainment is considered to impart some of the microtextural features observed on glacial grains (Krinsley & Marshall, 1987), Mahaney et al. (1988, 1991) studied a suite of samples recovered from supraglacial drift. The majority of grains recovered were mechanically weathered and mass wasted onto the glacier surface, so the microtextural record includes release from source rocks and gravitational transport.

The microtextures frequently encountered in this study show that fracture faces dominate on grains mechanically weathered on bedrock. Other microtextures encountered show that minor fractures also occur, perhaps as a result of collisions during gravitational transport. Considering the several forms of mass wasting, from viscous masses of sliding and flowing material in landslides and debris flows, to free-falling masses in rock avalanche debris, one might expect primarily abrasion in the former, and broken and shattered grains in the latter. Together with chemical etching, present on three-quarters of the grains studied, quartz grains mechanically weathered and gravitationally transported carry a microtexture record of fresh fracture faces that is very different from other geologic agents.

Venezuelan Andes

Prominent talus cones in the Québrada de Mifafi of the western Cordillera of the Venezuelan Andes (western flank of the Pico el Aguila 1:25,000 quadrangle; Mahaney & Kalm, 1996), a tributary of the Rió Chama which drains the eastern and western Andean cordillera of Venezuela, were sampled with the intent of analyzing the microtextures on quartz grains making up the bulk of the matrix material in the slide rock. Talus cones rise above the valley floor to an altitude of approximately 100 m, with area diameters of 30–60 m at the base, rising to an apex of the cone intersecting prominent bedrock ledges, some of which are fed by active avalanche chutes from massive granite bedrock outcrops above. The median particle size within the cones is fine pebble (6–8 mm), and the mode particle diameter is 1–3 mm (coarse sand to granule), which means that the most frequently occurring particle sizes are suitable for study by SEM.

The source rock is partly weathered and presently undergoing granular disintegration caused by a combination of physical and chemical weathering processes. These outcrops, overrun by Pleistocene ice originating from Páramo de Piedre Blanca, may well contain sands left behind by the ice, which could become mixed with grains originating from weathering of the source rock. With this in mind, the source rock was sampled to recover grains dislodged as a result of granular disintegration, followed by sampling at the head of the talus, in mid section and along the base. The objective was to determine the fractography that might be produced by physicochemical release at the source followed by gravitational transport through the slide rock pile.

As shown in figure 8.4, a number of key microfeatures result from the release of source rock. First, grains at source are well coated with precipitates of Si and

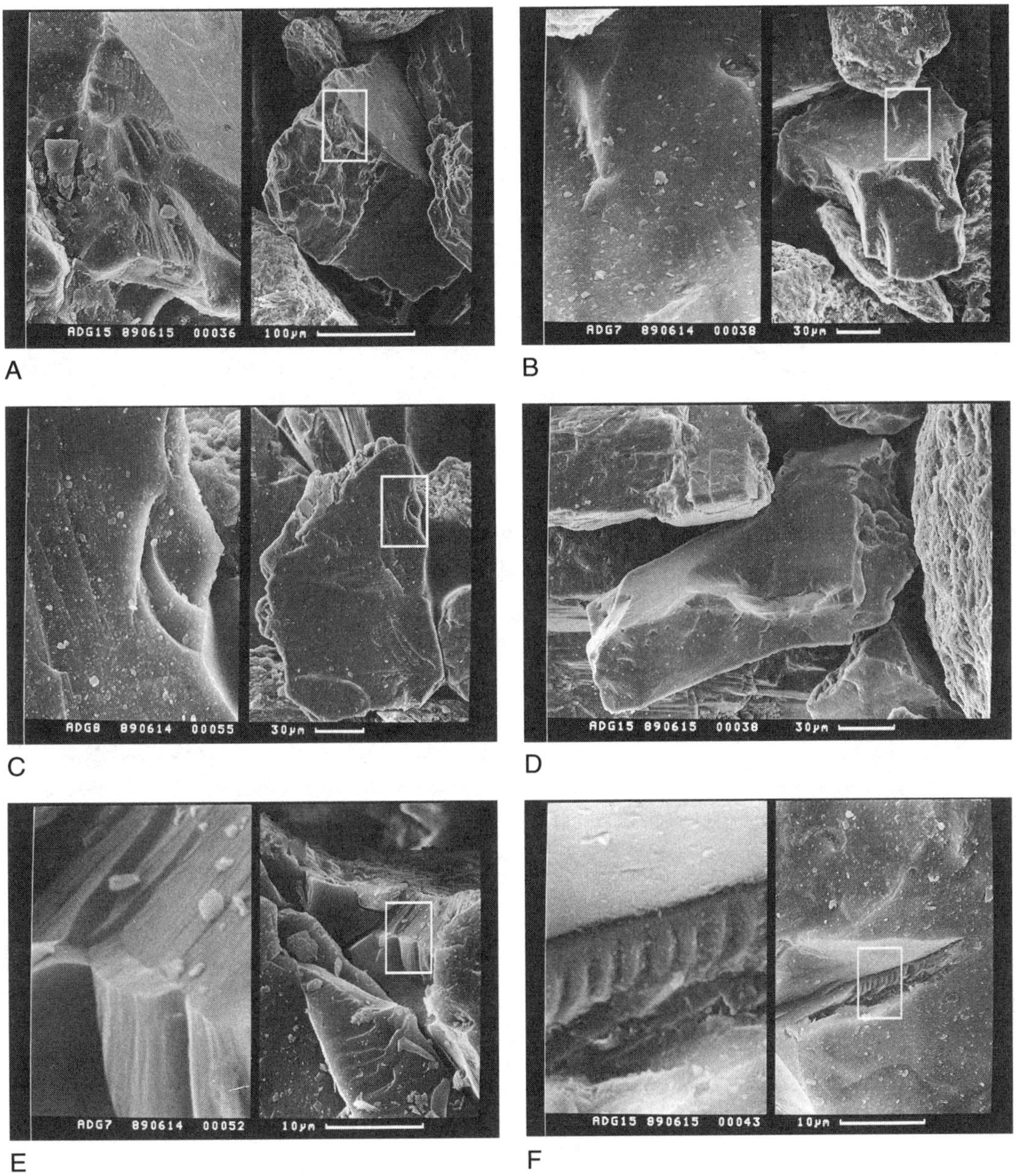

Figure 8.3 Grains from rockfall debris on glacier surface (Mahaney et al., 1991). (A) Quartz with weathered surfaces and fracture faces from ADG15 (enlargement ×5). (B) Quartz center with fracture faces. The enlargement (×5) shows a fresh fracture face with small, fresh craters in ADG7. (C) Large fracture face with fresh face, minor conchoidal fractures and chemical alteration on the obverse side (enlargement ×5). (D) Fresh fracture faces and a contrasting dissolution surface in ADG15 with some minor rounding. (E) Fresh fracture faces and surface separation either from mechanical weathering, or impact on the glacier surface in ADG7 (enlargement ×5). (F) Enlargement (×5) of a groove in a grain from sample ADG15 that could result from frost riving in bedrock.

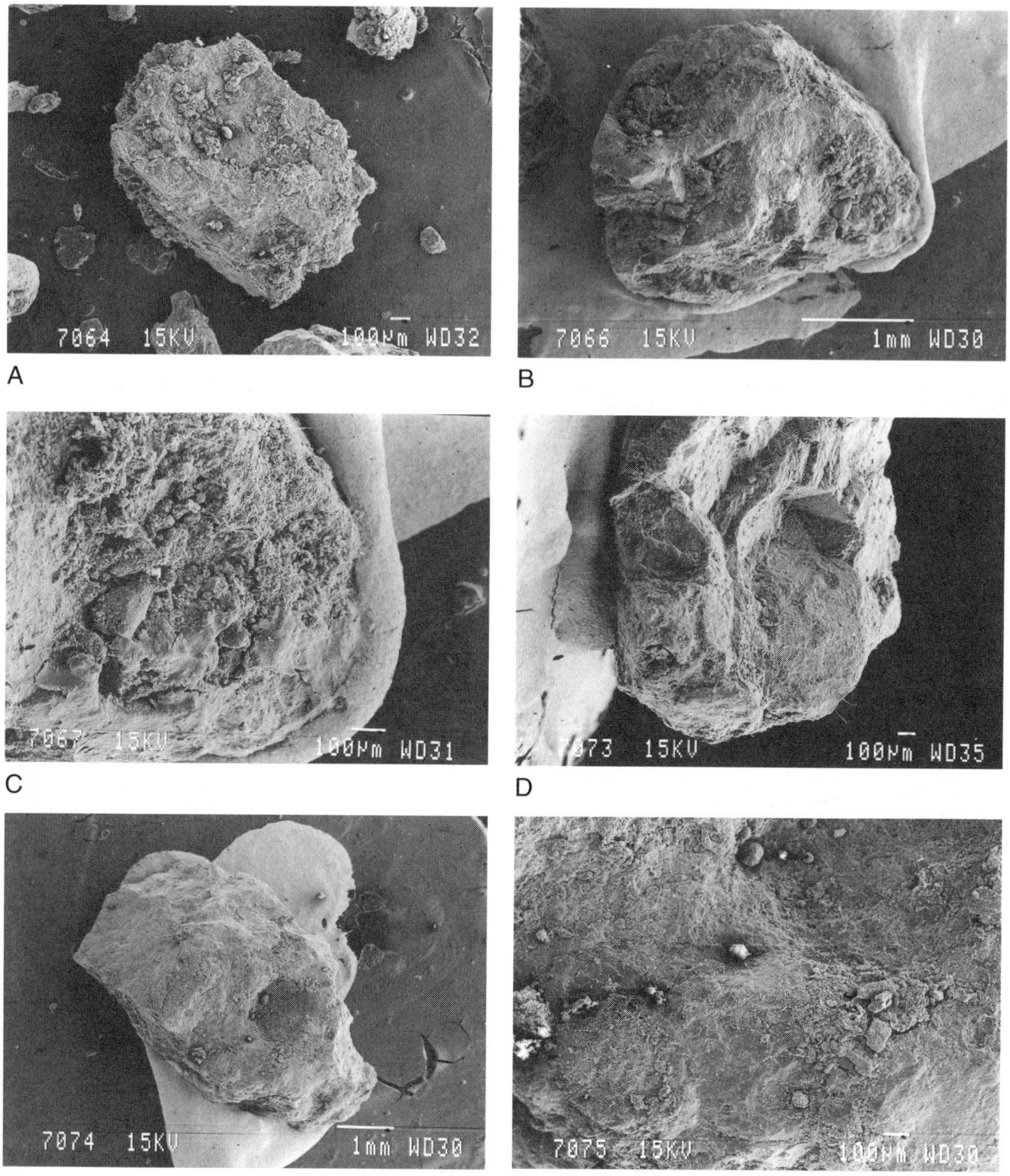

Figure 8.4 (A) General frame showing the shapes and sizes of coated grains among the sand-size particles in dislodged source rock. Notice the adhering particles packed into recessions and craters. (B) Very coarse sand with craters filled with packed silt- and clay-sized grains and coated with Si and Fe. (C) Enlargement of the grains packed in the crater on the right of B. (D) Quartz coated with Si in mid-stream talus. Small, coated fracture faces in mid grain. (E) Quartz covered with adhering particles and a coating of Si and Fe (Si:Fe = 4:1); (F) Enlargement showing the packed grains in craters and the coatings in E.

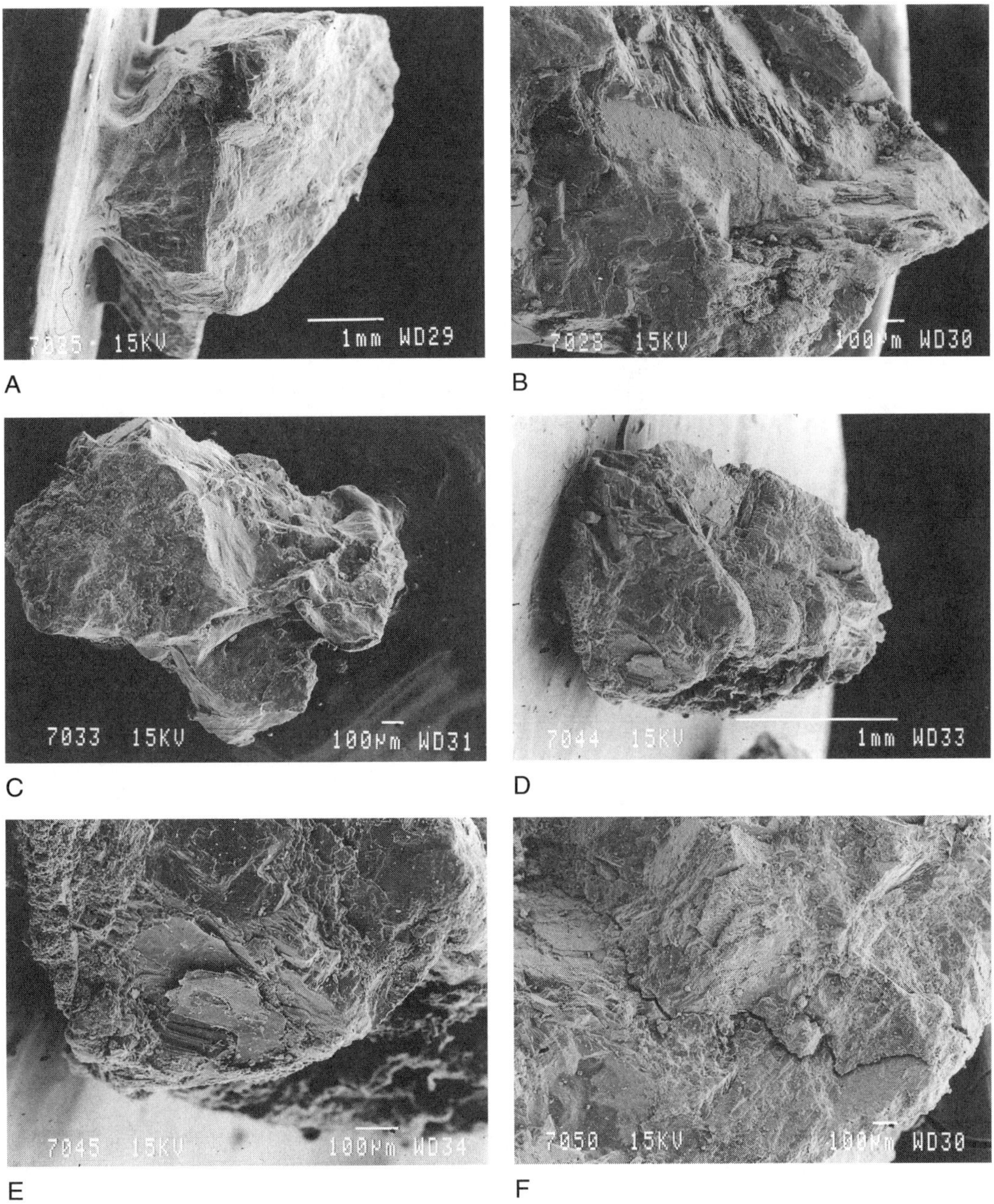

Figure 8.5 (A) Large granule in the Mifafi talus with a fracture face at the base. (B) Fractures on the edge of quartz in released source rock. (C) Fractures in the lower edges of quartz covered with precipitated silica. (D) Quartz with an Si–Fe coating, packed grains in craters and slickensided mica plates in the lower left. (E) Enlargement of the slickensided micas shown in D. (F) A long jagged crack in quartz, with splays extending to the left.

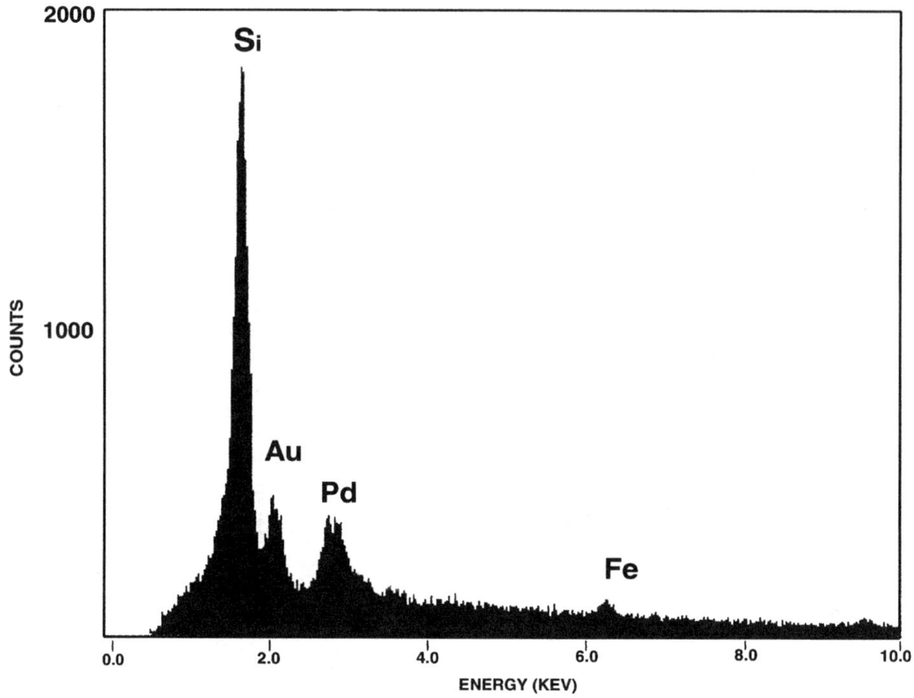

Figure 8.6 EDS spectrum showing the relative peak heights of Si and Fe on a coated portion of a quartz grain (figure 8.5A) in Mifafi talus C. When compared with fresh quartz, the Si peak height shown here is 10 percent higher, indicating a thin Si coating combined with a similar amount of Fe. The Au and Pd peaks are from the coating.

Fe, raising the possibility that the outcrop source rock was weathered prior to late Pleistocene glaciation and somehow escaped the passage of ice down valley. Second, source rock particles as large as granule size (2–4 mm) show, along with sand, the unmistakable imprint of fracture face development. Even individual sands are riddled with fracture faces that apparently originate either as a result of ice crystal growth or as a consequence of a snapping off process when chemical weathering removes the support of overhanging rock, resulting in the release of a small or large mass of source rock. Dislodgement by ice crystal growth may be impossible in the present climate where freezing only rarely occurs; freezing depths of only a few centimeters in rock could hardly be considered to provide a closed system with sufficient compressive strength to overcome the tensile strength of the rock. A minor amount of surface fracturing results in superficial breakage in grain skins, most often occurring across fracture face surfaces. Finally, on some grains the unmistakable imprint of small roots raises the possibility of root dislodgement along the source rock face.

Within the talus itself, quartz grains are only slightly abraded from gravitational transport. Cracks, and sometimes multiple cracks within a single grain, dominate the suite of grains studied. These multiple cracks, different from fracture faces produced by the release of grains upon weathering, presumably result from the release of kinetic energy caused by low-velocity collisions with larger particles, or the compression caused by overloading of debris within the talus cone. At the base of the cone where grain fracture is most severe it is impossible to determine the pathway followed by grains from the cone apex downward in the system. Grains may reside on the cone surface, or follow a subsurface route to emerge at the base of the cone. The grains reported here come from matrix material at the surface of the cone.

Most grains retained their coatings after transport in the talus (figure 8.6); only occasionally were coatings dislodged by contact with other grains. Some fracture faces might be produced in the talus by grain-to-grain contact but given the available data, this is difficult to determine. The statistical data indicate a similar number of fracture faces in the talus at the apex of the cone as at the base, so it seems implausible that movement within the talus is sufficiently forceful to produce fracture faces. Likewise the absence of fractures on grain surface areas with removed or scraped-off coatings suggests insufficient stress to fracture the surface. Perhaps grain relaxation in the talus is sufficient for stick–slip and other allied processes so prevalent in glaciers not to operate.

Grain packing as shown in figure 8.5D is a normal process in talus, as indicated by the presence of small fine silts and clay packed into craters and crevices, including some cross-grain fractures, in the talus sands. This is similar to grain packing in loess described by Assallay et al. (1998). More than 50 percent of grains analyzed in the Mifafi talus exhibited grain packing, presumably a product of meteoric water washing finer material into pore space in the talus, evaporation of pore water, and adherence of fines to the coarser particles.

Edge rounding, which might be expected to occur in talus, also seems inoperative. This might relate to poor pivotability amongst the population of finer grains resident in talus, but it follows from transport pathways that grain coatings are preserved intact with little modification. Also, the absence of striated surfaces seems to indicate insufficient stress to groove and abrade quartz under gravitational transport. In sum, the outstanding feature of grain transport in talus is the production of cracked grains, often with cracks extending completely through the mineral and coated surface.

Because cracked, coated grains are not present in the released source rock, they must be fractured during transport in the talus system.

Cape Breton

To the north of the entrance to Cape Breton Highlands National Park, on the St. Lawrence side of Cape Breton in eastern Canada, outcrops of Carboniferous strata contain quartz-rich conglomerate that provides source rock for talus cones with heights of 50–150 m stretching all the way to sea level. North of Grand Étang, subformations of the Riversdale Group of sandstone, siltstone, shale, and coal, also contain conglomerate, which provides source rock for prominent talus cones.

A relatively small 30 m high talus cone was sampled at source—at the top, in mid-section, and at the base—for matrix material to analyze the microtextures

Figure 8.7 (A) Quartz grain in the source rock of the Cape Breton talus showing root channels oriented right to left. (B) Partially fresh quartz with a crack presumably due to frost riving. (C) Enlargement of the crack in B and adhering particles. (D) The edge of a quartz grain with fracture faces and minor conchoidal fractures. (E) Angular quartz with multiple fracture faces, some adhering particles, and clay coatings in the mid-talus stream. (F) Crack in a quartz grain (arrow) with branching splays, from the lower talus.

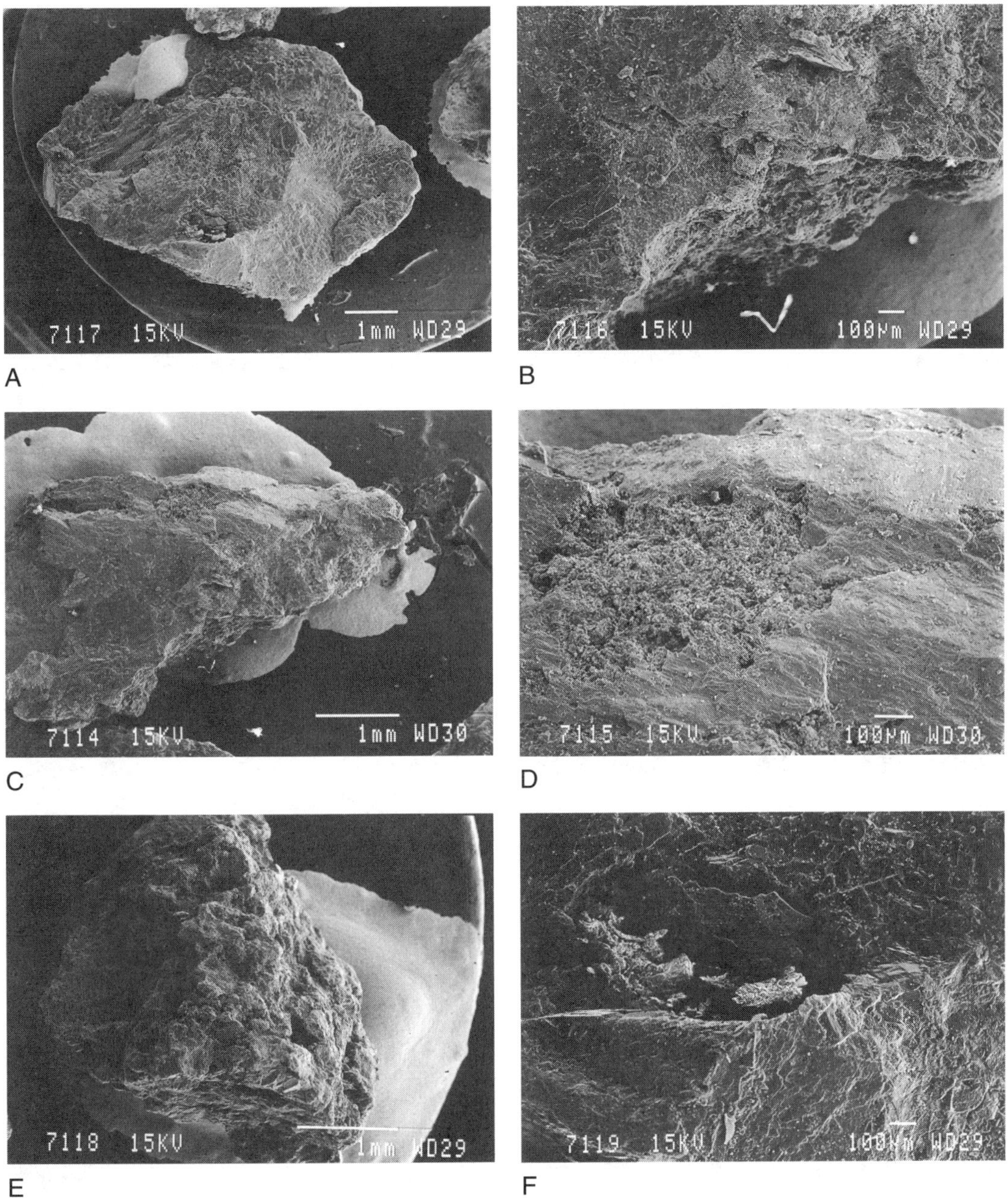

Figure 8.8 Quartz in lower talus, Cape Breton. (A) Whole quartz grain with a pothole in the bottom (center). (B) Upturned plates on a coated quartz surface. (C) Flat quartz with fracture faces (top). (D) Clay coating on quartz with Si:Al = 2:1 indicating vermiculite or smectite. (E) Orthoclase well coated with clay. (F) Enlargement of a pothole filled mainly with quartz.

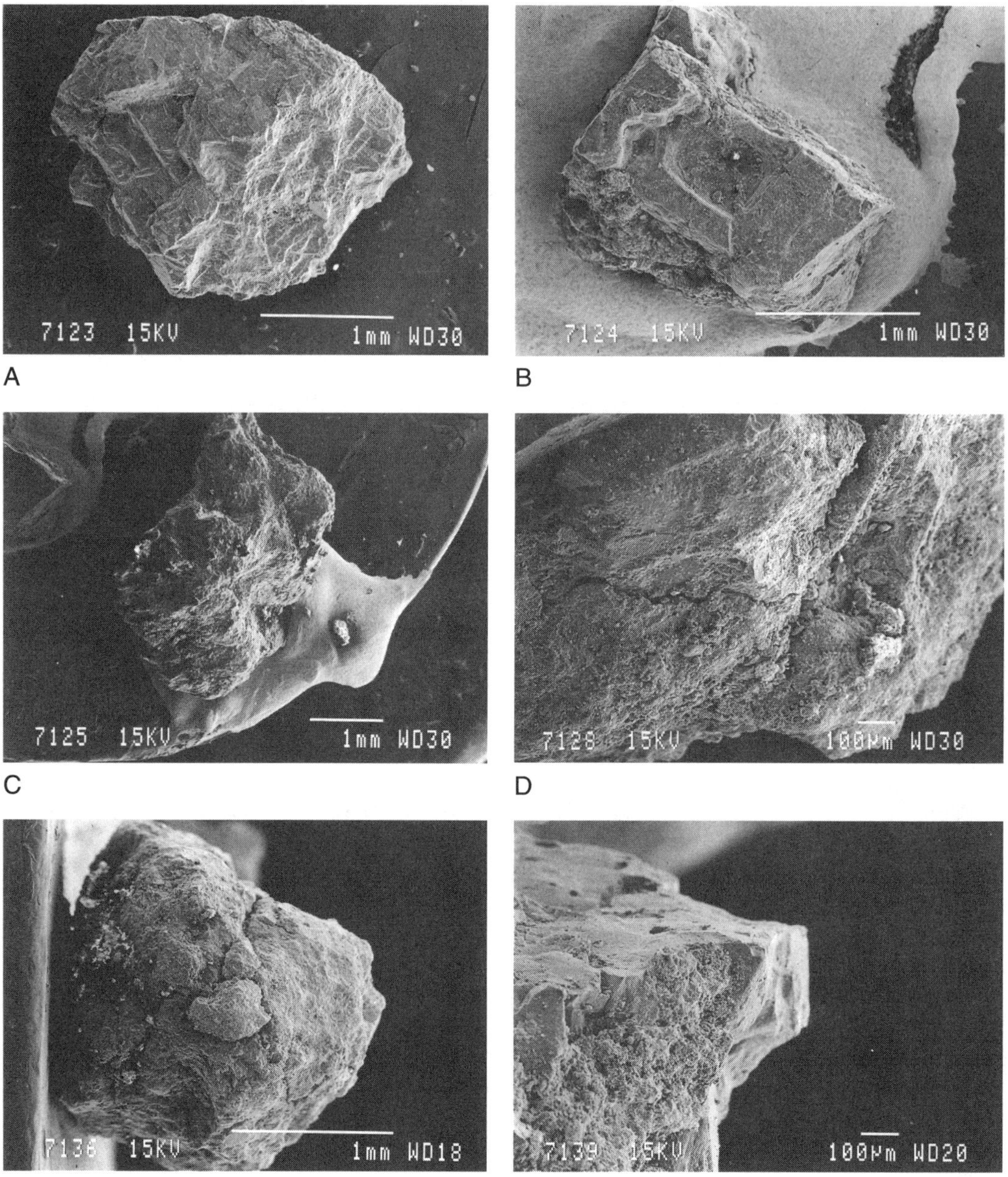

Figure 8.9 (A) Ca-plagioclase, relatively unweathered, with a spalled surface, in S39 talus (Oregon). (B) Partly weathered orthoclase in S2 talus. (C) Very well weathered quartz in S2 talus. (D) Quartz with cracks normal to each other through part of the grain in S2 talus. (E) Weathered amphibole with cracks, in S2 talus. (F) Fresh versus weathered faces on quartz with Si:Al = 1:1 indicating kaolin clay minerals.

produced on release of bedrock and in transport through the talus pile (figures 8.7 and 8.8).

In this environment, ice crystal growth is a real possibility, as winter temperatures are severe, with a mean annual temperature of +5°C and extremes of at least 30°C. Such temperatures make the fracture-prone conglomerate the ideal rock for physical weathering. With multi-mineralic material, and therefore different tensile strengths, the growth of ice from meteoric water invading rock crevices would rely only on the expulsion of gases and the development of an ice cap, thus creating a closed system. Physical weathering would probably be extreme here and certainly stronger than in the Venezuelan highlands. One might expect, for example, a higher production of fracture faces and less coating on grain surfaces.

Oregon Coastal Range

Sands in mass wasted debris in the Oregon coastal range (figure 8.9) yielded a number of sand and small pebble-size clasts with broken surfaces sometimes extending to considerable grain depth, and upturned plates, often coated with clay and occasionally with silica and Fe oxides. Even though this area was beyond the range of glaciation, there is a striking resemblance between these grains and felsic clasts studied in the Cape Breton Highlands of eastern Canada and the high Venezuelan Andes.

Conclusions

Mass wasting, as an active geologic process, is a fertile area for SEM–EDS microtextural investigations. When SEM–EDS is combined with source rock analysis, it becomes evident that granular disintegration is the major process in bedrock weathering. This is accompanied by the large-scale production of fracture faces. These ubiquitous, fresh fractured planes vary in size from almost whole grain areas to small $5 \times 5\,\mu m$ patches on individual grains. Often coated with Si and clay (Si and Al), and occasionally with Fe, these grains have fresh fracture faces that indicate the grain areas where separation from bedrock occurred. In some cases, the absence of coatings on fracture faces indicates that separation is relatively recent.

Relatively few dislodged grains were fractured in the source areas, as compared to the active talus piles. In the active talus piles, as many as 50 percent of grains studied contained multiple fracture faces penetrating nearly entire grains. In these cases, over 90 percent of grains studied contained fractures showing through the coatings, indicating they had occurred after dislodgement from bedrock and entrainment in the talus. As one of the outstanding microfeatures of slide rock transport, grain microfracturing could be considered diagnostic of gravitational transport. Only more detailed studies will show if it is a universal feature on quartz in different environments, especially in nonglacial localities. In glacial localities, overriding with ice might fracture grains in bedrock, giving rise to some of the microfeatures observed in talus.

9 Glacial Grains

Glacial grains probably carry the greatest range of microtextures when compared with grains affected by other geologic agents. This includes the full range of fractures, grooves and abrasion microfeatures listed in chapter 5, as well as solution–precipitation microfeatures and other coatings that may pre-date or post-date a glacial episode. At the same time, some grains make the glacial trip unscathed by their sojourn in the ice. These fragments may carry a record of release from bedrock with unremarkable fracture faces, and without the usual grain-to-grain contact that produces triangular-faceted, sharp-edged grains with moderate to high relief. The latter are usually well abraded with a multitude of fractures, grooves, and well-worn abrasion features.

As will be discussed in this chapter, quartz is the ubiquitous and well-studied recorder of glaciation in many environmental settings (Krinsley & Doornkamp, 1973; Whalley & Krinsley, 1974; Perttunen & Hirvas, 1982; Helland & Diffendal, 1993). But other minerals, including all the plagioclases (Mahaney, 1990a), many heavy minerals (Mahaney & Milner, 1998), and gold (Mahaney, 1992b; Loen, 1994; Kinnunen, 1996), to mention only the major mineral suites, may also be used. In this chapter the discussion will concentrate on quartz as the prime environmental recorder.

The relative hardness of quartz (7.0 on Mohs hardness scale) and its widespread occurrence in nearly every environment make it an ideal candidate for microtextural study. It is often unblemished, existing as colorless, hexagonal crystals, although sometimes colored by impurities. With an absence of cleavage, especially in the coarse fractions (2000–50 μm), it is composed exclusively of silica–oxygen tetrahedra, with oxygens joined with silicon into a strong geometrical network (Frondel, 1962; Bloss & Gibbs, 1963). Unlike weaker minerals it is capable of transiting the glacier system many times, and as will be seen later on, it is possible to examine complex quartz grains that have had many sojourns in the ice. Because quartz is subject to slow dissolution and precipitation, it is also an ideal mineral to record subaerial and diagenetic weathering events; moreover, it is possible to observe preweathering features overprinted with the fresh scars of glacial transport, followed by more recent weathering events.

Mechanical experiments designed to simulate the glacial environment are uncommon. However, experiments have been accomplished, using either quartz grains on a quartz plate with hand pressure, or uniaxial stress applied to silica spheres of fine gravel size encased in ice. This experimental data and the imagery related to it will be discussed in chapter 15.

Glacial Load

The load transported by ice ranges from clay to boulder and block-size material, including any minerals provided by source rocks. It also includes a minor amount of eolian-delivered grains as well as mass wasted material derived from nunataks and cirque walls. The supraglacial load may be light to heavy, depending on the nature of the source rocks and the climatic setting (tropical mountains to Arctic or Antarctic environments), and consist of material delivered by airfall influx and gravity to the glacier surface. There is grain-to-grain contact to be sure, but at pressures close to 1 atmosphere and generally lacking the confined stress of the grain-to-grain contact that occurs in the basal ice. The englacial environment is relatively free of clastic load, although subsurface channels may move large volumes of meltwater that may occasionally carry grains in turbulent sluiceways. It is these grains that suffer heavy abrasion, causing overprints that partially mask previously inherited weathering (e.g., preweathering) and other fractures and/or abrasion inflicted during the initial stage of ice transport.

The basal ice, containing the bulk of the clastic load, together with variable amounts of water provides the environment with the highest probability of grain-to-grain contact. Increasing water content, tends to create hydrostatic pressures close to one bar and nearly equal in all directions. With decreasing water content, shear stress rises somewhat but falls far below the pressures that result from stick–slip processes (up to ~200 bars) where sudden slippage generates vibrations—the sudden convergence of kinetic energy to elastic energy. The vibrations produced by glacier movement are manifest in the fractures seen in great numbers across the glacial grains. No other geologic agent can generate the strong vibrations that produce deeply embedded microfeatures in clasts.

There are many pathways by which mineral grains may enter, move through, and exit a glacier. Theoretically, grains of any description may sink into the ice from the glacial surface, be carried by surface and subsurface meltwater streams to exit the terminal areas, be forced to the surface by thrusting, and/or be entrained from underlying bedrock and/or till. This begs the question of where to sample till, and whether it is best to concentrate on subglacial deposits where glacial grinding is apt to be at a maximum. Alternatively, one could look at end or lateral moraines, which might likely contain a mix of material deposited directly in contact with ice or from meltwater flow.

It is not the purpose of this atlas to go into till lithogenesis in any detail, but suffice it to say that basal tills of whatever origin (lodgement, meltout, flowtill, deformation) provide ideal settings in which to study the effects of glacial comminution recorded on quartz and other grains. This is especially the case where till is associated with other sediments emplaced by fluvial and/or eolian action.

Unlike eolian and fluvial grains, glacial grains are generally held in bondage by the ice or by other grains in rigid suspension, so the ultimate amount of grinding occurs with movement down glacier. Eolian and fluvial grains, as previously discussed, are capable of considerable random movement, and while they are capable of high-energy collisions, they are not subjected to the sustained and concentrated high energy inflicted on glacial grains. While eolian grains may produce scars somewhat reflecting microfeatures resulting from glaciation, the damage is far less deeply embedded in eolian grain surfaces.

As suggested by Barcilon & MacAyeal (1993) and Mahaney (1995b), stick–slip motions at the base of glaciers tend to fracture and abrade particles in ice. Particle shape (defined principally by $c\sqrt{ab}$) is governed partly by bedrock release (frost riving; Coudé-Gaussen & Balescu, 1987), partly by the preweathered characteristics of grains (Mahaney et al., 1991), and partly by transport via different geologic agents. Within the ice body, quartz grains form inhomogeneous elastic inclusions with different strengths, and respond to variable stresses by forming fracture and abrasion microfeatures. If the particles are already weakened by chemical preweathering, and if they have fracture faces (Mahaney, 1995b) derived from frost riving (Mahaney et al., 1991), they may easily comminute into smaller particles. This is particularly the case if they come into contact with one another in the basal ice layer where stick–slip processes produce high basal shear stresses and high strain rates (Barcilon & MacAyeal, 1993). Basal till, with meltwater lubrication, glides over bedrock and occasionally surges, subjecting particles carried as load to high fracture and abrasive stress, engendering low-velocity impacts which impart microtextures onto quartz sand surfaces. While there are similar shallow, conchoidal, and linear fractures on quartz from eolian (Mahaney & Andres, 1996) and fluvial environments, the angular shape, deep entrenchment of conchoidal and linear fractures, and frequent directionality of troughs and grooves (striations) are unique to grains emplaced by glacial transport.

Microtexture Recognition

Almost all the microfeatures listed in chapter 5 can be found on glacial grains. The exceptions include multiple fracture faces, v-shaped percussion cracks, and shock-melted textures which emanate from mass wastage, fluvial transport, and asteroid impact, respectively. All dissolution, weathering, preweathering, and dissolution features resulting from subaerial weathering and/or diagenesis are presumably inherited from interstadial or interglacial climates having survived the damage inflicted on a large proportion of grains moving through glaciers.

Taking the array of microtextures shown on figure 5.1 as possible candidates for microfeatures likely to be observed on glacial grains, the researcher is faced with the question of how many grains to analyze and at what resolution. All microtextures should be logged by their frequency of occurrence, and as discussed in chapter 4, it is necessary to analyze several samples and several fractions of each sample in order to arrive at a meaningful interpretation. A quick overview of each sample can be obtained by analysis of the fine and very fine sand fraction taking into account that angularity is often extreme in the finer fractions. This should be followed by analysis of at least 20 grains in the medium sand fraction (500–250 µm) and 20 grains in the coarse to very coarse fractions (2000–500 µm). These grains should be collected as randomly as possible (see chapter 4) and studied under the light microscope prior to using the SEM.

The degree of relief is one of the first assessments to make when analyzing a sample suite. Because of heavy fracturing and abrasion, high relief often dominates on glacial grains. This needs to be assessed and given a percent frequency of occurrence. Next, the degree of angularity, edge rounding, and all fracture and abrasion microfeatures need to be identified, and the frequency of occurrence tabulated. Lastly, all weathering, precipitation, and dissolution features should be assessed in a similar fashion. Usually glacial grains will be compared and contrasted with grains resulting from the action of other geological agents, perhaps in the same section or between sections.

As a test of how to carry out and assess the analysis, two examples will be used. The first will be from the Virunga Mountains of north-western Rwanda, with basalt bedrock of late Pleistocene age, overlain with weathered residual regolith, fluvial, and eolian deposits. Analysis of the sand fraction through this section reveals information on a variety of levels, from low-resolution recorders such as volcanic feldspars, to higher resolution grains of volcanic quartz in the lower section, and the sudden appearance of glacially-crushed quartz in the eolian beds.

A second test will invoke the study of indurated but non-cemented sandstones of Devonian age in the Aruküla Formation in Estonia from which material was entrained into Weichselian ice during the Last Glacial Maximum (Mahaney & Kalm, 1995, 2000). The sandstone was collected without the aid of chisel and hammer, thus ensuring the absence of artifacts produced in the collection process. The sands were loose enough so that it proved possible to analyze the grain size distributions in great detail. The overlying reddish till, which had inherited its color from the underlying sandstone, contained many grains that appeared to be overprinted with a glacial signature. As you will see, the two data sets produced quite different microtexture signatures.

In undertaking the Estonian study it proved possible to include a small sample of meltwater-emplaced grains. These glaciolacustrine and glaciofluvial samples, while not large enough to provide a good statistical sample, nevertheless produced a third signature contrasting greatly with the fluvial and glacial sample suites.

Virunga Mountains

Microscopic analysis of weathered sands from a sequence of volcaniclastic, fluviatile, and eolian deposits (figure 9.1) in the Virunga Mountains of north-western Rwanda (1° 30′ S, 29° 30′ E) unexpectedly revealed what appeared to be glacially crushed quartz in the eolian and fluvial beds (Mahaney, 1990c). Because the Virunga Mountains are unglaciated and the bedrock is basalt, replicate samples of quartz sand were studied throughout the section. The objective was to determine the degree of grain modification and range of microtextures on statistically significant populations of quartz in each paleosol horizon. The nearest ice center is the Ruwenzori Mountains, 150 km to the north, almost directly in the path of the north-easterly trade winds that dominate in winter.

The Karisoke area in which the KAR1 section is located lies in the saddle between Visoke (3711 m a.s.l.) and Karisimbi (4507 m a.s.l.) volcanoes on the Zaire–Rwanda border. The bedrock is basalt of late Quaternary age covered with residual regolith weathered during the Early Middle Wisconsinan/ Weichselian interstade. The lowermost paleosol is, in turn, overlain with weathered tuff, loess, and alluvium weathered partly during middle to late Wisconsinan time and on into the Holocene. The 2Ab horizon complex, radiocarbon dated at 6330 ± 200 yr BP, is probably slightly contaminated and therefore older than this radiocarbon date suggests. The ~40 cm thick 2Ab horizon complex of black peat-rich material is considered to have formed during the Late Glacial Maximum (~18 ka) when grass and heath vegetation of the Afroalpine zone replaced the forest of bamboo and *Podocarpus* (Mahaney, 1990a,c).

Glacially crushed quartz sand in the eolian and fluvial units suggests transport from an ice center, presumably the nearby Ruwenzori Mountains, followed in some cases by reworking in streams. The micrographs shown in figure 9.2A–D are representative of the quartz sand that makes up 10–15 percent of grains in the pedostratigraphic sequence above the 4Ahb horizon. The grains shown in figure 9.2A include glass and volcanic plagioclase mostly coated with gibbsite and metahalloysite. The frame outlines an angular, coarse quartz sand grain with an uncoated surface partly covered with adhering particles that are mostly glass. The grain immediately above is also quartz, as is the larger grain in the lower right. The enlarged grain has medium relief, sharp edges, craters of unknown origin, and minor dissolution features. It does not belong to the country rock and is similar to other quartz grains that have been glacially crushed.

The grain shown in figure 9.2B is a triangular-faceted quartz grain of very fine sand, well fragmented, which may represent a spall from a larger grain. It has sharp edges, irregular grooves on the upper end, and subparallel fractures that are partly abraded. It is uncoated, which sets it apart from other volcanic grains in the same horizon, although it carries an abundance of adhering particles. While this is a prototype of what most people think a glacial grain looks like, there are many other shapes and sizes that appear in tills and in redistributed eolian sediment.

The quartz sand grain (3Coxb horizon of KAR1) shown in figure 9.2C is most certainly wind blown and probably glacial in origin. Showing some minor edge rounding and fractures, it is partly coated with a mix of gibbsite and metahalloysite. An upturned plate at the bottom of the grain could be eolian or glacial. The enlargement shows the clay coating as the nodular form of gibbsite. A quartz grain recovered from the 3Coxb horizon in KAR1 (figure 9.2D) shows partial details of the weathered outer skin, and the fresh interior broken with a clean fracture face and ridge extending toward the toe of the grain can also be observed. The outer skin contains weathered fractures showing extreme dissolution. Numerous craters litter the fresh front end of

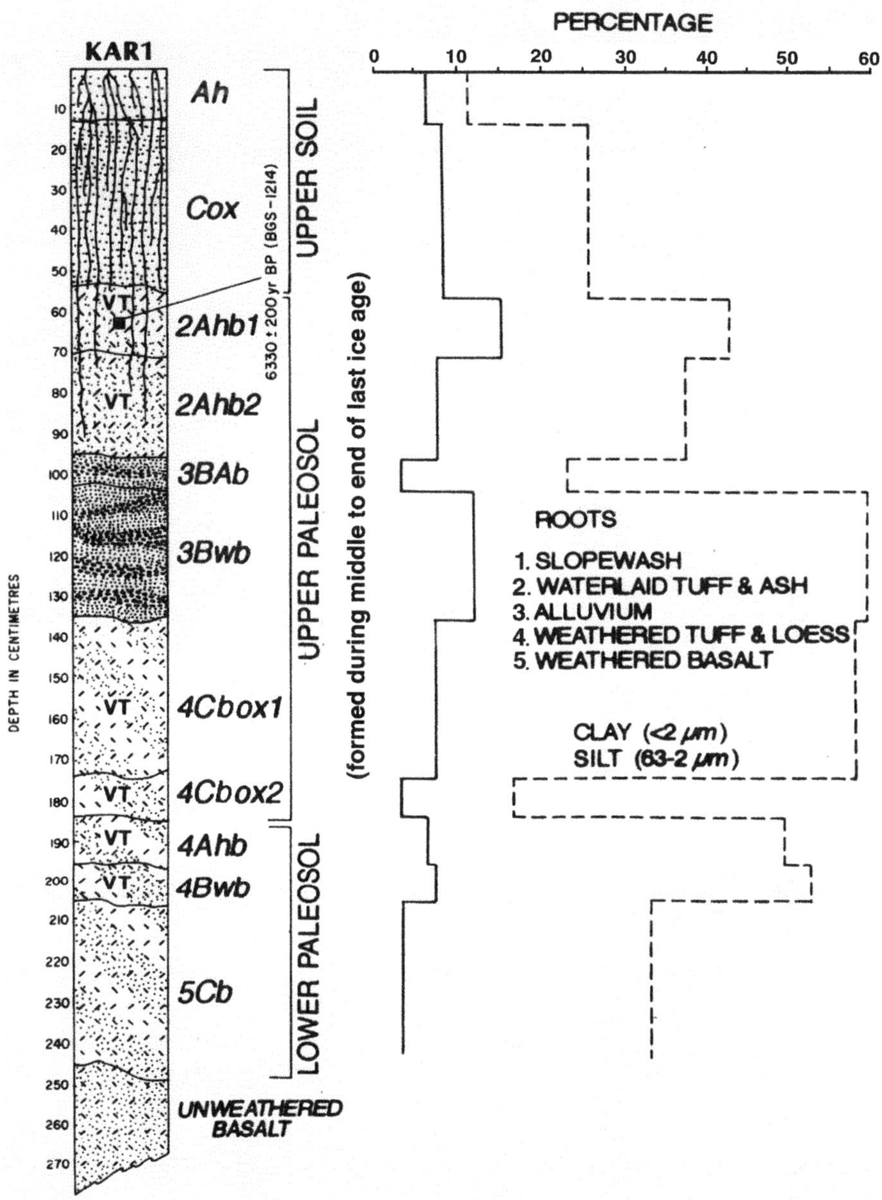

Figure 9.1 Stratigraphy at KAR1, Virunga Mountains, north-western Rwanda. Adapted from Mahaney (1990c) by permission of the publisher, Taylor & Francis AS.

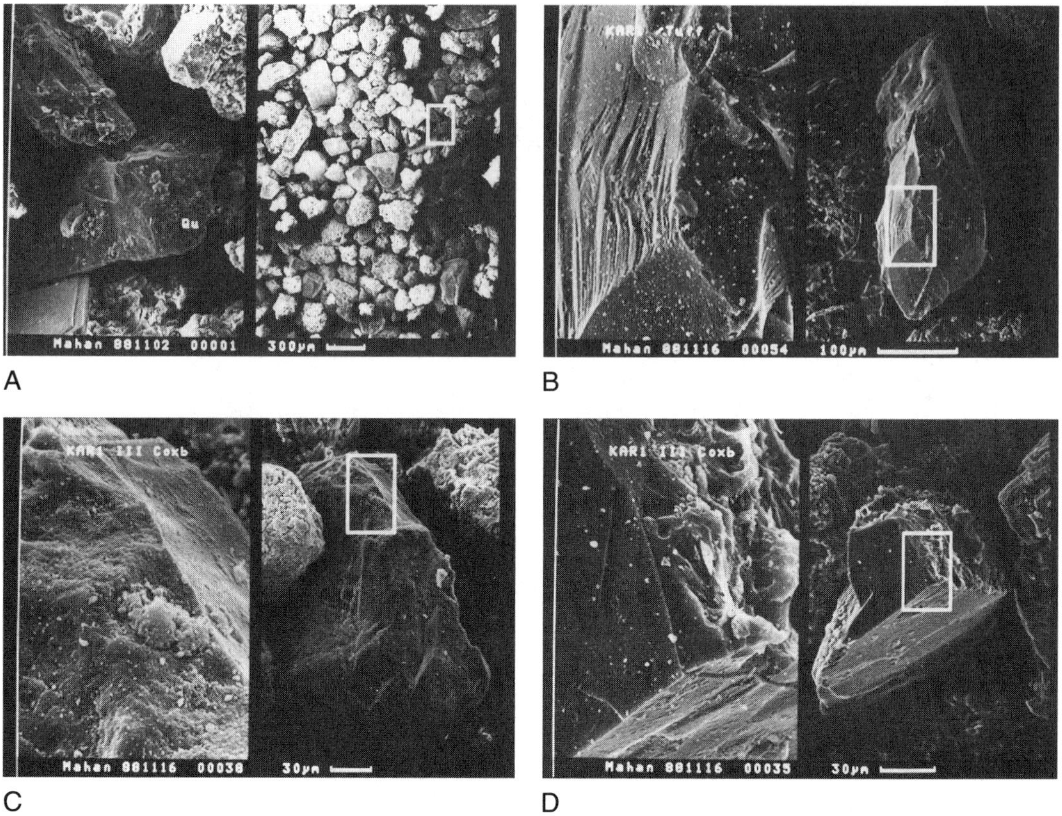

Figure 9.2 (A) General frame of grains in the 3Bwb horizon consisting mainly of amphibole and volcanic feldspars with quartz along the right margin. The enlargement (×10) contains two angular, sharp edged, and relatively clean quartz particles of fine sand size. (B) Triangular-shaped quartz grain in the 4Ahb horizon, with partly abraded subparallel fractures shown in the enlargement (×5). (C) Relatively clean quartz grain in the 3Bwb horizon surrounded by coated volcanic feldspars. The enlargement (×5) shows the coated portion of the quartz particle. (D) Quartz with a pronounced fracture face (bottom) and preweathered remnant (top).

the grain and some appear to be v-shaped percussion scars caused by impact in turbulent water.

In conclusion, this is one occurrence where a study of weathering of volcanic sediments including mainly sanidine, analbite, and K-analbite yielded unexpected eolian quartz unrelated to the country rock. Routine scanning of the entire section to establish differences in weathering that could be used to establish relative age, turned up a small percentage of quartz sand that shed light on glaciation and eolian transport from the nearby Ruwenzori Massif.

Estonian Sandstone–Till Test

Many workers who have recognized various depositional environments from SEM analysis of quartz have relied on the analysis of sediments located close to glacial environments (see Setlow & Karpovich, 1972). Thus, many grains could be reworked from glacial environments and mixed with eolian (see Mahaney & Andres, 1991; Mahaney, 1992a,b), fluvial, or coastal sediments. To avoid contamination with grains that could be affected by glaciation, indurated and non-cemented samples of Middle Devonian (Eifel) Aruküla Stage sandstone, emplaced in a fluvial environment (Kleesment 1994, 1997) in Estonia and Latvia (figure 9.3A), were subjected to SEM analysis. Comparing these samples with sands from basal till, deposited under thick ice (see Mahaney (1995b) for a discussion of thick versus thin ice and its effect on the range of recovered microtextures), it proved possible to study contrasting effects of water and ice transport on quartz grains. Similarities and differences in microfeatures were used to provide a means of identifying quartz grains originating from glacial and fluvial depositional environments (Mahaney & Kalm, 2000).

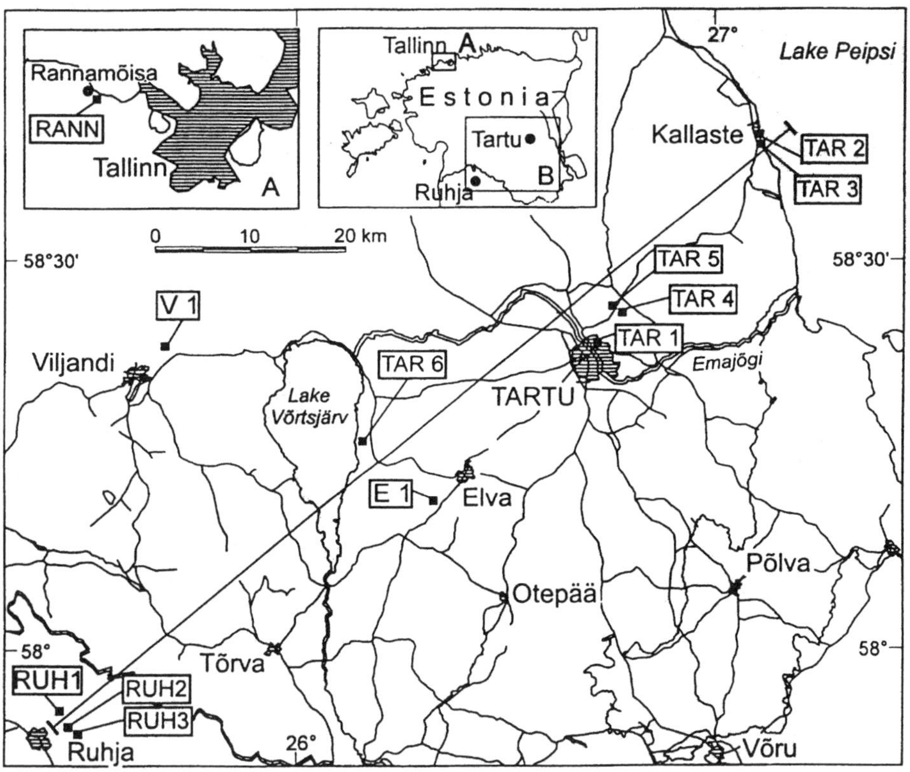

Figure 9.3 (A) Location of Estonian till and fluvial samples. Reprinted from Mahaney & Kalm (2000) by permission of the publisher, Taylor & Francis AS.

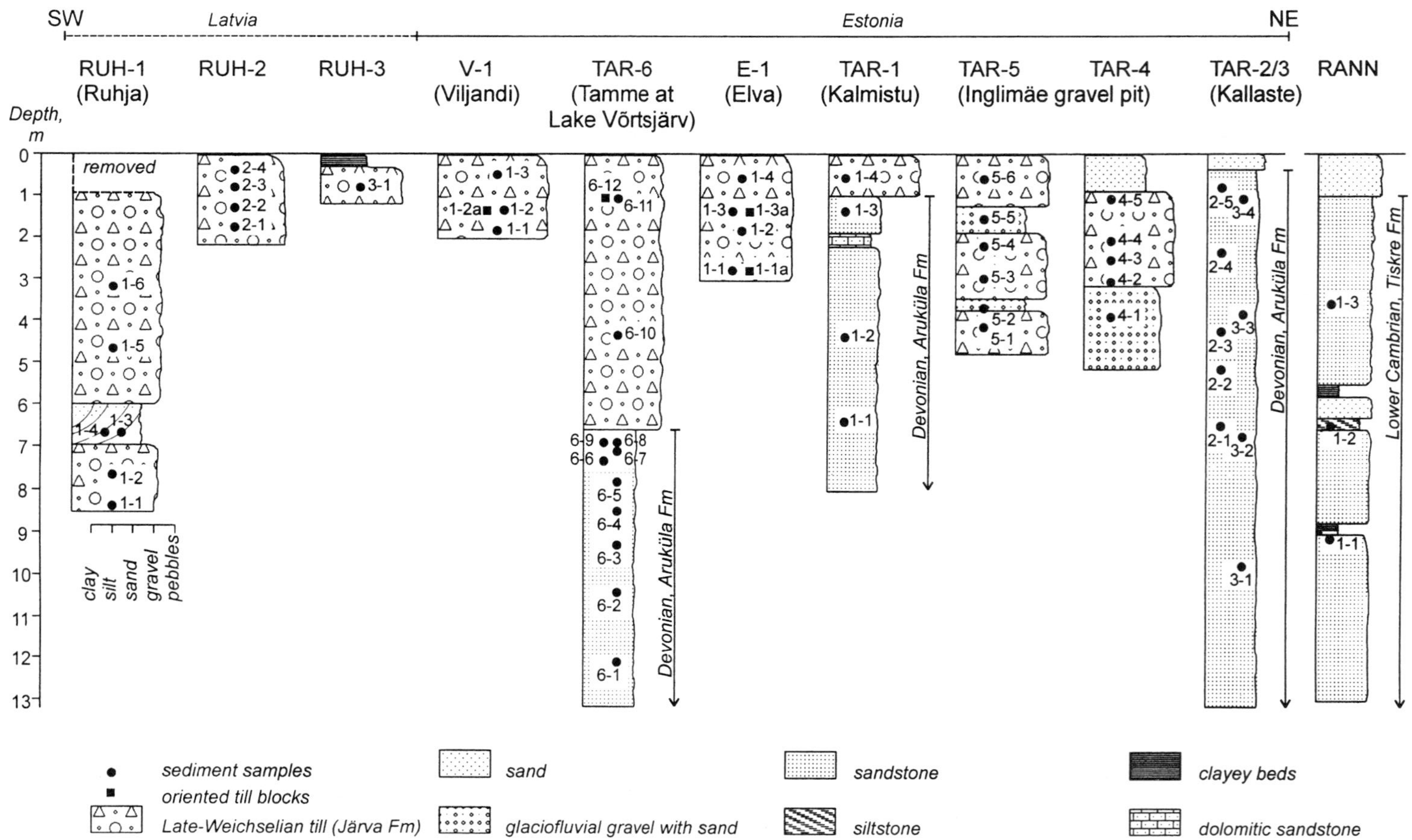

Figure 9.3 (B) Stratigraphy of Estonian and Latvian sections on a NE–SW transect. Reprinted from Mahaney & Kalm (2000) by permission of the publisher, Taylor & Francis AS.

The tills belong to the Vôrtsjärv or Latvia (Estonian and Latvian terminology, respectively) subformations and are correlative with late Weichselian glaciation (Kondratiene, 1993; Kirismäe & Kalm, 1995; Raukas & Kajak 1995, 1997). The glacial deposits (for site locations, see figure 9.3A, comprising ground moraine, were emplaced by subglacial deposition of ice in direct contact with the underlying sandstones (Raukas, 1995). Devonian sandstones (Eifel and Givet), particularly of Aruküla and Burtniek stages, underlying Pleistocene deposits in the study area, are quartz-rich (70–90 percent), slightly indurated, and non-cemented (Viiding et al., 1981; Kleesment, 1994, 1995). Emplacement of the Aruküla Formation in a continental water basin close to sea level (Kleesment, 1994) and well away from any glacial source (Torsvik et al., 1992) provides an ideal setting in which to assess overprinting of the glacial signature on water transported material.

The stratigraphy of the different sections is shown in figure 9.3B. The TAR-1 (figure 9.3) section was sampled near Tartu (for the location, see figure 9.3A), where a reddish-brown pebbly till overlies cross-bedded sandstone. Sections TAR-2 and 3 (figure 9.3) are from closely spaced exposures of Aruküla (Eifel) sandstone located 20 m apart. Section TAR-4 (figure 9.3C) consists of glaciofluvial beds overlain with late Weichselian till, and TAR-5 (figure 9.3C) consists of till with inclusions. The TAR-6 (figure 9.3B) section includes exposures of Aruküla (Eifel) sandstone and late Weichselian till. The E-1 and V-1 sections (figure 9.3) at Elva and Viljandi are both till exposures with a high proportion of carbonate pebbles. The Latvia Formation till (late Weichselian) exposure at RUH-1 contains a mix of basal till and outwash sand and gravel. The sections of RUH-2 and 3 contain basal till and laminated flow till.

The origin of quartz in South Estonian tills is principally from Precambrian igneous and metamorphic rocks of the Fennoscandian Shield, and from Cambrian and Ordovician sandstone exposures along the southern coast of the Gulf of Finland. Thus, the transport distance for more than 50 percent of the quartz grains studied here is more than 100 km (Mahaney & Kalm, 1995, 2000).

The Sandstones

The Devonian sandstone is composed of a mix of quartz, feldspar, and mica, with quartz as the dominant mineral. A representative suite of samples is shown in figure 9.4A–F. Some grains show minor dissolution, most probably from diagenetic weathering. The quartz grains show pronounced rounding on edges and over 50 percent of all the grains analyzed are round to subround, with multiple v-shaped percussion cracks (figure 9.4B, C). Less than 3 percent of all the quartz grains are subangular, with minor dissolution effects and minor conchoidal and linear fractures (figure 9.4D). Abundant mica (see the biotite grain in figure 9.4E), which is a particularly weak mineral subject to brittle fracture in the Devonian sands, indicates a short transport distance and absence of crushing effects. The deep troughs in figure 9.4F, while common in the glacial grains, are very rare (<0.5 percent) in the Devonian sandstone. As common abrasion features, troughs of any size and magnitude require considerable and sustained pressure to form, which is lacking in the fluvial environment.

Fresh microtextures are well preserved on the suite of grains studied. However, many grains also contain weathered and preweathered surfaces (the left side of figure 9.4B, C, D, and F). As in previous studies (Mahaney & Kalm, 1995, 2000), "a preweathered grain is defined as one on which dissolution etching has been partly obliterated by fresh abrasion or fractures. Some grains show clear evidence of fracturing followed by soft to severe dissolution etching, whereas others are greatly affected by dissolution weathering. Given the porous nature of the sandstone and its proximity to meteoric water at the surface, it is probable that diagenetic weathering was prevalent."

Within the population of fluvial sands, the dominant shapes are subangular to round (figure 9.5A). The weathering in this population ranges from <10 percent preweathered to approximately 90 percent fresh. Quartz is the dominant mineral, with lesser amounts of feldspar and mica. The well-preserved mica shown in figure 9.4E indicates either a short transport distance, or low current velocity and/or rock fragment armoring, which would protect such fragile grains from destruction.

Quartz grains in TAR 2-3 exhibit pronounced subangularity to subrounding (figure 4C, center and lower right of frame) and edge rounding (figure 9.5C), hallmarks of long distance transport. Fewer quartz grains have step features (figure 9.5D) compared with glacial grains, and these microfeatures are only occasionally deeply inscribed in the grain surface. In addition to well-preserved mica, weathered and well-preserved orthoclase grains are rare in the sandstone samples. Also, occasional intergrowths of various grains are present including quartz cores armored with orthoclase shells.

In TAR 3-3 a representative mix of subangular to rounded grains shows at least 50 percent fresh to weathered particles (figure 9.5A). The round quartz grain in the coarse fraction, bearing v-shaped percussion cracks in the center of figure 9.5A (enlarged in figure 9.5B), is the extreme rounded form observed in water-transported sands. The quartz grain to the right bears a high frequency of micro-dissolution features, the result of diagenesis. Whereas fractured grains occur in the fluvial sands (figure 9.5C), they are relatively uncommon (<10 percent).

Glacial Grains

In till TAR 4-5, some quartz grains show both extensive fracturing and abrasion with pronounced directional features including curved and linear troughs (figure 9.5D–F). Occasionally, some crushed grains show pronounced dissolution effects, which suggest they were weathered after the last glaciation (i.e., during diagenesis).

Fractured quartz assumes many shapes (as shown on figures 9.6A, D, E). These suggest various degrees of glacial crushing and rounding by abrasion in meltwater. The frame in figure 9.6B shows the quartz grain in the center of figure 9.6A along with some perfectly preserved grains of mica. Occasionally quartz occurs as a cleaved form, extensively fractured (figure 9.6C). Still other quartz grains are coated with Si (figure 9.6F), which may have resulted from grinding in water. The grain in figure 9.6F may have inherited its Si coating from diagenetic processes in the sandstone followed by crushing effects (see the lower fractured area on figure 9.6F).

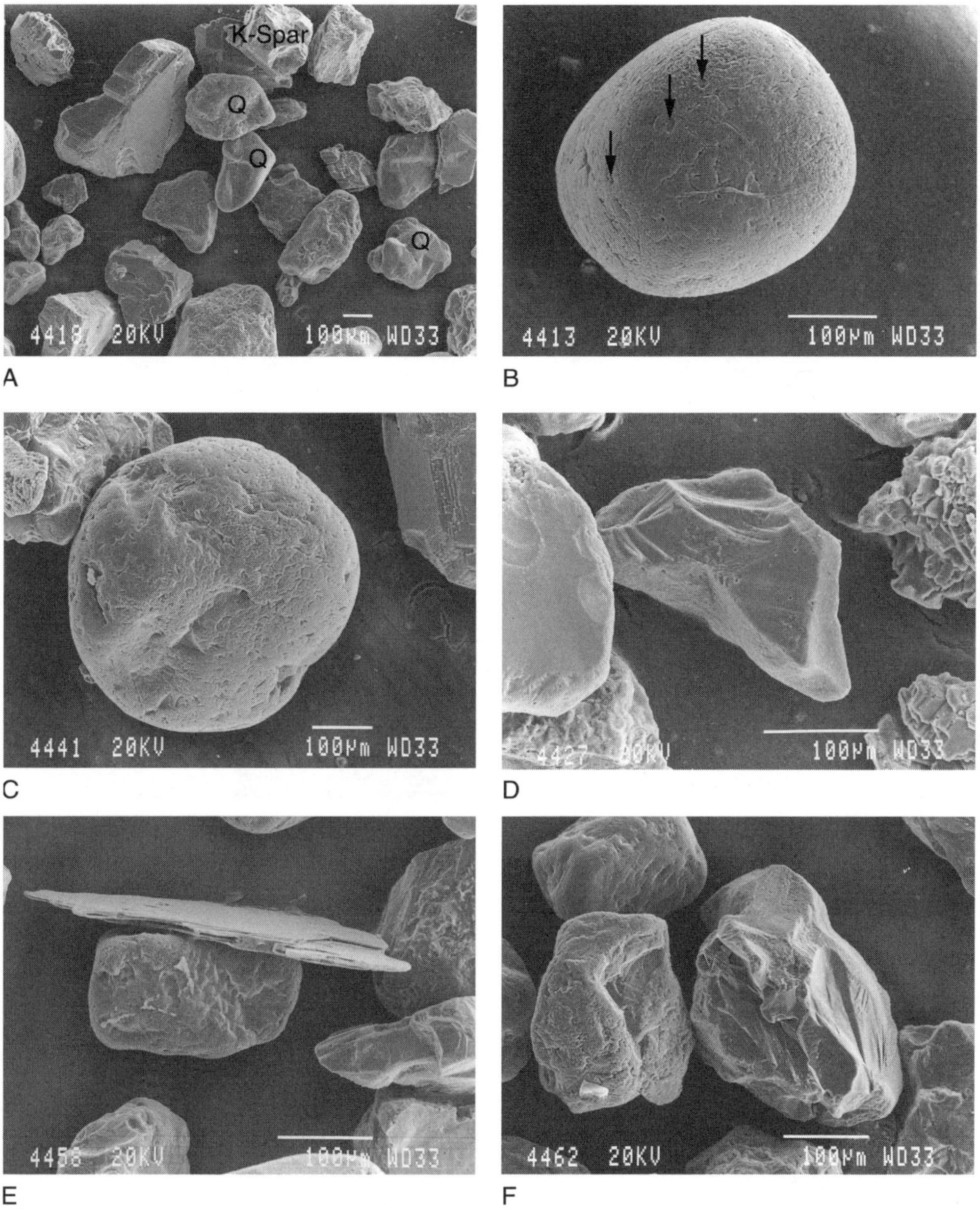

Figure 9.4 (A) General frame of water-transported grains in sample TAR 2-2 showing subrounded to subangular particles. The well-weathered grain at the top center is K-feldspar. Quartz (Q) dominates and many particles have bulbous edges. (B) Round quartz grain with v-shaped percussion cracks (arrows) and dissolution microfeatures. (C) Rounded quartz with multiple v-shaped fractures in sample TAR 2-3. Rounded quartz accounts for more than 50 percent of the grains in this sample. (D) Slightly weathered and fractured quartz in sample TAR 2-1. (E) Biotite and quartz grains in sample TAR 2-1. (F) Two subangular quartz grains with deep (approximately 30 µm) curved and linear troughs in sample TAR 2-1. The grain to the left is preweathered. Reprinted from Fig. 3, Mahaney & Kalm (2000) by permission of the publisher, Taylor & Francis AS.

Figure 9.5 (A) General frame of grains in sandstone from sample TAR 3-3 showing a mix of rounded to subangular grains. (B) Round quartz grain in center of A with pronounced v-shaped percussion cracks. (C) Extensively abraded and fractured quartz, atypical for fluvial sands, with a preweathered surface (top) in TAR 3-3. (D) In contrast, grains in till show well-fractured, striated and preweathered microfeatures (TAR 4-5). (E) Extensively crushed and preweathered quartz grain with a pronounced trough (arrow) 15 µm wide and 20 µm deep. (F) Extensively abraded quartz grain in sample TAR 4-5. Note the minor dissolution effects on the abraded surface. Reprinted from Fig. 5, Mahaney & Kalm (2000) by permission of the publisher, Taylor & Francis AS.

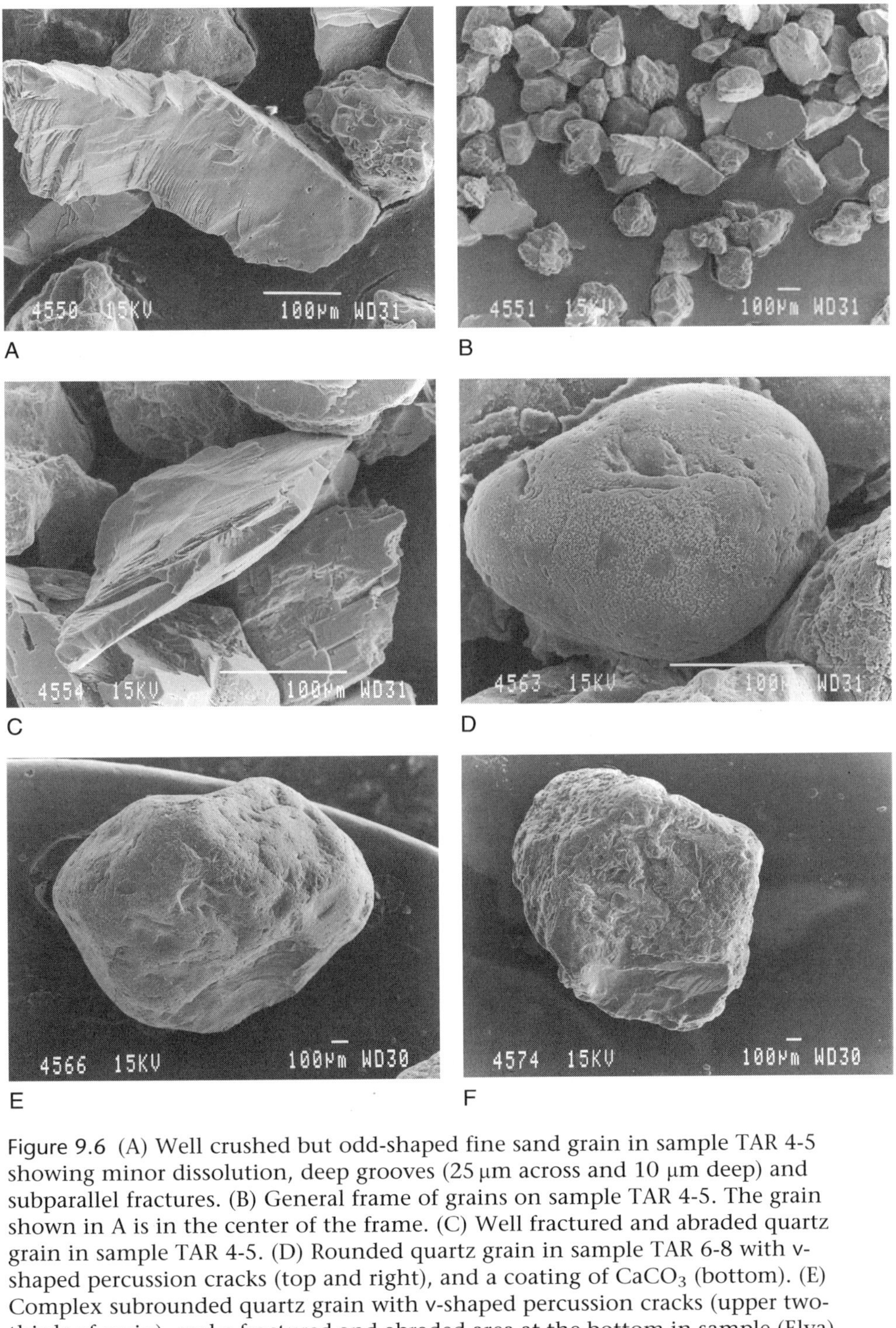

Figure 9.6 (A) Well crushed but odd-shaped fine sand grain in sample TAR 4-5 showing minor dissolution, deep grooves (25 µm across and 10 µm deep) and subparallel fractures. (B) General frame of grains on sample TAR 4-5. The grain shown in A is in the center of the frame. (C) Well fractured and abraded quartz grain in sample TAR 4-5. (D) Rounded quartz grain in sample TAR 6-8 with v-shaped percussion cracks (top and right), and a coating of $CaCO_3$ (bottom). (E) Complex subrounded quartz grain with v-shaped percussion cracks (upper two-thirds of grain), and a fractured and abraded area at the bottom in sample (Elva) E1-1a. (F) Quartz grain that is partially fractured (upper three-quarters of grain) in sample (Elva) E1-1a with thick Si coating. The area at the bottom is freshly fractured. Reprinted from Fig. 7, Mahaney & Kalm (2000) by permission of the publisher, Taylor & Francis AS.

The subsample of grains in indurated sandstone that survived fluvial transport during the Devonian (without a glacial source) showed unequivocally the overall effect of transport in water. While some of the microtextures on these grains might be inherited from source rock, certainly the abrasion microfeatures, v-shaped percussion cracks, and grain roundness are the result of transport in streams, which is discussed at length in chapter 7.

Water transport adds another dimension to the use of SEM microtextures, to isolate a range of microtextures unique to fluvial processes. Since meltwater grains have a glacial source, it is likely they will carry a record of the glacial environment into the fluvial system. This makes the isolation of quartz grains transported solely by water with no glacial source, an ideal test to determine what microtextures fluvial processes might produce on quartz. While one may assume grains in the continental sandstone were transported by water over gentle gradients of a few degrees, their state following release from source rocks is unknown. Certainly there was no frost riving, but chemical preweathering is quite likely to have occurred.

The grain shapes presented and discussed indicate a wide range of microtextures in tills, often with variable edge rounding and sharpness (even on the same grain). From the quantitative summary that follows, the ultimate aim is to extract and analyze the key microtextures representative of the glacial environment—either as microfeatures peculiar only to the glacial environment or as those overlapping it.

Summary

The Phanerozoic stratigraphic setting in southern Estonia and Latvia, where Weichselian tills and glaciofluvial deposits are underlain by indurated Devonian sandstone, provides an ideal situation in which to study contrasting SEM microtextures on sand-size clasts. While some glacial grains are reworked, complete with an original set of microtextures pre-dating the Pleistocene, the quantitative data (figure 9.7A–C) from microscopic analysis show some important differences among the three suites of samples.

The major differences between the late Weichselian tills and the Devonian sands include a marked increase in subparallel linear and conchoidal fractures in the glacial grain suite, as predicted by Krinsley & Doornkamp (1973) and Mahaney et al. (1988b). In direct contrast to the glacial grains, grooves and sharp angular microfeatures are nearly completely absent on the Devonian sand samples and low relief dominates. The major glacial-crushing features include linear steps, deep troughs, and mechanically upturned plates, all of which are completely absent on the Devonian grains. Lattice shattering and arc-shaped steps are <5% on glacial grains.

Grains from the Devonian deposits have a higher incidence of v-shaped percussion cracks, which are common in fluvial sediments and are caused by grain-to-grain collisions in suspension (Krinsley & Doornkamp, 1973). This microtexture, plus edge rounding, occurs on over 50 percent of the Devonian sandstone and glaciofluvial deposits, but in only 20 percent of the late Weichselian tills. Clearly, while there may be some water transport on, within, and underneath glaciers, its effect is greater in terrestrial streams.

Quartz grain weathering is complex and open to different interpretations. There is a considerable range and intensity of preweathered and weathered surfaces present, and dissolution (etched) surfaces and overprinted grains have been observed in this suite of samples. "Weathered surfaces" refers to grains that are completely, or almost completely weathered, without any glacial crushing microfeatures. Preweathered grains show some weathered surfaces in juxtaposition to fresh glacial crushing microtextures. When present in tills the former group is considered to have been preserved in the ice during transport (Mahaney, 1992a), without any grain-to-grain contact; whereas the latter group consists of grains that were weathered prior to glaciation and later crushed during glacial transport.

Precipitation features refer to coatings of Al, Si, and Fe on grains that sometimes survive glacial transport (presumably without glacial crushing) (see Mahaney, 1995a,b). Overprinted grains are those with weathering microfeatures that also carry fresh glacially crushed microtextures. The Devonian sandstones carry higher percentages of preweathered, weathered and precipitation coatings compared with glacial sands. Within the glacial sample suite, there are high percentages of grains (up to 25 percent) that exhibit a wealth of information about weathering histories prior to glaciation. It is apparent that careful scrutiny of such grains has the potential to yield important information about Quaternary interglacial climates and even pre-Quaternary weathering in general.

A comparison of Devonian sands with glacial grains from basal tills shows a range of fresh to weathered round grains with v-shaped percussion cracks dominating in the Devonian fluvial sample. Many grains carried bulbous edges suggesting eolian transport prior to fluvial abrasion. About 50 percent of the Devonian grains observed were preweathered relative to <20 percent in the glacial suite.

Within the suite of glacial grains studied, the *dominant* microtextures reported by Mahaney and Kalm (2000) include: subparallel linear and conchoidal fractures, edge rounding, step features, linear and curved troughs, extensive abrasion (sometimes with v-shaped percussion features), multiple adhering particles, upturned plates, maximum relief and preweathered coatings of various kinds. The data indicate a mix of sharp-edged and edge-rounded grains within all the tills; the sharp-edged grains dominate.

This study, with a large population of samples, makes it possible to define a glacial grain as one with medium to high relief showing abrasion across as much as 40 percent of the observed grain surface. Sharp edges dominate, but approximately one in five grains show some edge rounding which, along with v-shaped percussion cracks, could be inherited from the underlying sandstone. Other microfeatures appear to be the result of glacial crushing. These include linear and curved troughs, steps, and upturned plates. The absence of crescentric gouges (chattermarks) is noteworthy in that, according to prevailing opinion (Folk, 1975; Bull et al., 1980), they dominate on glacial grains. But chattermarks were not observed among the quartz grains in this study and they are exceedingly rare in other glacial stratigraphic studies (Mahaney 1990a, 1995b). While some of these microfeatures are also known to occur on grains transported by eolian and fluvial processes, the extent of crushing tends to reshape grain surfaces with many abrasion and fracture microfeatures; over

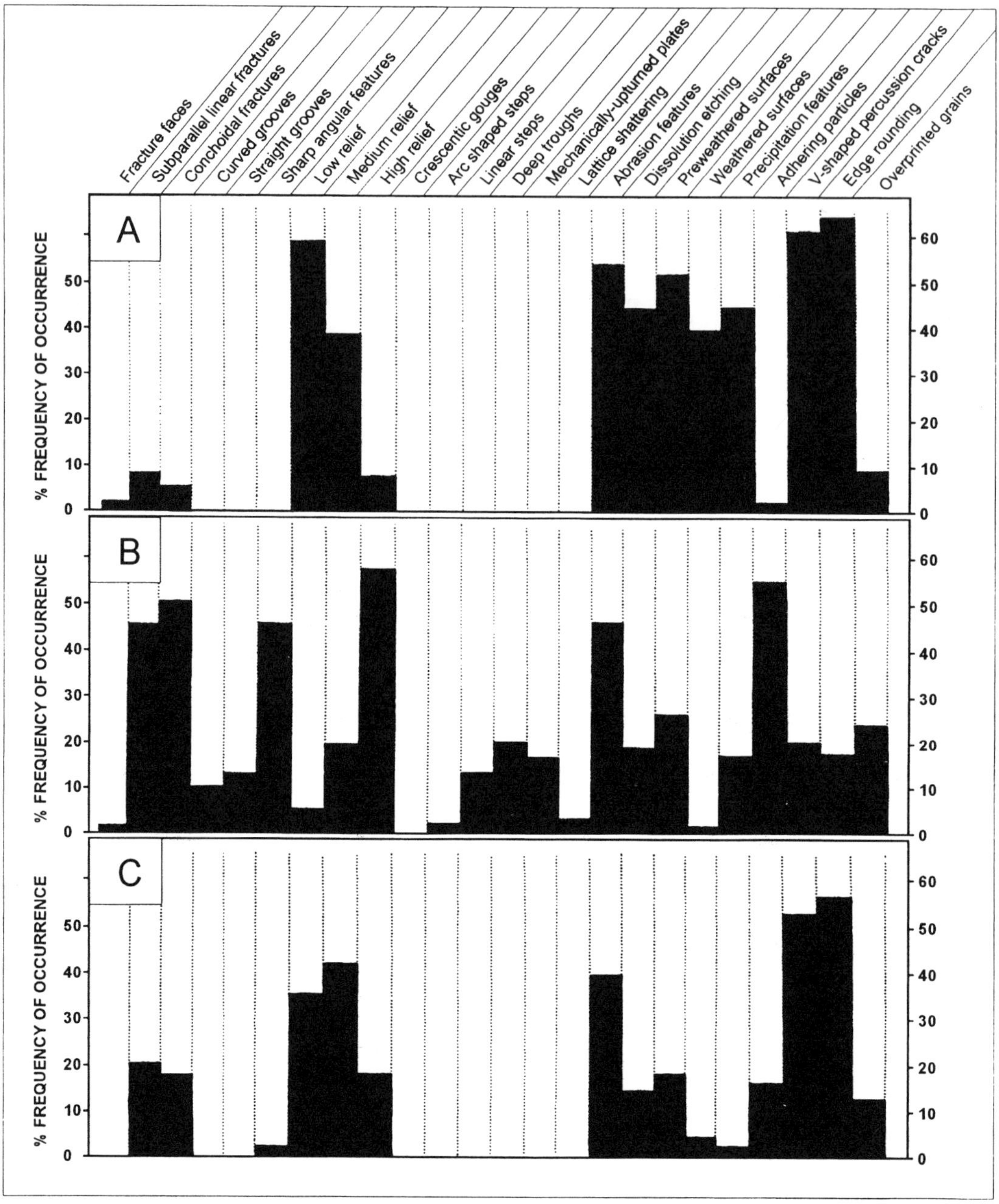

Figure 9.7. Quantitative summary of grains from (A) sandstone, (B) till, and (C) glaciofluvial beds.

60 percent of linear and conchoidal fractures present on glacial grains are deeply inscribed. Because some preweathered grains (~ 15 percent) are found uncrushed, they must be perfectly protected from grain-to-grain contact during their sojourn in the ice.

Continental Glaciation

Wisconsinan and pre-Wisconsinan tills provide a wealth of information (Mahaney, 1990b) on glacial processes and preweathering that occurred prior to glacial entrainment. The middle group of tills (sample WCC2; Wildcat Creek, Indiana), of Illinoian age, shown in figure 9.8A exhibit fresh surfaces, with a range of fracture features including step-like features and severe abrasion that in some cases has likely obliterated fractures. The surrounding grains show a range of dissolution microfeatures resulting from weathering, presumably during a pre-existing interglacial. The most extremely preweathered grains are to the right and left; the grain to the right carries a thin Fe coating on the upper half.

One of the most striking features on grains from thick and extensive valley glaciers and continental glaciers are the straight and curved grooves resulting from severe abrasion. These are unidirectional features, as shown in figure 9.8, that are as much an indisputable proof of the glacial environment as striations on rock. Whether these grains are frozen into the ice and dragged along the surface or are affected by ice drag on particles that eventually break off from bedrock, they are clearly the product of ice contact and are similar to macro-features that have been invoked to prove glaciation for nearly two centuries.

The enlargement in figure 9.8B shows some of the v-shaped percussion features that indicate the effect of water transport, presumably meltwater flow in this case. Also, some rounding along the edge of the grain appears to have occurred when this grain was incorporated into till. The grain in the background is an edge rounded particle, nearly equant, and carrying surface scars of meltwater transport in somewhat greater abundance than the grain in the foreground.

The grain in figure 9.8C (the grain in the lower right of figure 9.8A) is shown in enlargement with directional grooves, step-like features across a cirque-like amphitheater, overprinted with superficial radiating fractures less deeply inscribed. The enlargement shows that a surface plate has been lifted out leaving the amphitheater-like feature, the radiating fractures of which suggest that plate removal may have resulted from a high-stress collision.

Upturned plates, while common on some high-velocity eolian grains, are also common on glacial grains, as shown on the quartz in the upper left of figure 9.8D. The upper part of this grain is weathered whereas the lower four-fifths of the grain are severely disrupted. The grains flanking the quartz grain in the center are all preweathered to varying degrees, presumably by interglacial weathering environments prior to the Illinoian Glaciation. The grain in the center is a representative triangular-faceted particle with directional troughs, fractures, and abrasion features common on glacial grains.

Many glacial grains undergo reshaping from larger fragments, which leaves protrusions in the middle or on the ends. In figure 9.8E, an assemblage of grains (fresh on top, weathered below) shows varying degrees of edge rounding with sharper edges on thinner portions of the grain. The grains at the top show pronounced edge rounding where grain bulk is more massive. Directional troughs and abrasion are pronounced on the grains at top. The enlargement shows nice clean fractures partly overprinted by abrasion, with a single dissolution cavity. The enlargement of the same grain in figure 9.8F shows

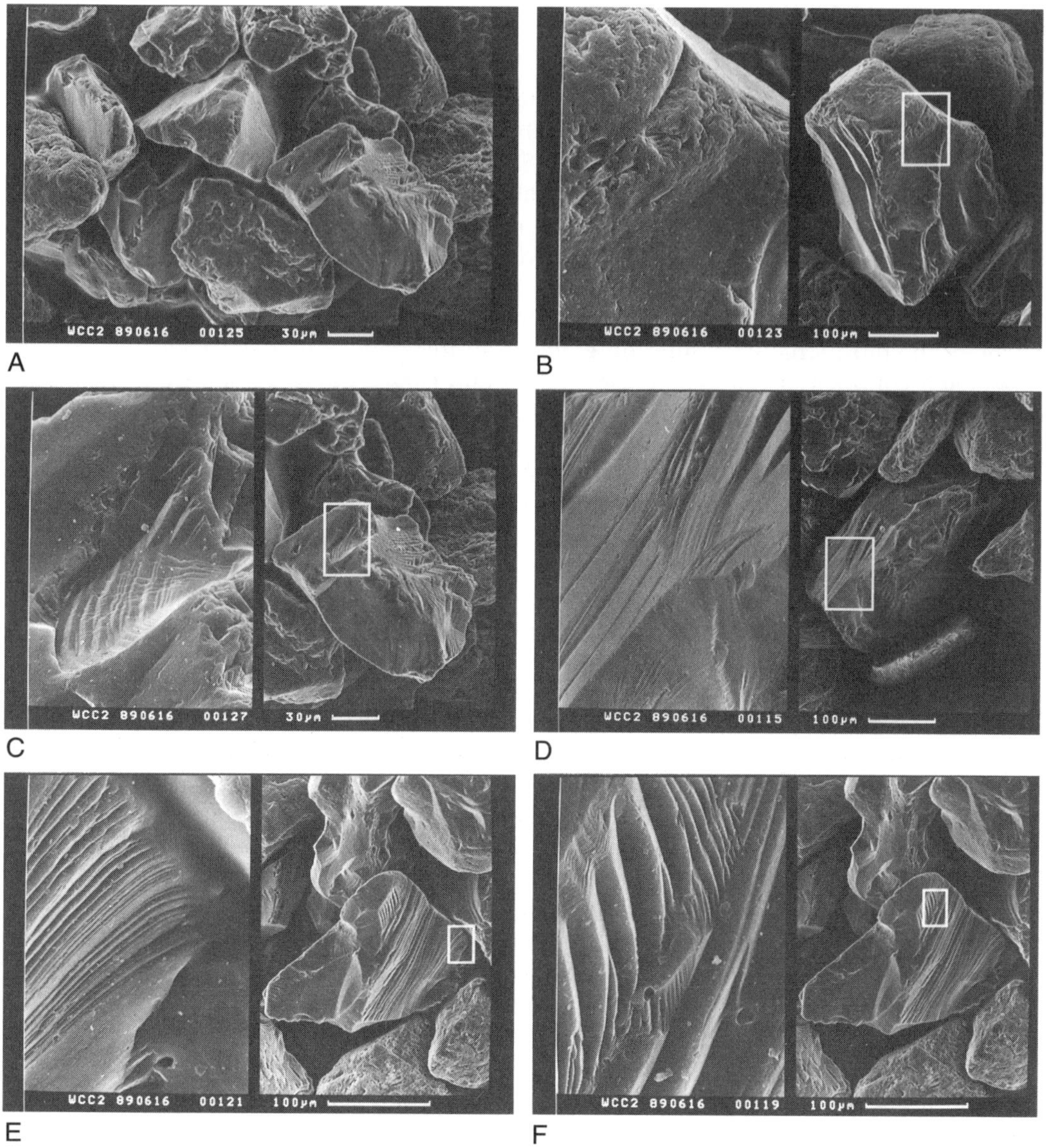

Figure 9.8 (A) Mix of preweathered and unweathered grains in till WCC2 of Illinoian age, Wildcat Creek, Indiana, USA. Fresh grains are clustered in the center, with well-abraded fractures and step-like features. Quartz grains to the top, bottom, left, and right show varying degrees of extreme dissolution which presumably occurred prior to glaciation and survived a sojourn in the ice. (B) Two quartz grains showing contrasting heavy fractures and striations (bottom) and edge rounding together with v-shaped percussion scars from water transport in till WCC2. The enlargement (×5) on the fractured grain shows some v-shaped percussion cracks. (C) Fractured and well-abraded quartz. The enlargement (×5) shows high- to low-frequency fractures merging into step features in the upper center.

two dissolution cavities on fresh fractures that are obviously not preweathered features.

Antarctic Tills

The grain shown in figure 9.9A is a heavily damaged grain representative of tills emplaced by continental ice. Refashioned into a triangular-shaped particle, it exhibits deep fractures, an upturned plate along the bottom of the grain, a somewhat curved directional trough from the top to the mid region, and step-like features at the top right; much of the grain is extensively abraded. The enlargement (×5) shows detail on the deeply embedded fractures. The grain at the bottom is partly preweathered while the grain to the upper left is almost exclusively covered with deep wedge-shaped craters flanking a long, nearly straight directional trough.

Detail on the grain in figure 9.9A is shown in figure 9.9B with an enlargement (×5). This detail shows, among other items, an irregular series of zig-zag fractures rarely observed on glacial grains, but known to occur on neotectonically released grains in caves (Mahaney & Sjöberg, 1993). Considered to be the product of high stress fields, such microfeatures might be expected in thick continental ice with a low water content where high local stick–slip movement might produce pressure capable of dislocating the lattice. Also in Till 42 (for the till location see Mahaney et al., 1996a), heavy abrasion has obliterated fractures (see figure 9.9C).

Till 134, near the MacKay Glacier (Mahaney et al., 1996a), displays grains with high degrees of abrasion and directional troughs. The grain in figure 9.9D shows a thin, narrow trough oriented north–south on the frame, with another and much larger trough oriented normal to it. Some preweathering is noticeable on the upper nose of the grain but the lower three-quarters is almost exclusively abrasional. The enlargement shows the detail of the large groove with subparallel fractures thinly embedded into the upper part of the groove.

Till 480L, near Prydz Bay (Mahaney et al., 1996a), also produced grains with a multitude of abraded surfaces. Faint vestiges of fractures were apparent on some of them, as shown in figure 9.9E along with sharp edges and subdued relief, similar in kind to grains reported from the West Antarctic Ice Sheet by Tulaczyk et al. (1998). While adhering particles are rather scarce on most of the sands shown here, they are more prevalent on the abraded grains.

Figure 9.8 (*cont.*) (D) Quartz cluster in till WCC2 showing heavy fractures and grooves on the center grain, upturned plates on the grain in the upper left, and preweathered grains in the top center, upper right, and right. The enlargement (×5) shows abraded fractures. (E) Quartz from the fine sand fraction of WCC2 showing well formed and partly abraded fractures, part of an abraded groove left of center, and an impact zone with radiating fractures near the top of the grain. Grains on the top are well fractured and abraded; grains toward the bottom show severe dissolution. The enlargement (×10) shows details of abraded fractures. (F) Detail of the deep fractures on the grain in E. The enlargement (×10) shows variations in high- to low-frequency fractures presumably caused by variable vibrations through the grain.

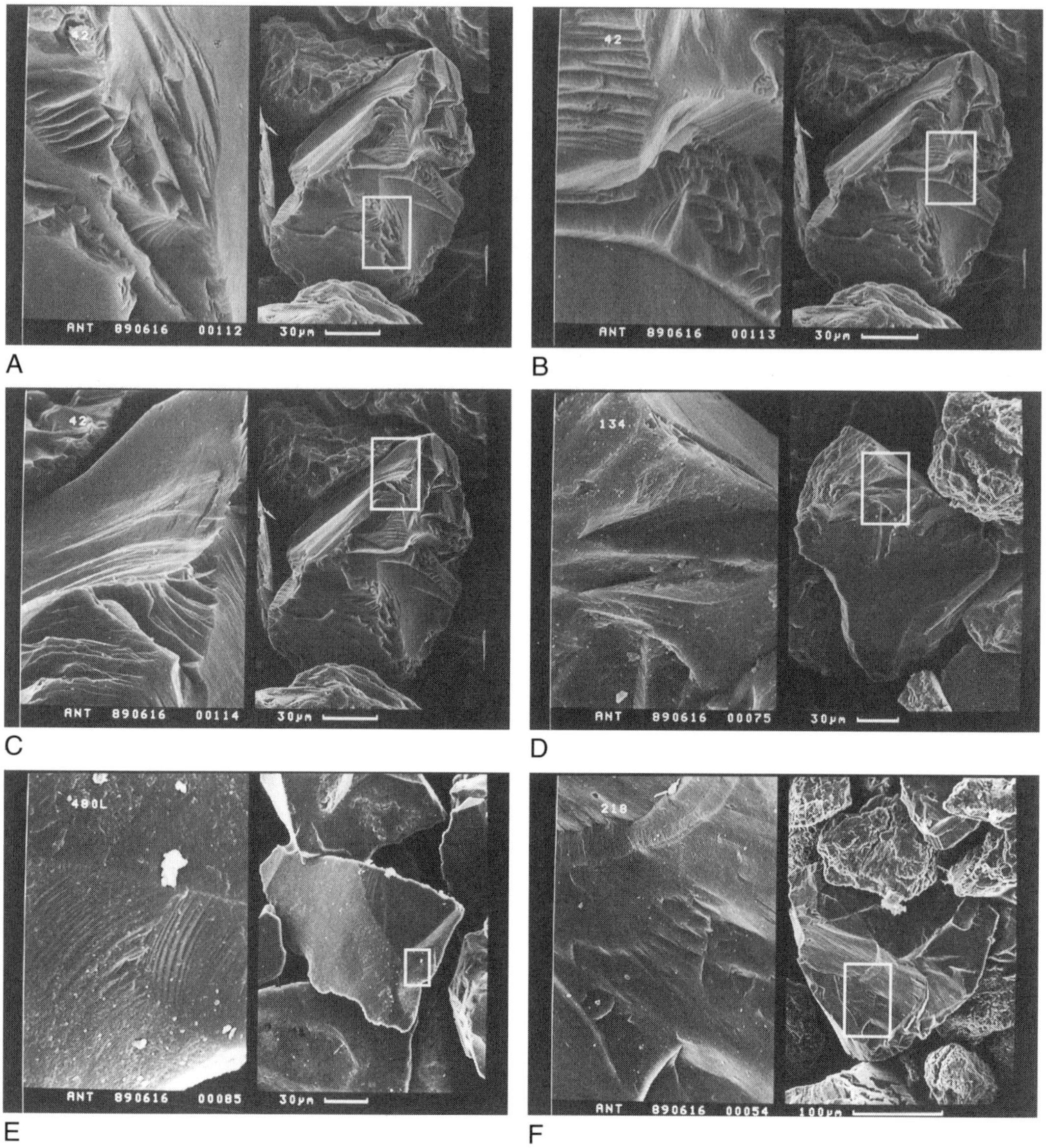

Figure 9.9 Tills from the Dry Valleys, Antarctica. (A) Quartz showing extreme fractures and abrasion microfeatures. The enlargement (×5) shows combinations of grooves and fractures rarely seen on middle-latitude and alpine grains. (B) The same grain as in A from till 42. The enlargement (×5) shows normal conchoidal fractures with zig-zag fractures normally seen only on grains affected by tectonic movement. (C) The same grain as in A. The enlargement (×5) shows the effect of abrasion on deep fractures. (D) Quartz in till 134 with sharp edges, severe abrasion, fractures, and a deep groove some 30 µm deep and 40 µm wide. The enlargement is ×5. (E) Sharp-edged quartz in till 480L. The enlargement (×10) shows fracture vestiges which survived abrasion. (F) Sharp-edged, fractured, and abraded quartz. The enlargement (×5) shows abraded multiple striations.

Quartz in Till 218, from the Scott Glacier, exhibits well-abraded steps and fractures with thinly embedded radiating fractures that could result from high stress impact with other grains in the ice (figure 9.9F). Note that this grain is surrounded by a multitude of preweathered grains transported by inland ice.

The most intense crushing comes from the thickest ice, as indicated by observations made by Mahaney et al. (1988a) following a thorough examination of samples from Mount Kenya and southern Canada. Quartz sand grains in thick ice, such as the Antarctic Ice Sheet, are presumably subjected to transport over long distances, with maximum stress generated by stick–slip mechanisms at the base of the ice.

Alpine Glaciation

Quartz and other mineral grains recovered from various tropical and middle latitude alpine areas show ranges of fracture microfeatures, striations, and abrasion that appear to be related to ice thickness and distance of transport (see Mahaney et al., 1988a). Quartz and other minerals surviving transport in ice in alpine cirques carry a range of fractured and abraded preweathered grains showing a low frequency of occurrence of subparallel linear fractures, conchoidal fractures, and abrasion that suggest low shear stress, low stress field conditions for stick–slip occurrences, and, of course, minimum transport distance. As previously discussed, many alpine subglacial tills are well lubricated with water giving them very low shear stress and high pore water pressure, and making it difficult to fracture grains carried as basal load. This condition must also affect stick–slip processes (Barcilon & MacAyeal, 1993; Mahaney, 1995b) creating a situation where, even if stick–slip were to occur, the resulting shear stress would not overcome the shear resistance of the grain.

Reworking of grains in the glacial environment, while hardly confined to alpine locales, is evidenced from several suites of samples from African massifs, the Andes, Tibet, the Rocky Mountains, and the European Alps. The greatest ratio of preweathered to fresh grains is often observed in alpine areas for both Neoglacial and late Pleistocene tills studied to date. There are many glacially crushed grains with extensive dissolution features, indicating they underwent glacial transport at some distant time and were later weathered in a subaerial weathering zone, and eventually re-entrained in an active glacier.

To illustrate this point, the grains shown in figure 9.10 show a multitude of representative fractured and abraded microfeatures on a single grain in Till HK6 from the Zillertal Alps of Austria. The large grain shown in figure 9.10A contains both new growth features (center), which give the appearance of slightly distorted hexagons, and large (~50 μm) craters, which may be confined to relatively young quartz (Molén, 1992). The craters may be related to new growth that survived collision within the ice, or it may be related to an unloading effect following the release of confining pressure. Whatever the origin of these crude cavities, they are confined to the glacial environment, and presumably result from grain-to-grain contact.

The smaller grains surrounding the fine sand in figure 9.10A are almost all quartz, and are all either preweathered like the larger grain, or they are abraded without much in the way of fracturing effects, as in the grain to the right. The enlargement in figure 9.10A shows the degree of preweathering. The same grain is shown in figure 9.10B; the enlargement shows only fresh fractures in the upper half with preweathered fractures below. A close-up of the same area in figure 9.10C illustrates the differences between soft preweathering and fresh fractures that occurred during Neoglaciation.

The grain shown in figure 9.10A is enlarged in D, with a new growth of quartz showing a slight distortion of the lattice. The genuine absence of fresh fractures on this grain together with minor v-shaped percussion scars suggests the latest transport involved mainly collisions with other grains in a fluid.

The quartz grains shown in figure 9.11 illustrate at various magnifications the range of microfeatures one might encounter with late Pleistocene valley glacia-

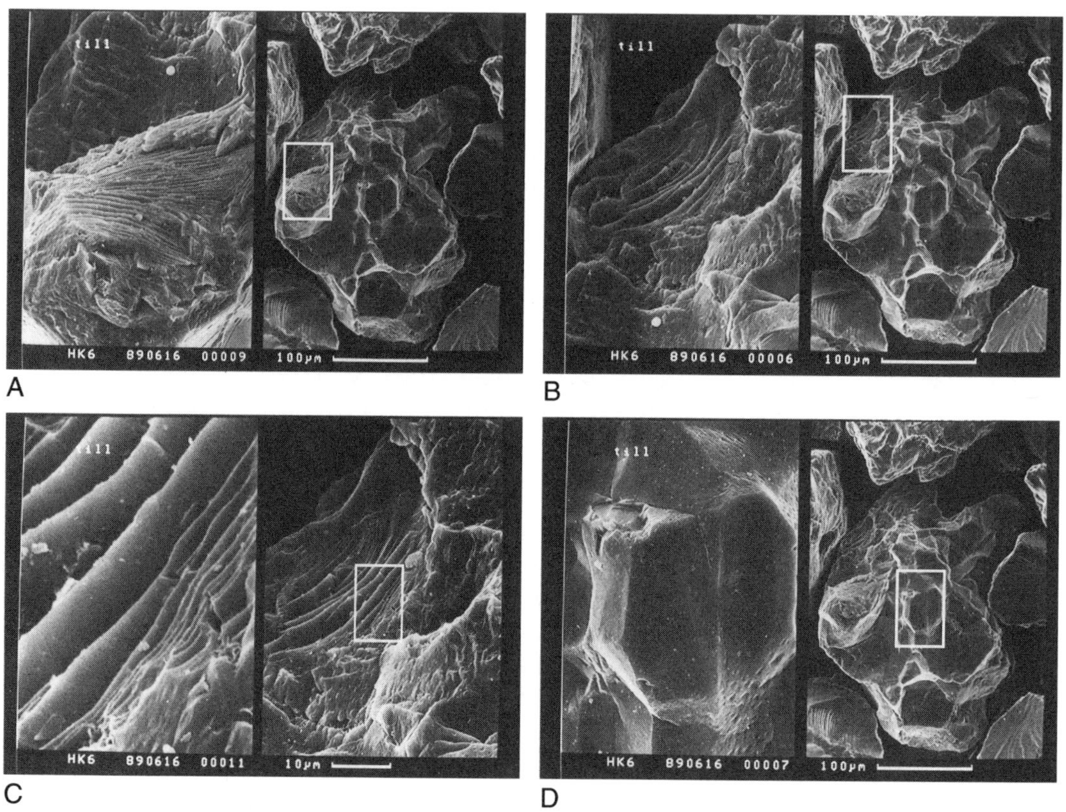

Figure 9.10 (A) Quartz sand in sample HK6, late Pleistocene till, Zillertal Alps, Austria. The grains at the bottom and top are quartz; the grain to the right is biotite. This is representative of quartz with extensive new growth. The enlargement (×5) shows rather well-preserved glacial crushing. (B) The same grain as in A showing somewhat lower frequency conchoidal fractures. The enlargement is ×5. (C) Conchoidal crushing on the grain in A showing fresh fracture planes. The enlargement is ×5. (D) Details of new growth of quartz on the grain shown in A. Minor v-shaped percussion scars are apparent but, for the most part, the new growth is unaffected by glaciation.

tion. The two fresh grains shown in figure 9.11A are well-faceted complex grains showing preweathered dissolution on their upper parts and minor fracturing and abrasion on the lower flanks. The fracture pattern on the lower right appears to be an impact structure with radiating fractures merging with a later formed microstriation some 6 μm in width and depth, which is oriented normal to the fractures. Higher magnification of the lower part of the grain in figure 9.11A, shown in B, illustrates the freshness of the radiating fractures and the step-like nature of the irregular groove. In figure 9.11C, the grain in figure 9.11B is shown at the same scale, with the enlargement showing the subparallel linear fractures partly abraded. The fractures on the edge just above the box in figure 9.11A are shown in figure 9.11D. The enlargement shows fresh fractures about 2.0 μm in width and about 3.0 μm deep on an abraded surface.

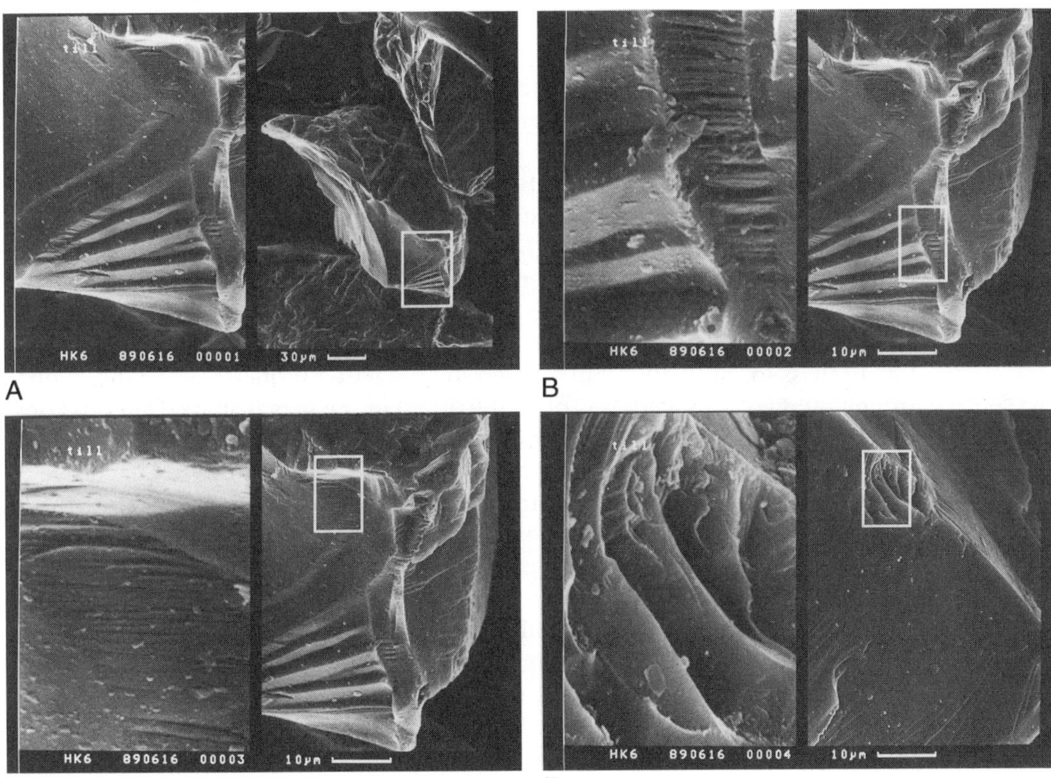

Figure 9.11 (A) Angular quartz in till from site HK6, Zillertal Alps, Austria. The enlargement (×5) to the left shows radiating grooves with widths and depths of 4-5 μm. (B) Close-up view of the striations in A showing subparallel fractures climbing vertically along the grain surface. The enlargement (×5) shows detail of the grooves and fractures all in pristine condition. (C) Fresh subparallel fractures on the top of the grain in A, with an enlargement (×5). (D) Fresh conchoidal fractures on a sharp edge of the grain in A with few adhering particles. The enlargement is ×5.

Conclusions

The glacial grain, depicted in Krinsley & Doornkamp (1973) as a triangular-faceted, refashioned quartz particle, with a multitude of fracture and abrasion features, is one of the common particles found in the sand fraction of tills. As shown here, it is by no means the only particle found in till, although depending on the ice thickness and distance of transport, it could amount to 40 or 50 percent of the grains observed in till. There are other grains, such as equant grains with minor damage, flake-shaped grains often with an abundance of abrasion and little fracture, partly streamlined particles with variable damage, and preweathered grains. The preweathered grains often reach 30 percent of the total population of grains in till observed under the SEM, which means that as many as one in three of the grains analyzed might carry important information about preglacial weathering. Indeed, in some instances, the degree of preweathering might yield important information about paleoleaching and the strength of interglacial climates (see Mahaney, 1990a).

Microtextures considered to be of unequivocal glacial origin include directional curved grooves or troughs in mutiple pairs. Occasional straight grooves could have a tectonic origin whereas multiple grooves are always glacial in origin. "Groove" is a useful term for lightly embedded striations, with "troughs" reserved for deeply embedded features, in the order of 5 μm or more. While lenear grooves are occasional byproducts of neotectonic release and movement within fault planes, all troughs and irregular curved grooves are as much the exclusive product of the glacial environment as the macrostriations that have been used for nearly two centuries as chief indicators of glaciation. In the micro world it may be that minerals harder than quartz (topaz, zircon, etc.), while rather low in mass in most tills, could, when in contact with quartz, leave directional troughs and grooves. Perhaps this is the origin of the pencil-shaped zircons commonly found in some tills (Mahaney & Milner, 1998).

Deeply inscribed fractures are common on glacial grains, but not exclusive to the glacial environment. Over a large population of grains, however, conchoidal and subparallel linear fractures would be expected to be one of the most frequently occurring forms. Only thick valley glaciers and glaciers of continental proportions would be expected to produce deeply embedded features of this type, although they are also found in rock glaciers (Lewis & Hanvey, 1993).

Sharp edges where grains have been worn away by abrasion are also a hallmark of the glacial environment. They are not unequivocally glacial, but taken as a proportion of the whole in a large population of samples they can be expected to have a frequent occurrence.

Relief is often considered to be at a maximum on glacial grains, but with extreme abrasion, which occurs with thick ice, the grain surface may be worn to produce minimum relief. With cirque ice and outlet glaciers, grains often have moderate to high relief, which may in fact mislead the investigator into thinking considerable grain modification and reformation has been achieved.

Crescentic gouges (Folk, 1975), often touted as the common glacial microtexture, are actually extremely rare (figures 5.22 and 9.8F)—so rare that it proved possible to recover only a few observations among the hundreds of thousands of analyzed grains.

Step-like features resulting from well placed and deeply embedded fractures are unequivocally glacial in origin.

The fact that so many preweathered grains make the glacial trip without damage means that not all glacial grains come into contact with one another. While they provide important information on interglacial climates, their coatings can range from carbonate and silica to iron and manganese; even clay coats have been documented from the Antarctic (Mahaney et al., 1996a).

As with other geologic agents, to prove a glacial origin requires a large number of samples from different size fractions (very coarse–coarse; medium; and fine–very fine sand) and also replicate samples.

10 Tectonic Microfeatures

Outside of the work of Kanaori et al. (1985, 1991) and Mahaney & Sjöberg (1993), little research has been undertaken to identify microtextures within fault planes, in shear zones adjacent to faults, and/or as a result of neotectonic release of source rocks (Kanaori et al., 1985). Additional studies on the use of quartz grains from fault gouges is warranted, to establish the age of faulting and the degree of grain modification.

Building on this initial work, with limited samples from the San Andreas Fault in California and the Boconó Fault in Venezuela, a tentative collection of frames has been selected to provide some possibilities for future work. Aside from the production of fine-grained material such as mylonite, in the mill of the fault and in the adjacent shear zones on either side, it is theoretically possible to produce rounded and stretched grains of metamorphic minerals of variable composition, and grain boundary microcracking (Kanaori et al., 1991). In, or immediately adjacent to the fault, it should be possible to easily overcome grain tensile strength, and perhaps also overcome the elasticity of grains such as quartz, resulting in tensional release producing strands of silica that might resemble a torn wire or cable. Sawtooth fractures produced from neotectonic release in caves, might also be produced in the fault plane. While not totally unknown in till, sawtooth fractures could be said to be diagnostic, if they were frequently encountered in any analysis of grains recovered from faults.

Spindle shapes and all types of rolled forms of quartz might also be expected from the movement of fault planes. Indeed, the prime product of low-energy fault movement might be excessive abrasion, where, if the grains are not totally reformed or obliterated in the fault, they might be expected to be abraded, polished, and perhaps slickensided to the point where they take on sheen easily recognizable on the SEM.

With these concepts in mind, I will examine, first, neotectonic release in caves in Sweden (Mahaney & Sjöberg, 1993); and second, a suite of samples available from two major fault zones: the San Andreas Fault along the California coast, and the Boconó Fault stretching from the Pacific Coast of Ecuador to the Caribbean Sea along the Venezuelan Coast.

Swedish Caves

The cave Trollhålet, situated close to the city of Lysekil in a granite hill at 13 m a.s.l. (see figure 10.1 for the location), was formed by marine abrasion along an approximately 0.35 m wide subvertical crevice of neotectonic origin striking N20°W–S20°E. The crevice, rounded by abrasion, has walls polished by cobbles moving in and out, powered by surf breaking currents. Some 70 examples of caves of this type are known in Sweden. With a height of 2 m and a width of 0.95–1.5 m, the passage is 18 m long. Formed originally during early Quaternary interglacials, and successively widened during later interglacial periods, caves of this type were unaffected by low sea-level stands during the last glacial. Exfoliation of quartz inside the cave is attributed to either neotectonic processes or hydration caused by meteoric water leaking in from overlying infill emplaced by industrial activity (Mahaney & Sjöberg, 1993).

Ulorna, a boulder cave in a collapsed mountainside (figure 10.1), is a huge collapse. It was first described by Hillefors (1970), who described it as "a huge cone of gigantic rocks at a steep fracture-zone, along which the mountainside has fallen out" (translated by Sjöberg in Mahaney & Sjöberg, 1993). The collapsed rocks are situated on a moraine outside of which two low, glacially striated roches moutonnées are found, just above the present shore. Hillefors' (1970) interpretation of this feature was that it formed by pressure release (e.g., unloading) along surface-parallel joints steeply dipping toward the fjord. He dated the collapse to either late glacial or postglacial time, and considered that it could have been triggered by an earthquake.

The site, investigated by Sjöberg in 1989, was found to contain, within the main rock mass, an interconnected system of cave passageways of several hundred meters' length (Sjöberg, 1989). Although similar to other neotectonic caves of type 2 (i.e., caves in collapsed mountain slopes, earlier described by Sjöberg, 1987), the Ulorna cave is characterized by the presence of abruptly broken striae on some of the boulders lying in a mass on the cave floor, which indicates a sudden collapse of the whole hillside. In one of the outer cave passages, a gigantic rock, forming the straight and vertical wall of the passage, is covered by a dense net of fresh, parallel striations. These striated surfaces appear absolutely fresh, as if they had been found at the front of a melting glacier. Thus, a sample of this wall was collected for SEM test, in order to determine, if possible, whether or not the striations had a glacial or a tectonic origin.

The SEM provides evidence for fracturing caused by exfoliation and striated surfaces that may have a neotectonic origin (figure 10.2A). More than 60 percent of the particles from Trollhålet cave observed on the SEM showed coatings of Fe and Al and dissolution etching (figure 10.2B), indicating that the exfoliation slab had been subjected to considerable chemical weathering prior to mechanical release. This supports the argument that the cave is old, possibly dating from the upper Neogene. A high percentage of adhering particles on larger sand grains are also coated with Fe and Al. Minor conchoidal fracturing on the edges of these sand particles (see figure 10.2A) may result from mechanical release from source rocks.

The deep and extensive fractures from neotectonic release in Ulorna cave shown on figure 10.2D are possibly due to lattice failure, although they could be rather unusual features resulting from mechanical release.

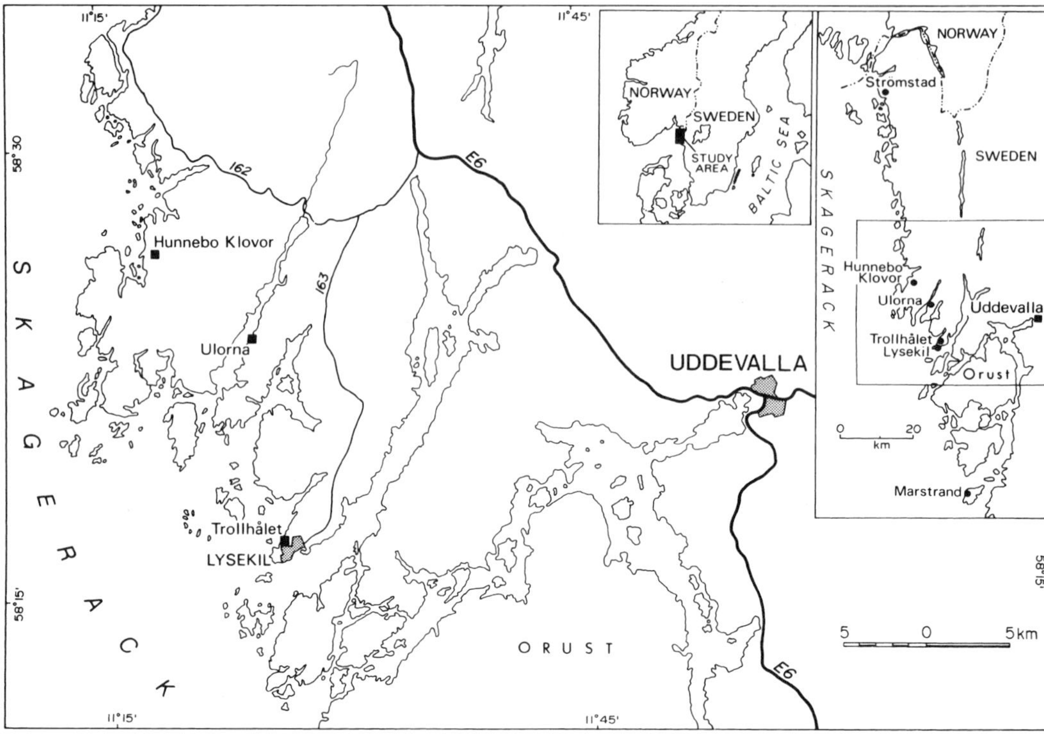

Figure 10.1 Location of caves in Bohuslän, southern Sweden. Trollhålet and Ulorna caves are described in the text. Weathered grains of probable upper Neogene age and glacial grains from the last glaciation in the Hunnebo Klovor are described in Mahaney & Sjöberg (1993). Reprinted from Mahaney & Sjöberg (1993) by permission of the publisher.

The Ulorna cave samples are derived from a striated boulder near the cave mouth that is considered to have a neotectonic origin. Edge fracturing and upturned plates (figure 10.2C) on these grains may result from high-velocity impact generated by neotectonic forces when the boulder was released from source rocks. The resultant conchoidal fractures leave irregular-shaped edges that are not found on glacial grains. Indeed, the prevalence of upturned plates (figure 10.2C) would be very unusual even for grains transported by continental glaciers (Mahaney et al., 1988a). The prevalence also of quartz dissolution and overall weathering of particle surfaces (figure 10.2E) suggest, as with the Trollhålet samples, that relatively strong chemical weathering occurred prior to release by neotectonic processes (see chapter 8). Multiple weathering episodes of a similar nature have been interpreted in other areas, such as on Mount Kenya (Mahaney, 1990a), on river terraces in Uganda (Krinsley & Doornkamp, 1973, plate 87), in tills emplaced by the Laurentide, Fenno-Scandian, and Antarctic ice sheets (Mahaney, 1995b), and in samples from subaqueous environments (Krinsley & Marshall, 1987).

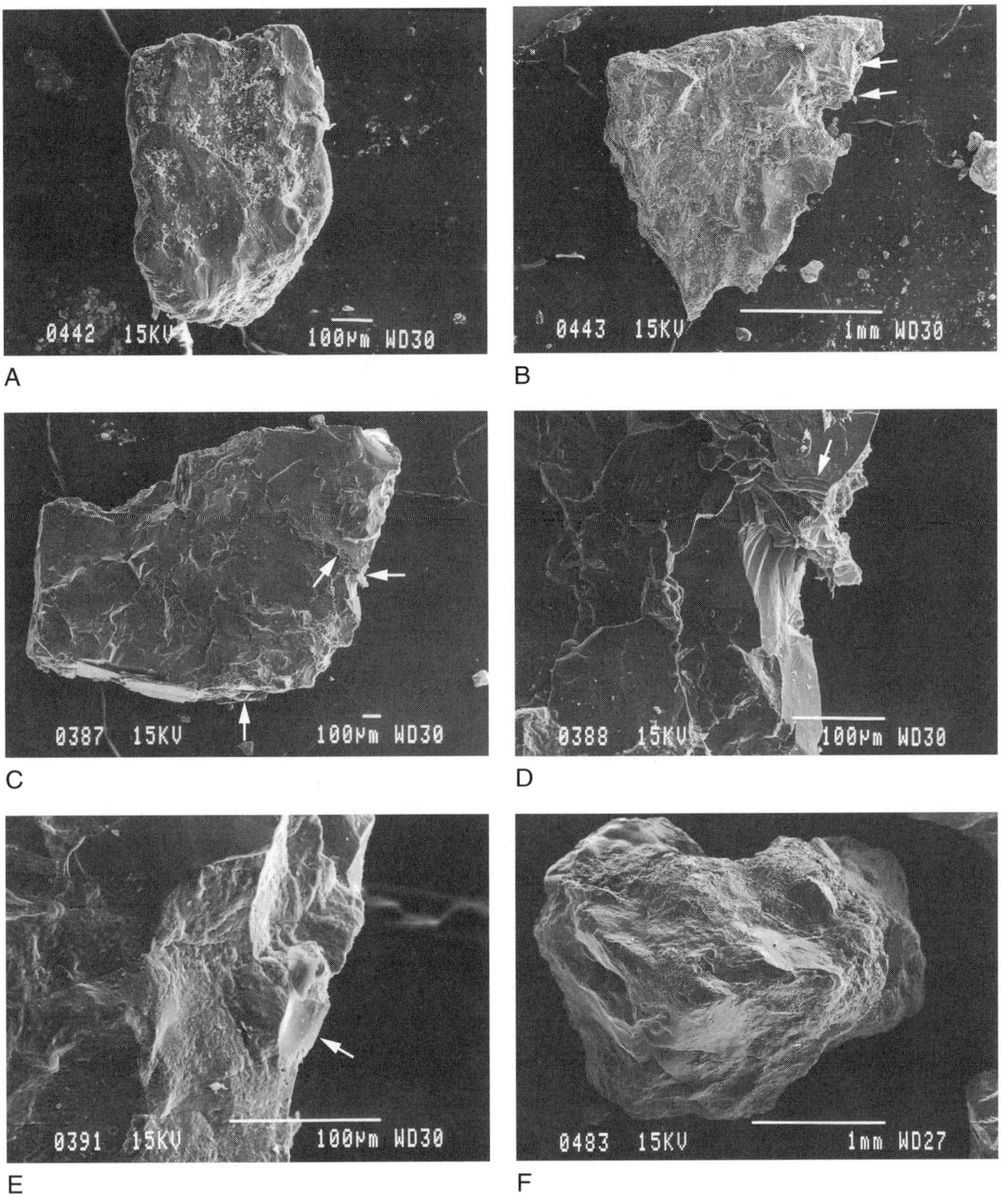

Figure 10.2 (A) Quartz from Trollhålet with numerous adhering particles, well-weathered conchoidal fractures, and a large concavity in lower left. (B) Quartz grain with adhering particles, coatings of Fe and Al, and deep, extensive exposure of sawtooth fractures (arrows) with lattice shattering or imperfections possibly caused by mechanical release. (C) Quartz grain from the Ulorna cave showing edge fractures and upturned plates (arrows) that may result from earthquake-generated mechanical release. (D) Enlargement of the grain in C showing fractures and upturned plates (arrow). (E) Enlargement of the upper right-hand part of C showing a fresh fracture face (arrow) where final breakage might have occurred. (F) Large quartz grain with fresh fracture faces (light areas, center) surrounded by weathered surface. Reprinted from Mahaney and Sjöberg (1993) by permission of the publisher.

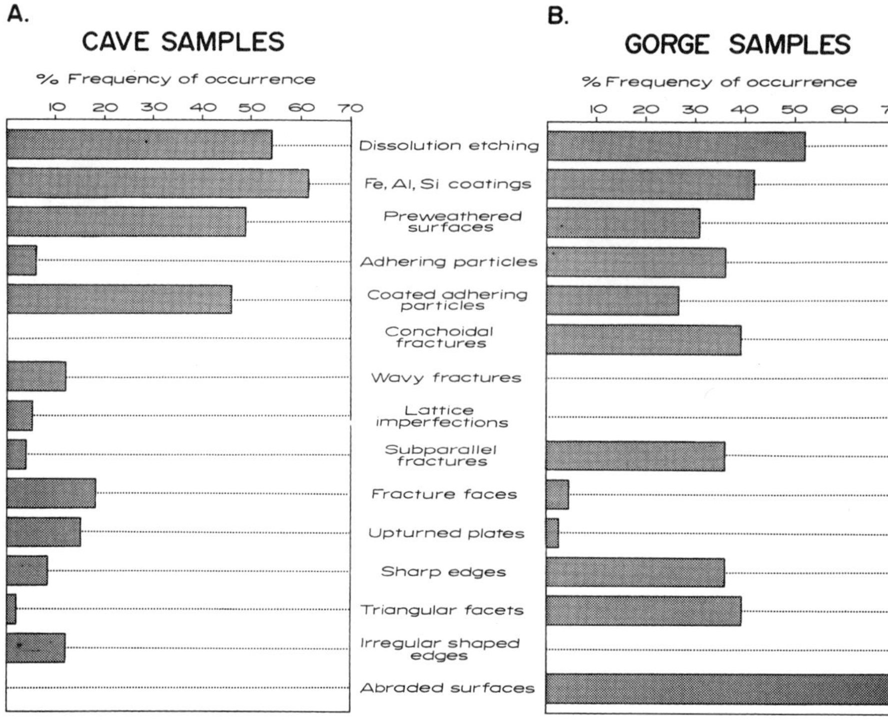

Figure 10.3 Quantitative summary of cave and gorge samples. Reprinted from Mahaney & Sjöberg (1993) by permission of the publisher.

A quantitative summary of quartz particles observed from the cave and gorge samples (discussed in Mahaney & Sjöberg, 1993) is shown in figure 10.3. When compared with nearby gorge samples affected by chemical weathering and glaciation, weathering microfeatures of quartz grains are somewhat more intense and severe within the cave samples. In particular, preweathered surfaces occurred more frequently among the cave samples. Most adhering particles are coated with various ratios of Fe and Al; some grains are coated primarily with Si and others with clay minerals. The sharp contrast in the frequency of occurrence of adhering particles on quartz grains in the cave and gorge samples suggests that most small particles found coating quartz sands originate from glacial grinding, as originally proposed by Smalley (1966). The contrast in the presence of conchoidal fractures on the gorge versus the cave samples is striking and illustrates that mechanical release of source rocks by frost in caves produces only a small number of fractures.

Wavy (sawtooth) fractures (figure 5.67), while restricted in occurrence on the cave grains, and not reported from the gorge samples, are also very rare on glacial grains (chapter 9). Lattice imperfections or shattering was not observed on the gorge samples, but occurred on less than 5 percent of the cave samples. Along with conchoidal fractures, subparallel linear fractures are more common on the gorge grains than on the cave samples. Fracture faces, more prevalent among the cave samples, presumably result from mechanical release of source rock material, as postulated by Mahaney et al. (1991) and discussed in chapter 8. Upturned plates, rather rare on glacial grains (Mahaney, 1995a,b), appear far more frequently in cave samples and may be due to high-velocity impact caused by neotectonic processes. Sharp edges and triangular facets are far more common on glacial grains than on cave grains.

Mechanical weathering and frost action in caves appear inefficient at producing angular grains. On the other hand, irregular sharp edges on quartz occurred predominately within the cave sample suite. Perhaps the most striking contrast occurred with respect to abrasion and, as expected, none of the cave samples showed abrasion surfaces.

Venezuelan Andes

The massive Boconó Fault, stretching from Ecuador to the Caribbean Coast of Venezuela, is one of the major fracture zones on Earth. In the Venezuelan Andes, in the vicinity of Sierra Nevada National Park, the fault gouge zone is found in Paleozoic gneiss and schist, which provide abundant quartz that can be studied for its microtextures. The images in figures 10.4 and 10.5 illustrate the breakage features and slickensided surfaces that can be encountered in fault gouge zones.

Polished surfaces are not as prevalent as one would expect but they can be found. Common among these grains are numerous fractures and thin slabs of quartz, often polished to a considerable degree.

In the area where Valera and Boconó faults converge, near Apartaderos on the outskirts of Sierra Nevada National Park, a sinistral wrenching motion produces a wide shear zone where rolled grains of quartz, zircon, and garnet provide a measure of the effect of extreme pressure and pervasive fluids on grain crystallography. The imagery shown in figure 10.6 represents some common features seen in grains recovered from bedrock in the area. For details of the location, see Mahaney & Kalm (1996).

In the shear zone of the Boconó Fault many grains are rolled, stretched, flattened, and cracked by the wrenching motion along the main fault line and either side of splays branching off to the north. As shown in figure 10.6, many grains are wrinkled and deformed first, followed by cracking which exposes ridges and cleavage in some instances (figure 10.7). Still other grains from the Boconó Fault become incorporated in talus. The primary features of these grains, as shown in figure 10.8, include their rolled appearance with realigned crystallography (figure 10.8F), cracked shell (figure 10.8C and D), and disrupted surface (figure 10.8C). The weathering features on these grains followed their modification in the shear zone of the fault system, and in some cases (figure 10.8F) occurred during fluvial and lacustrine fluvial transport.

In the area of Pedregal, grains of different lithologies—quartz, sillimanite, kyanite, and mica—in proglacial deposits subjected to tectonic readjustment following deposition, show evidence of short throw displacement along fault lines. In these zones, where tectonic adjustment along short throw faults of <100 cm occur, quartz grains tend to exist as thin wafers or shavings (figure

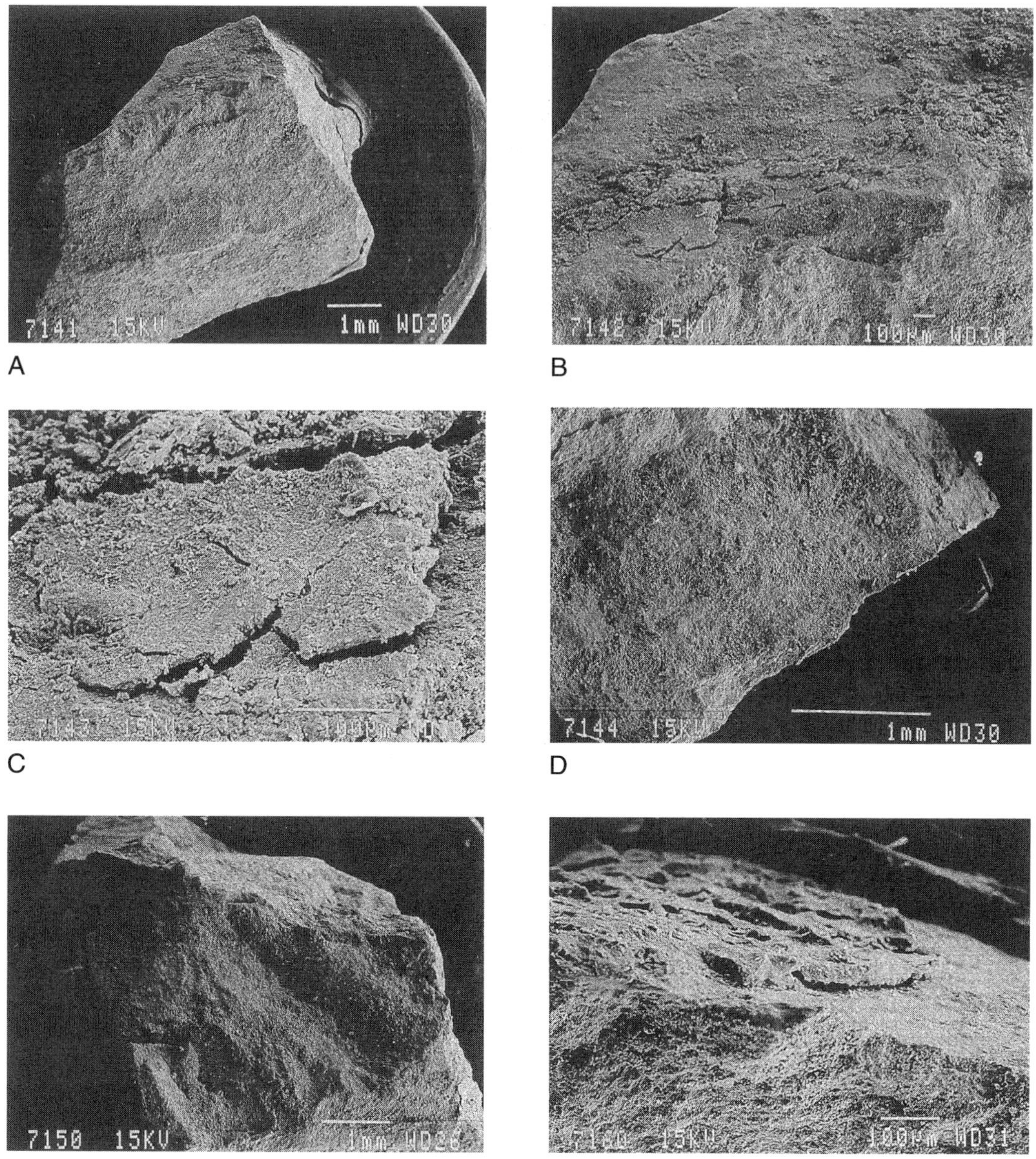

Figure 10.4 (A) A 6 mm diameter block of quartz with minor coating of Fe and Ti, block removal in the upper third and a strong slickensided surface in the lower third. Minor cracking is apparent on the upper edge of the grain. (B) Slickensided, fractured slivers of quartz on the upper edge of the grain in A. (C) An enlargement of a quartz slab fractured from the top of the grain in A. (D) Surface of a gouge plane in the fault, with minor fractures and rubbed and abraded grains. (E) Quartz from the gouge zone with deep grooves and small craters. (F) Thin plates of quartz dislodged in the gouge zone.

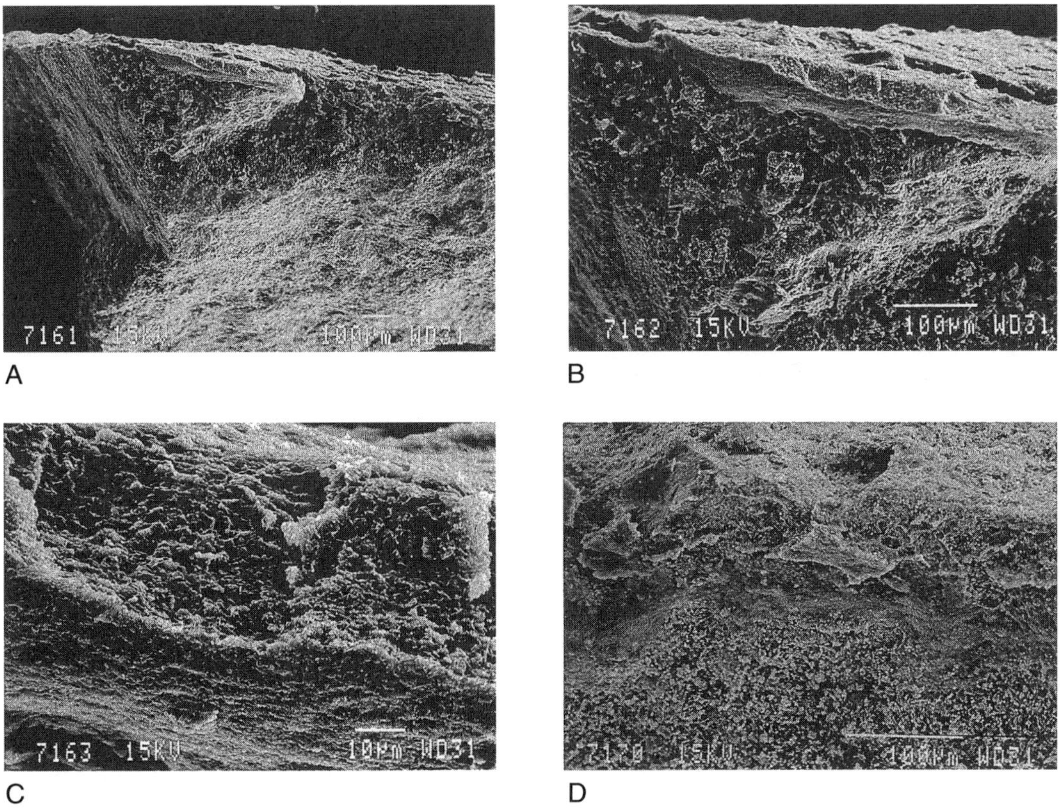

Figure 10.5 (A) A portion of quartz block from a gouge zone in the Cotamille Fault, a splay of the Boconó Fault, showing a polished (slickensided) surface. (B) Enlargement of the polished surface in A. (C) Enlargement of the area shown in B. (D) Enlargement of a gouge zone showing slickensided and rubbed grains flattened by movement along the slip planes, as well as some minor cracking.

10.9A), with pronounced fracture faces polished to form slickensided surfaces (figure 10.9B) and/or fresh surfaces with pronounced linear grooves juxtaposed against an original crystallographically realigned face probably formed in the stress field of the nearby Boconó Fault (figure 10.9C). Very angular pieces with minor subparallel fractures (figure 10.9D) are common on some grains from the fault trace. In figure 10.9D, a triangular edge remains intact on the whole fine sand grain, whereas in the samples studied many small triangular-shaped forms are common. While somewhat similar to glacitectonite grains, formed from glacial tectonic activity, these grains might be called proglacitectonite.

San Gabriel Fault, California

The San Gabriel Fault is part of the San Andreas System, the intensively studied, several hundred kilometer long boundary that separates the North

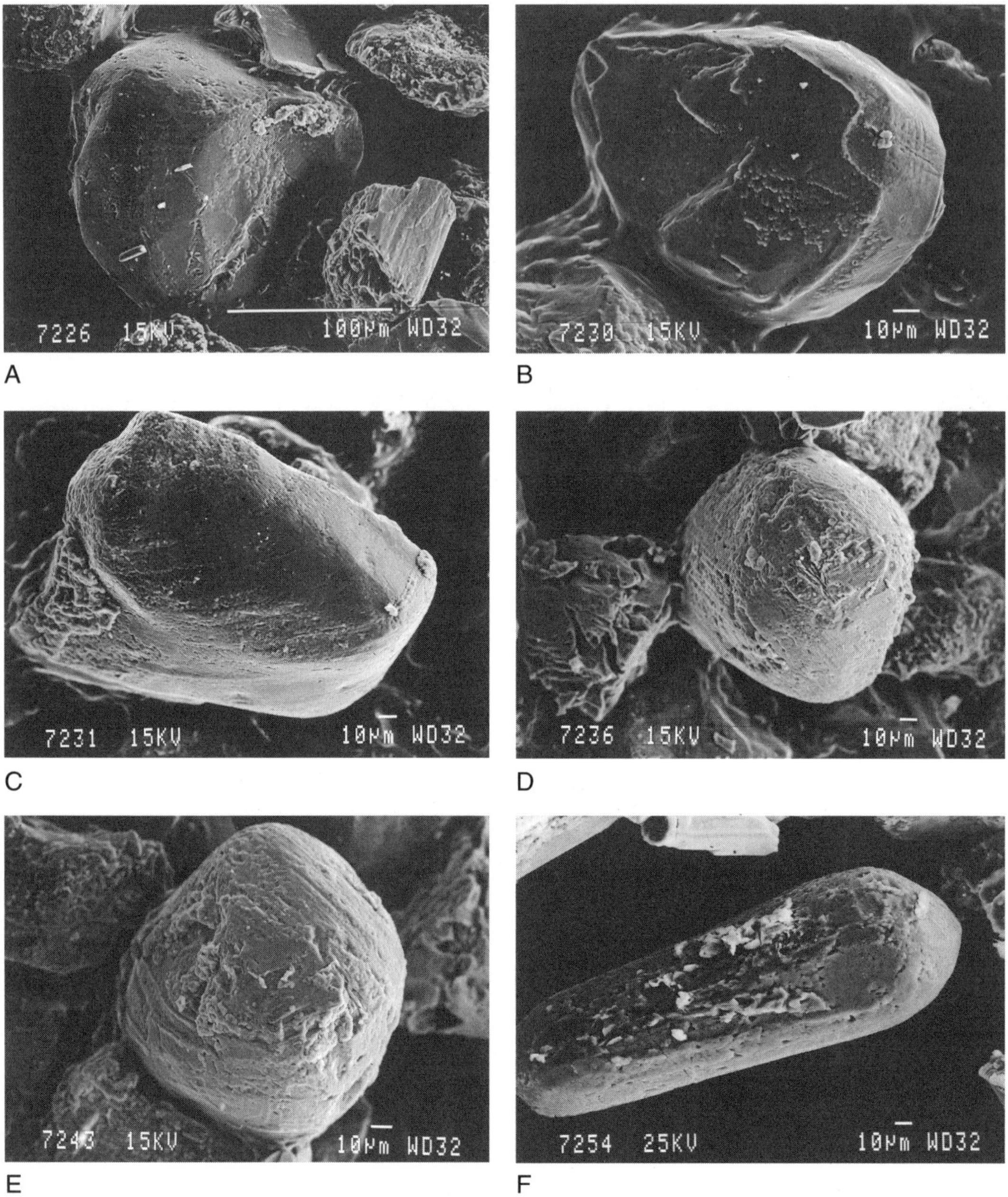

Figure 10.6 (A) Round, rolled quartz with a "wrinkled" form caused by pressure in the shear zone of the Boconó Fault. (B) Greatly deformed rolled quartz grain. (C) Rolled and stretched quartz grain. (D) Rolled quartz (center) with weathered K-feldspar (to left). (E) Intersecting ridges and cleavage in quartz with K-feldspar adhering particles. (F) Rolled and stretched quartz with adhering particles of Fe-stained muscovite.

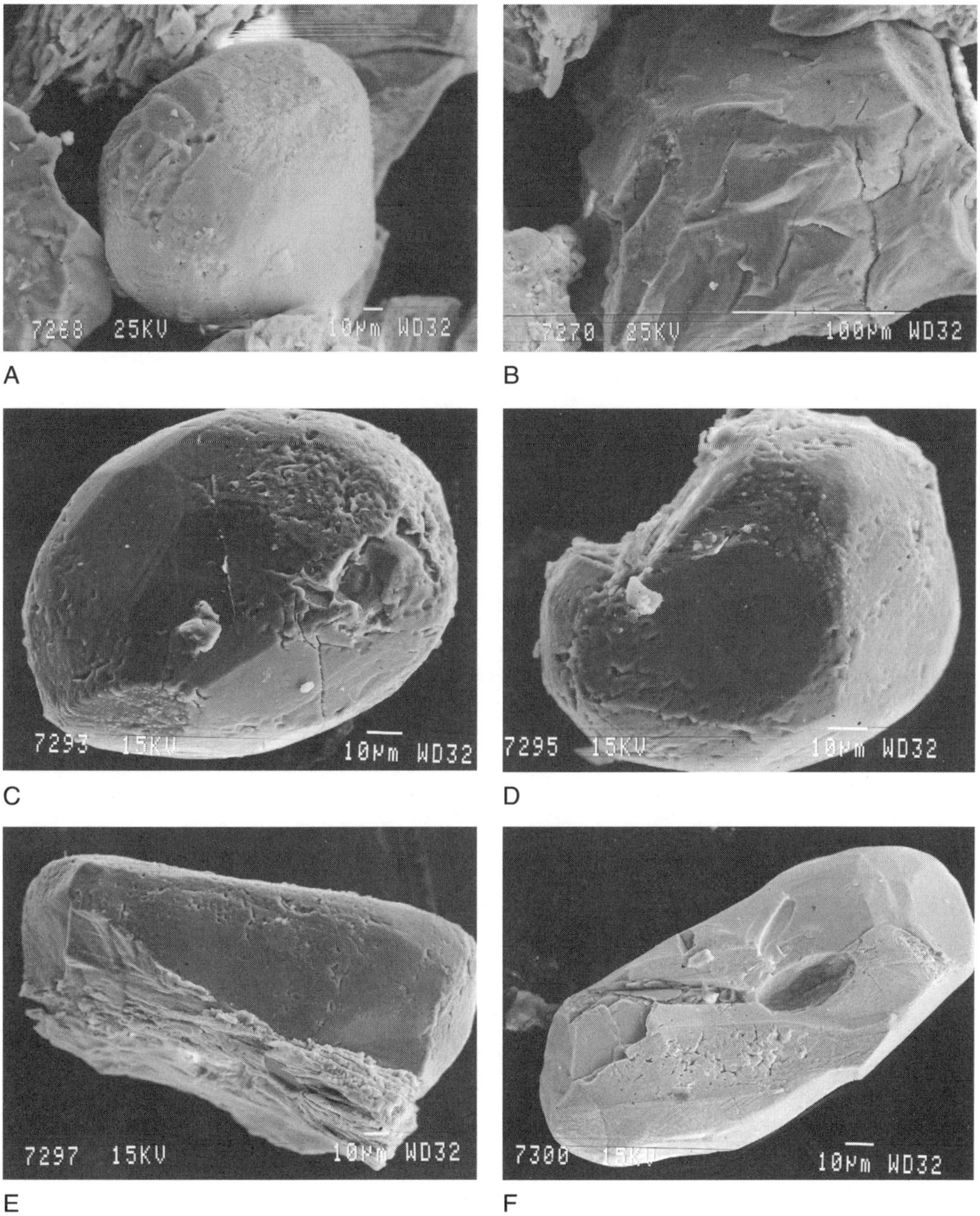

Figure 10.7 (A) Partly rounded, rolled quartz from the Boconó Fault with some dissolution in a crack above the center. (B) Cracked quartz with multiple fractures. The strong solution features on the quartz probably signify that garnets have been weathered. (C) Rolled quartz with a cracked and damaged surface. (D) Quartz rolled and reformed, then cracked. (E) Well-damaged quartz intergrown with muscovite. Some ornamentation on the quartz. (F) Monazite, with normal Ce and excessive La in the lanthamide series, flattened and cracked.

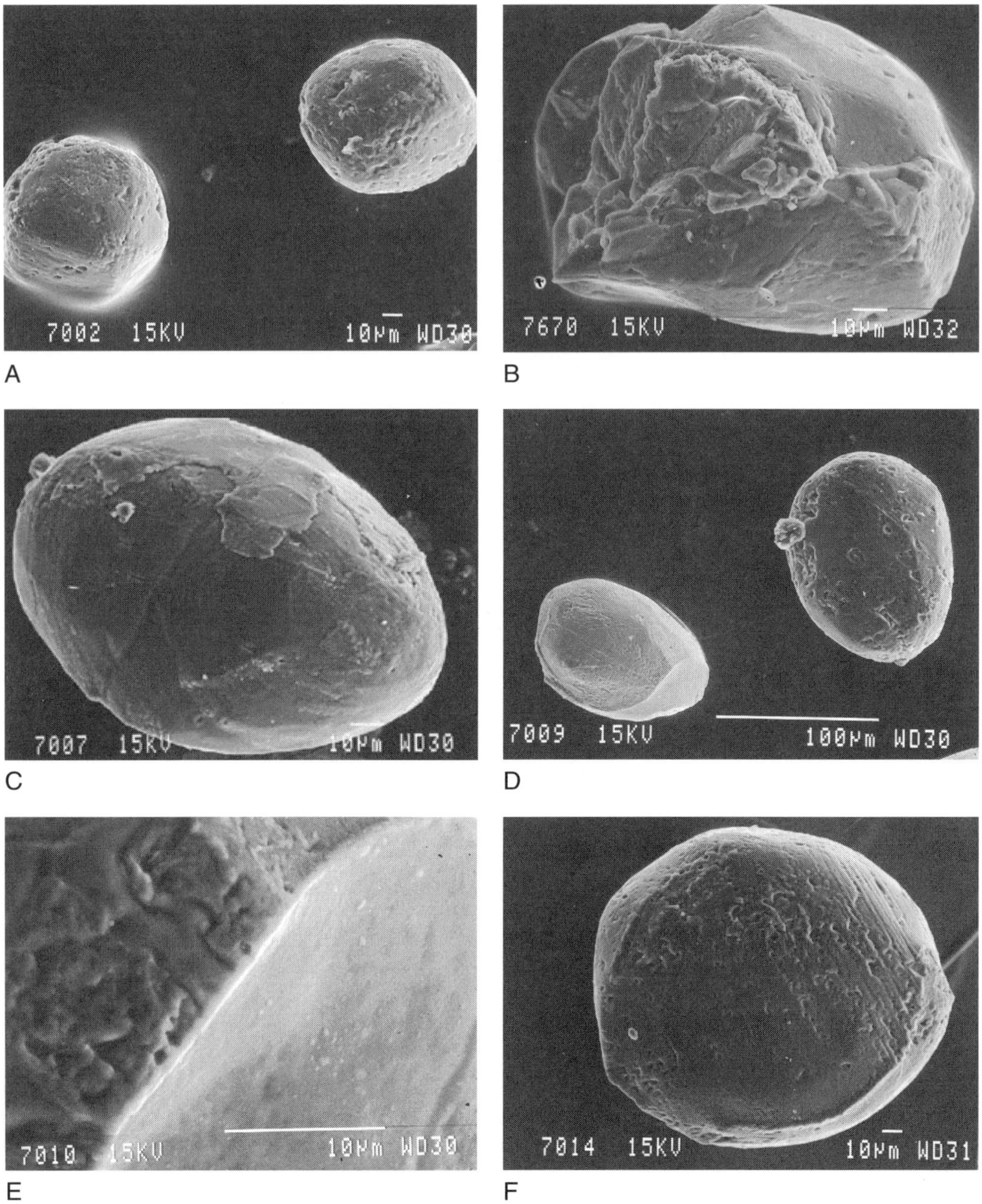

Figure 10.8 Rolled quartz from the Boconó Fault in talus of the Mifafi Catchment, northern Venezuela. (A) Coated and weathered, rolled quartz grains in talus of Holocene age, Mifafi Catchment, Venezuelan Andes. (B) Preservation of old etching and a new polished surface on the grain to the left in A. (C) Cracked rolled quartz with dislodged skin in Mifafi talus. (D) Rolled quartz grain from the Boconó Fault. The grain to the right is partly cracked. (E) Enlargement of the lower left grain in D, with a fresh surface to the right and a preweathered surface to the left. (F) Complex grain of rolled quartz with a fluvial history interpreted from the v-shaped percussion cracks.

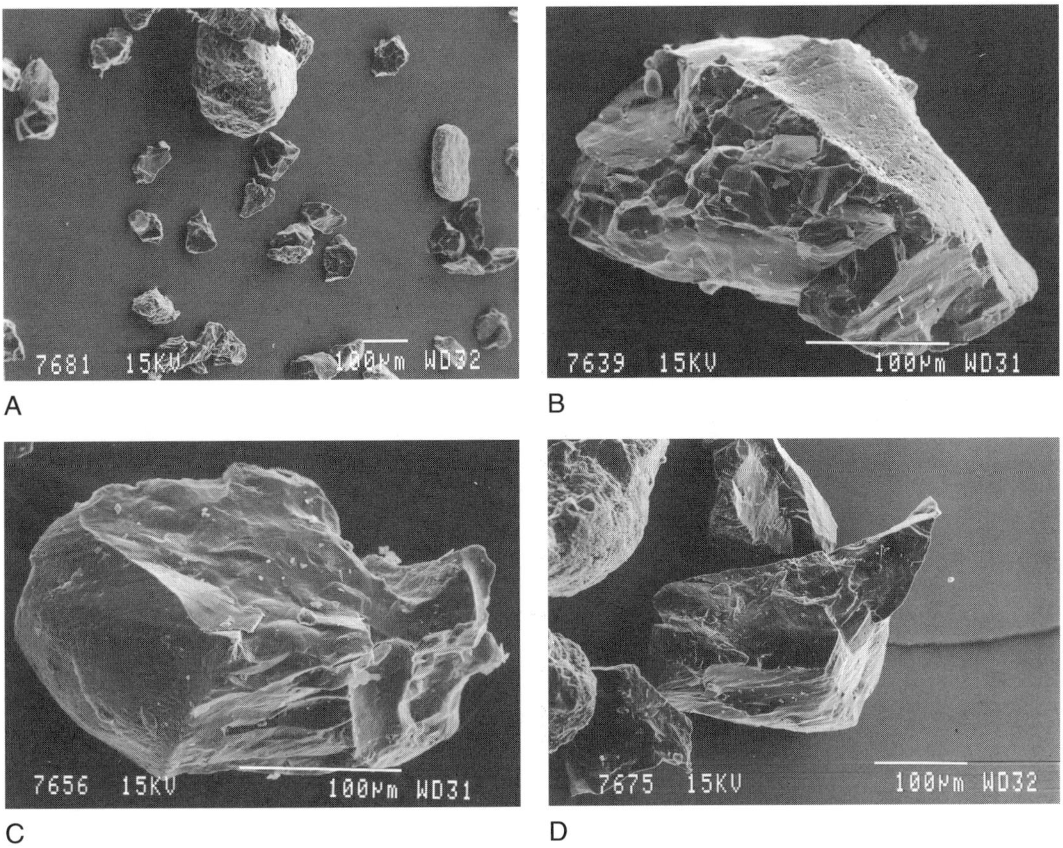

Figure 10.9 (A) General frame of thin wafers and shavings of grains subjected to proglacial tectonic stresses. (B) Scallop-shaped fractures and slickensides (bottom, upper right and upper left) on quartz. Slickensided surfaces (upper right) with minor lattice failure (lower left). (C) Fresh fractures (middle and right) overprinted on crystallographically realigned quartz (left). (D) Quartz segments with multiple shallow subparallel fractures and triangular edge.

American and Pacific plates. Most of the movement is right-lateral, strike-slip, although minor valleys and hills are common where a local change in strike results in vertical movement (transpression and transtension). The samples, taken from a "sag" or small enclosed depression caused by differential movement within the fault zone, show a rolled form similar to grains described above from the Boconó Fault. The San Gabriel and many other strike-slip faults in California have been extensively trenched in order to determine fault recurrence and time of last displacement. Such trenches, often 10 m deep, provide excellent exposures from which to collect the samples such as those shown in figure 10.10.

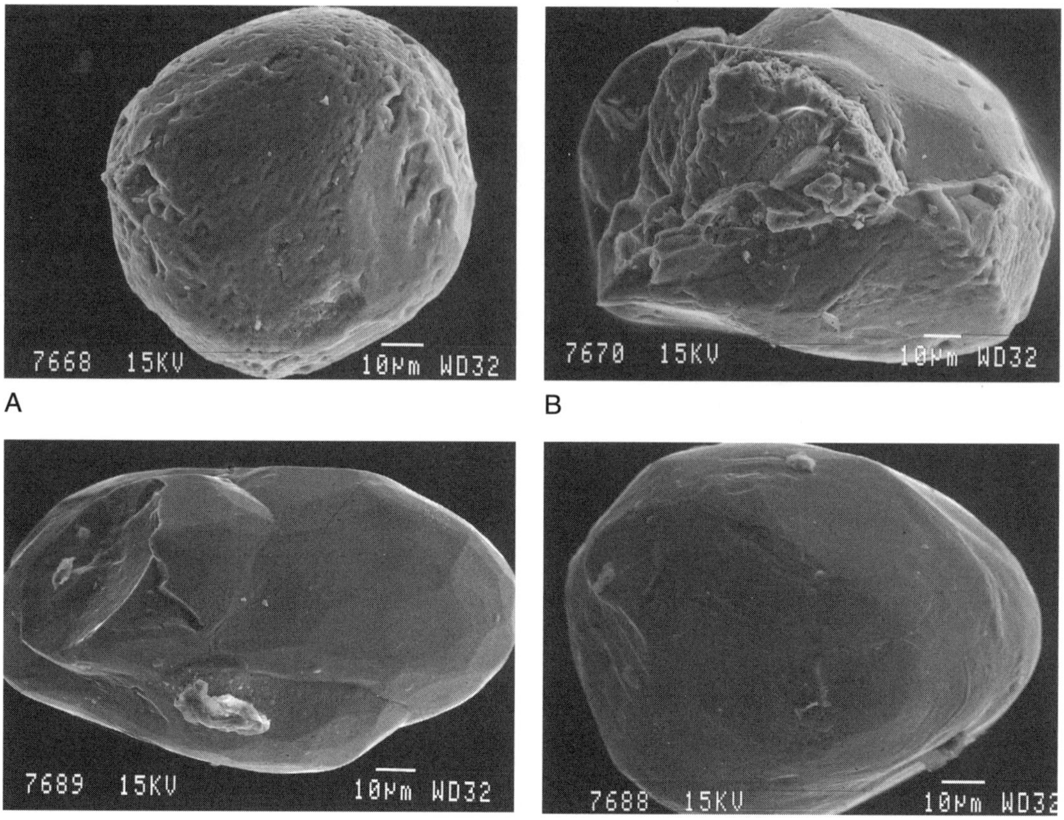

A B C D

Figure 10.10 Grains from the San Gabriel Fault, California. (A) Rolled quartz with a smooth, slickensided right margin. (B) Rolled quartz with thin Ca-coating and intense breakage (center and left). The upper and lower limbs are partly slickensided. (C) Rolled zircon with some exfoliation on the left, including a sawtooth edge. (D) Twined euhedral zircon with intense breakage (center and left). Samples courtesy of R.J. Shlemon, Newport Beach, California.

Conclusions

Quartz grains in Trollhålet and Ulorna caves carry a unique fractography, with markings including irregular conchoidal fractures that are pseudo-glacial in appearance, but not widespread on particle surfaces. In addition, sharp irregular-shaped fractures could be lattice imperfections, but are likely the result of neotectonic activity. The presence of upturned plates on many grains in Ulorna Cave are most likely neotectonic features signalling rapid release from the cave wall. Overall, microtextures observed on quartz particles from the two sites support the available geomorphological evidence signalling rapid release from cave walls as a result of seismic activity.

Small blocks of quartz recovered from the gouge zone of the Boconó Fault show that cracking and fractures seem to be the dominant microtextures in the high stress zone of the fault. Along with these microfeatures, polished and slickensided grains are less common than theory would predict. There are also numerous craters and abraded areas in the grains which result from movement along the fault.

In the shear zone between two faults, and in an area where wrenching of rock is likely to create high stress, rolled quartz, zircon, and garnet have been moderately reformed, giving rise to a distinctive grain morphology that is easy to recognize.

11 Asteroid and Cometary Impacts

Asteroid (bolide) or comet impacts begin with a nearly instantaneous conversion of kinetic energy of movement at several kilometers per second into heat from impaction. Surface impaction produces a temperature of several thousand degrees Celsius and the energy emitted is several million times greater than that emitted by the Sun. With radiation clustering at the 100 Å wavelength, in the soft X-ray section of the electromagnetic spectrum, the area engulfed by heat from the exploding asteroid or comet is consumed by melting rock, almost instantaneously.

The impactor drives a powerful shock wave radially outward in all directions—a wave that carries fractured minerals of all types (many with planar deformation features (PDFs) (shock lamellae)), partly shock melted from extreme heat, high into the atmosphere to altitudes where they may stay aloft for days or weeks until the hot gases cool to temperatures that allow subsidence. Only grains ejected outward from the surface blast survive; silicates exposed directly to a blast will be evaporated. For high-magnitude impactors, greater than 5 km in diameter, sands and silts raised by the impact are entrained by rising high-temperature fireballs, to return to Earth as shock-melted and fractured grains or even glass.

A representative impact crater on Earth removes a mass of earth about a hundred times the mass of the impactor. Using Meteor Crater in Arizona, for example, which is 1300 m in diameter, the size suggests the impactor had a diameter of ~130 m. To put this in a nuclear perspective, the impactor probably produced an explosive yield of 10–15 megatons of energy (Hartmann, 1999).

When friction in the Earth's atmosphere fails to burn up incoming meteorites they impact and explode with an energy estimated at 100 times their mass in TNT (Morrison, 1995). Meteorites greater than 50 m in diameter entering the Earth's atmosphere are subjected to increasing stress as atmospheric density increases, causing them to fragment, break up, and explode. If they explode above 15 km, at the normal upper altitude of the troposphere in equatorial latitudes (the troposphere is thinner at the poles), there is little damage to the surface. Stony asteroids must be larger than 50 m in diameter to damage the surface, but nevertheless a 50 m stone or nickel–iron meteorite strikes with about 10 megatons of energy, which is comparable to a large nuclear bomb. The 1908 Tunguska event in Siberia was such an impactor, and most probably an iron meteorite that broke up at altitude (Morrison, 1995).

Moving up the megaton scale, a 100 m diameter iron meteorite could devastate an entire region (part of a continent) on Earth with a 100 megaton explosion. An impactor 1 km in diameter could obliterate an entire hemisphere and release 100 gigatons of energy. A great impactor (extinctor) of 10 km diameter, similar to the Chicxulub impact (Hildebrand, 1993), is capable of unleashing 100 million megatons of energy and subjecting the entire Earth to a nuclear winter. An impactor landing in the sea would produce towering tidal waves that would devastate large coastal areas on several continents.

Catastrophism, long repugnant to Earth scientists, has been growing in favor over the last decade or so, as scientists realize that extreme events occur on Earth far more often than they had previously believed (Raup, 1986). Studies carried out by Oberbeck et al. (1993) indicate that great numbers of impacts have occurred over the last 2 billion years of Earth's history, leaving a great number of impact craters. Oberbeck et al. argue that textures of impact deposits are closely correlated with tillites and diamictites of Precambrian and younger age, and more importantly that some or most tillites might have an impact origin. If their hypothesis is correct, then tillites should contain impact ejecta with shock-melted quartz in great quantity. So far, none have been documented.

Quartz grains, shocked by meteoric impact, contain features that distinguish them from grains in other types of environments. The term "planar microstructure" refers to two kinds of shock lamellae in quartz: planar fractures (PFs) and planar deformation features (PDFs) (Stoeffler & Langenhorst, 1994). According to these authors, planar fractures (PFs) form sets of parallel open fissures, with a spacing of approximately 20 μm, oriented along crystallographic planes.

PDFs occur as multiple sets of parallel, planar optical discontinuities or as planes spaced 2–10 μm apart, some of which may be decorated with tiny vugs (decorated PDFs). Gratz et al. (1996) report that PDFs appear as narrow (0.05–0.5 μm) lamellae filled with amorphous glass (diaplectic glass). They are visible under the petrographic microscope; however, because most of these features are narrower than 1 μm, optical microscopy is not ideally suited for their identification. Gratz et al. (1996) further report that etching samples with HF, followed by examination with an SEM in the secondary mode, allows unambiguous distinction between etched-out glass-filled PDFs and glass-free tectonic deformation arrays in quartz. The TEM is also used to identify shocked quartz (e.g., Goltrant et al., 1991; Cordier & Gratz, 1995). PDFs appear on TEM micrographs as straight and very narrow (<0.1 μm) bands, either glassy or crystalline. TEM reveals many more PDFs than are optically detected (Goltrant et al., 1991). It should be noted here that most of the work with shocked quartz has been done in thin section. Many of the samples from consolidated rocks are difficult to disaggregate, and thus thin sections are easier to prepare than attempting to remove individual grains from rocks. The following micrographs, however, show secondary pictures of the surfaces of shocked quartz grains, which have not been altered chemically or physically. The SEM secondary photographs of shocked quartz grains in this chapter display definite PFs, and possibly PDFs. If sand grains could be removed from rocks with very little alteration of their surface features, the secondary electron examination of surface features might serve as an additional tool to identify shocked quartz.

K/T Mega Impact Ejecta

The source crater for the K/T impact ejecta is the 180 km wide Chicxulub Crater on the Yucatan Peninsula of Mexico (Hildebrand,1993). The Chicxulub impact, caused either by an asteroid or a comet, produced a massive high-energy flux of devolatized CO_2 and SO_2 from target carbonate and sulphate rocks, which led to a sulphurous rain and long-term greenhouse warming of at least 20 °C. The K/T boundary clay consists of two layers: a distal, globally distributed, nearly uniform 3 mm thick layer produced by the fireball, and a proximal, ballistically ejected layer of coarser debris found near the crater. The impacting body "skipped bombed" its way to the north and west impacting several localities in the Caribbean, western United States, and Canada. Samples recovered from Clear Creek, Colorado, by G. Izett of the College of William and Mary were provided for study as examples of shock-melted quartz found in the proximal ejecta of the second impact. The second suite of samples comes from the distal K/T clay layer at Gubbio, Italy. These samples were provided by Juri Plado, formerly of the Institute of Geology, Tartu University, Tartu, Estonia, and presently at the Geological Survey of Finland, Espoo, Finland. These proximal and distal facies are discussed below.

Clear Creek Colorado

The Clear Creek, Colorado, sample suite of K/T ejecta grains (Izett, 1990) was analyzed by SEM–EDS to verify composition and to determine the extent to which impacted grains differ from those produced by the operation of normal geological processes. The grains shown in figure 11.1A represent angular to subrounded forms commonly seen, not only as impact ejecta, but also as volcanic ejecta and glacially transported sands. All the grains in this frame are fractured and shock melted; some are abraded presumably from mid-air collisions or from rubbing when settling in a cloud of debris. An enlargement of the grain in the center of figure 11.1A, in C, shows a high density of melted quartz with many melted channels all interconnected—something never observed with dissolution microfeatures in quartz.

Extreme melting of quartz is shown in figure 11.1D, where much of the grain fabric has been melted or vaporized away; the lower part of the grain is fractured horizontally in thin layers (e.g., thin PDFs). Two Ca-plagioclases in figure 11.1B carry ellipsoidal melted craters and deep melted straight or curved trenches that are unlike forms seen on grains transported by normal functioning geological processes. Horizontally fractured and melted microfeatures on quartz are shown in figure 11.1E where the fractured surface appears superficial but with complete melt-out to the left of center. Melted fold-over structures on fractures are common on the suite of samples studied (figure 11.1F).

Microfractures abound on shock-melted quartz. A second general frame (figure 11.2A) shows angular to subangular grains with extreme, deep fractures and dislocated plates. Extreme lattice failure is seen on the bottom of the quartz grain in figure 11.2B along with a concave melt-out cavity partly recoated with Si; the grain to the right is partly melted and an upturned plate is apparent to the right. Extreme melting and fracturing with melting are common on some grains (figure 11.2C). Thin and wide fractures on quartz (figure 11.2D) might originate from different velocities on ejection or from collisions in the ejecta cloud. Fracturing and upturned plates, shown in figure 11.2E and F, are common on all grains studied.

Gubbio K/T Boundary Layer

Analysis of the 1 cm thick K/T boundary layer at Gubbio, Italy, provided an important corpus of information that illustrates how important the SEM–EDS can be in assessing asteroid/cometary impacts. A cross-section of grains in the 25–100 μm fraction yielded interesting results (figure 11.3). A mix of light and heavy minerals included tabular quartz with solution cavities that could come from a SO_4-rich atmosphere; quartz with a coating of Sn and Ir; quartz with a graphitic texture; quartz with a negative crystal imprint (i.e., the core removed by high-velocity impact), and v-shaped microtextures resulting from crystallographic failure (not from impact). While not conclusive evidence, as yet, for impact microtextures, these samples indicate that with a number of replicated observations, the imagery might provide important data supporting other stratigraphic evidence of a tremendous explosion at the end of the Cretaceous.

In addition to the quartz grains encountered in the boundary zone sample, particles of framboidal pyrite, which might be related to a SO_4-rich environment produced by S settling out of a dust cloud were observed along with rutile–anatase–brookite, volcanic quartz shards, and pyroxene crystals that most probably have an origin in nearby Italian volcanoes. There is enough tephra present to argue for coeval volcanism possibly caused by the Yucatan impact.

Sedimentary Dikes at Neugrund Crater

The submerged 7 km diameter Neugrund Crater is located 10 km off the north-western coast of Estonia (Puura, 2000; Suuroja & Suuroja, 2000). The carbonate rocks of the Kunda Stage of the Ordovician contain various size sand-filled sedimentary dikes found across an area of thousands of square kilometers in north-western Estonia. The largest of these cracks associated with strong brecciation are on the island of Osmussaar, about 5–6 km from Neugrund Crater (Suuroja et al., 1999). These cracks were earlier interpreted as the result of earthquakes (e.g., Öpik, 1927), but after the discovery of the Neugrund Crater the first interpretation, which provided a straightforward explanation for the sand-filled cracks and brecciation, relied on ejecta from an impact explosion

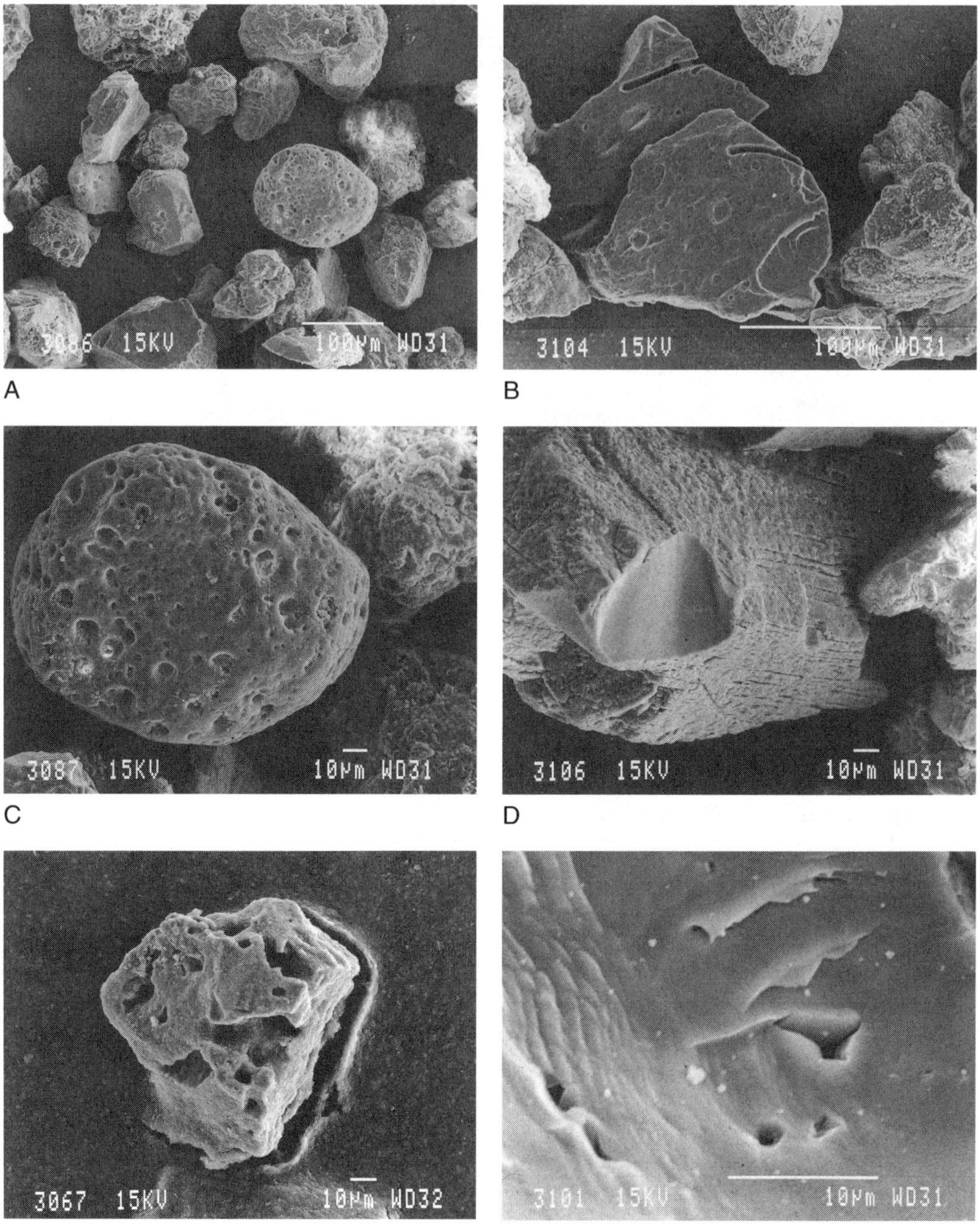

Figure 11.1 (A) Angular to subrounded grains from Clear Creek, Colorado ejecta. The fine sand size grain to right of center is shock-melted with interconnecting melt cavities. (B) Two partially melted Ca-plagioclases with ellipsoidal and lenticular shaped cavities. (C) Enlargement of center grain in A showing annealed smaller grains in some cavities. (D) Shock lamellae on quartz partially melted and annealed. (E) Microfractured and partially melted quartz. (F) Melted over fractures on quartz.

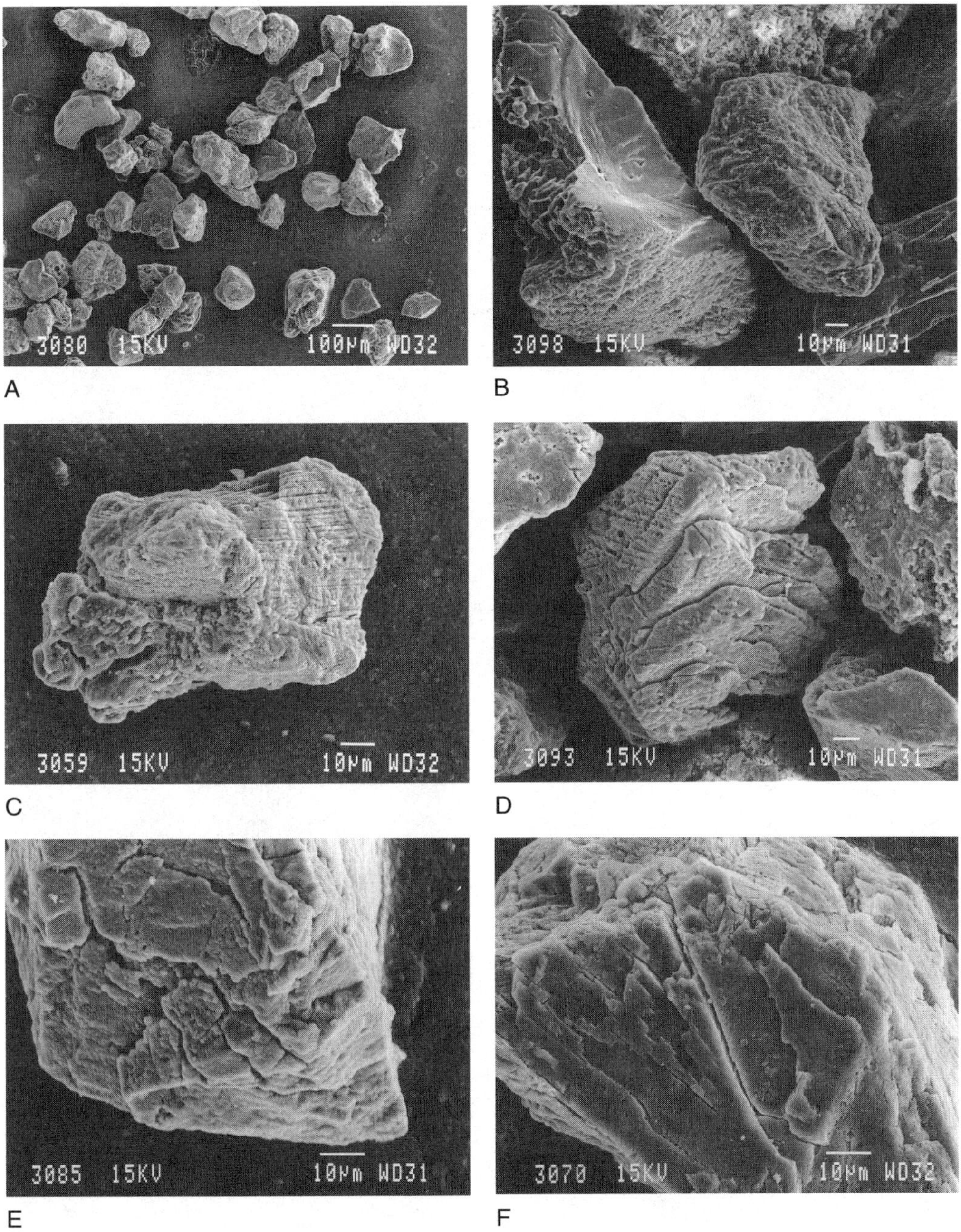

Figure 11.2 (A) Representative grains from a second replicate sample of Clear Creek, Colorado. These grains are all angular to subangular and carry multiple melted and fractured features. (B) The quartz on the left with a lattice-shattered lower area and a large melt-out feature on the top across from a shock-melted quartz grain with one upturned plate right of center. (C) Shock-melted quartz (lower left) with thinly fractured surface to the top and right. (D) Deep multiple fractures cutting thin horizontal fractures (shock lamellae) with some melting. (E) Deeply fractured plates on quartz. (F) Superficial fracture slabs of quartz with melted edges.

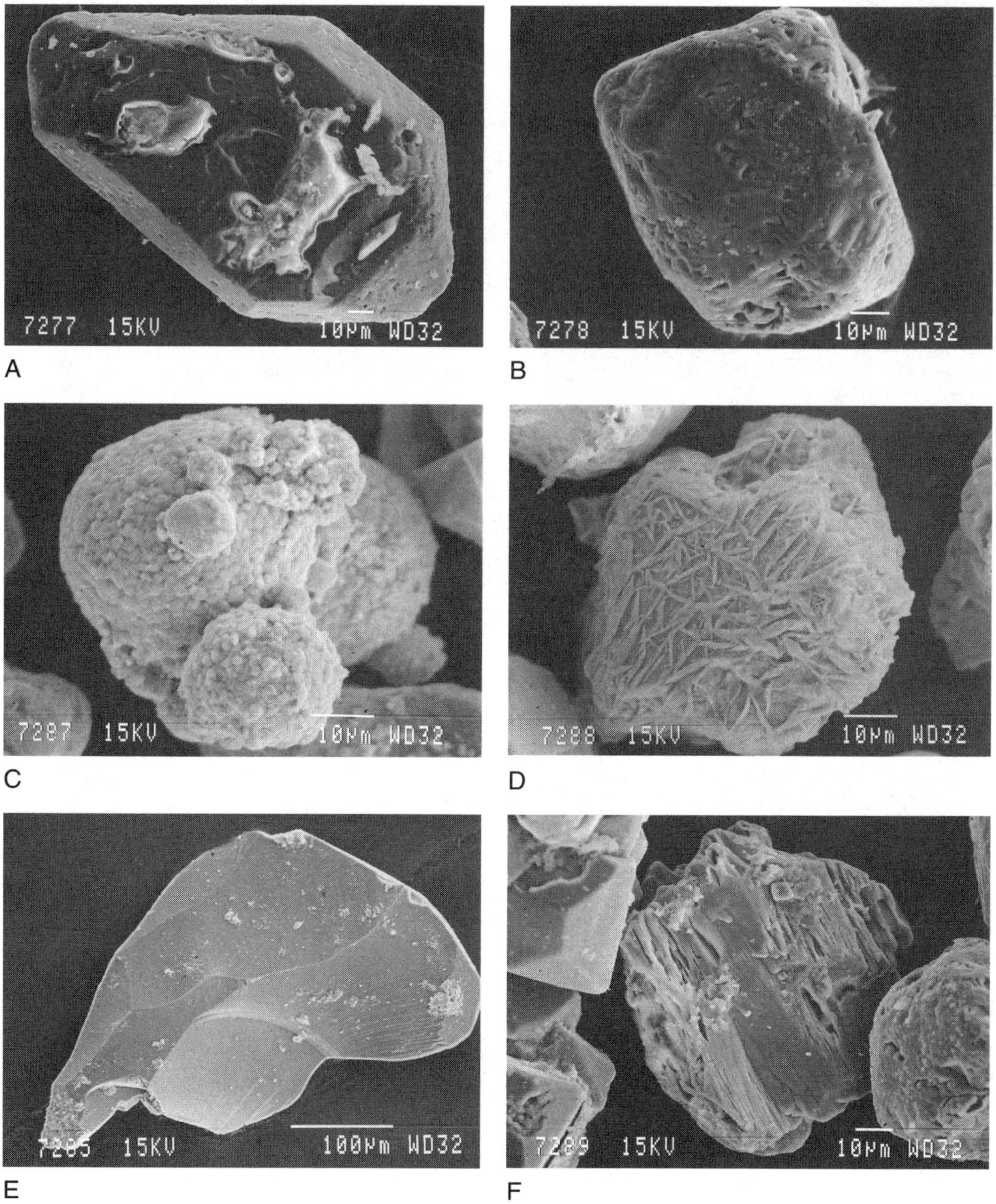

Figure 11.3 (A) Tabular quartz that could have formed in a SO_4^{-2}-rich atmosphere. (B) Quartz, partially melted, with small amounts of Sn and Ir on the surface. (C) Quartz with crystallographic limbs that produce v-shaped cracks. (D) Framboidal pyrite which could be related to a SO_4^{-2} cloud or to organic matter decay due to reducing conditions. (E) Quartz shard with some adhering particles. (F) Pyroxene, possibly hedenbergite.

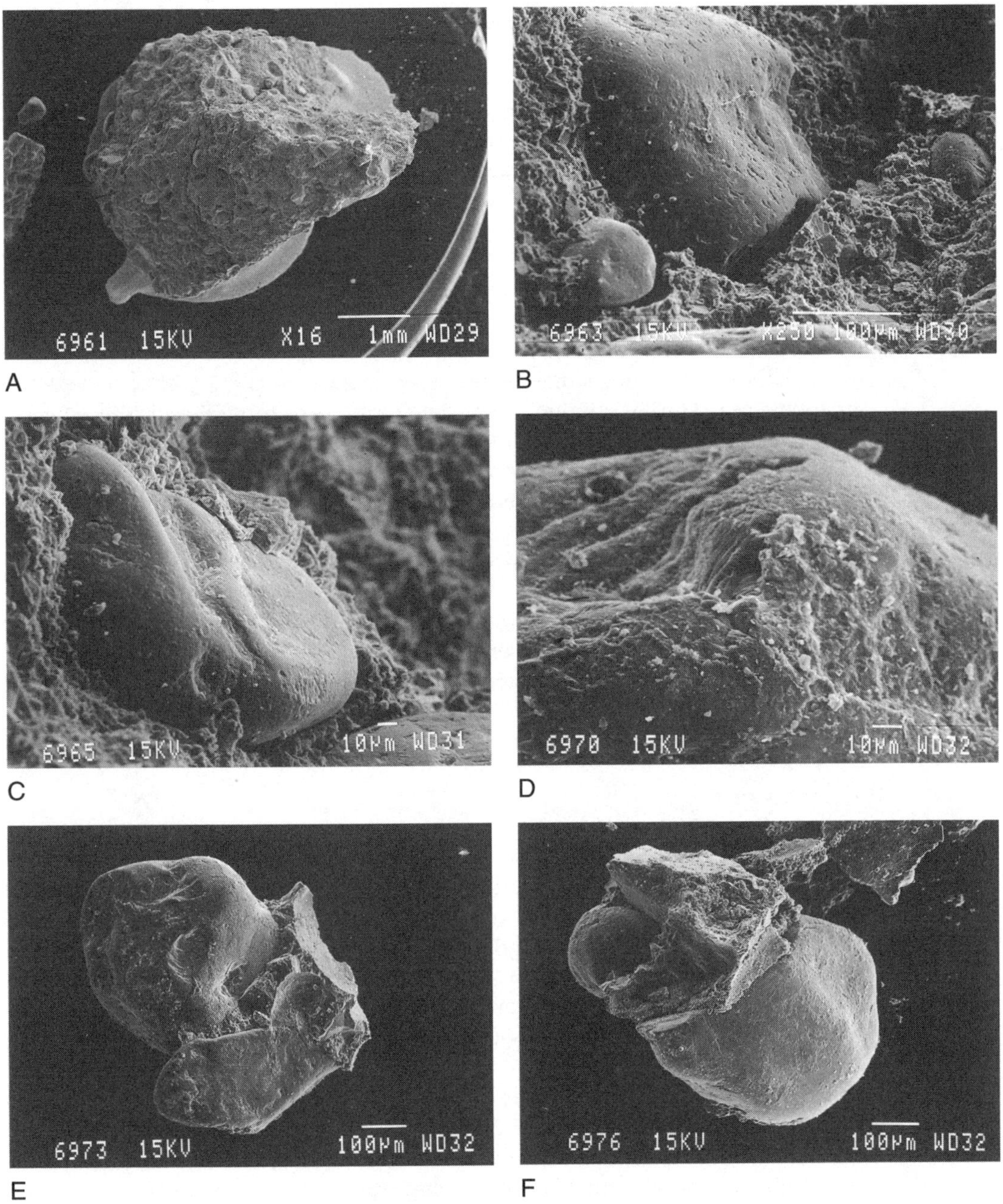

Figure 11.4 (A) Large granule of carbonate with quartz inclusions and a cracked surface. (B) Enlargement of quartz embedded in the carbonate granule in A. Note the v-shaped percussion cracks. (C) Rounded quartz with bulbous edges surrounded by carbonate matrix. (D) Long irregular shock-melted area on quartz grain. (E) Bulbous-edged quartz with multiple tear structures from collision with other particles. (F) Quartz grain with a thin coating of carbonate welded to carbonate that was melted and reprecipitated.

(Suuroja & Saadre, 1995). Originally considered to be of Ordovician age, the event was placed at 470–480 Ma. However, more recent exploratory work revealed that the Neugrund Crater is infilled with older, Cambrian sediments, and is most probably of Early Cambrian age (~540 Ma) (Suuroja & Suuroja, 2000). Consequently, the origin of these Ordovician sand-filled sedimentary dikes remains unclear. The grains of the coarse silt and fine sand fractions from the dikes on Osmussaar Island (around 5–6 km west of Neugrund crater) and from Pakri Peninsula (around 35 km east of the crater) were examined to determine if identifiable microtextures could help in understanding the geological events responsible for the formation of these dikes.

The results of SEM investigation show the shocked nature of the quartz grains in cracks in the Ordovician carbonate rocks from both Osmussaar Island and Pakri Peninsula (figure 11.4).

So far, two hypotheses may be proposed to explain the presence of shocked grains in the samples studied. First, the cracks were formed and filled during regional uplift and/or eustatic sea-level fall during the Kunda Stage, with the sandy sediment deposited in cracks eroded from raised ring walls of the Neugrund Crater emerged to erosion level at this time. Alternatively, the brecciation and sandy infilling of the cracks is related to a distant impact explosion in the present Baltic Sea to the west or north-west of Osmussaar Island. No such event is known to have occurred in this area. Although the age of the Granby Crater in central Sweden falls into the Kunda Stage (Granh et al., 1996), it is too small (~4 km in diameter) and too far away (>500 km) to account for the deformation structures and distant ejecta in north-western Estonia.

Conclusions

As with microtectonic features on sands, there is little published imagery on grains dispersed as a result of asteroid or cometary impact. A nickel–iron asteroid impact, occurring with a velocity of several tens of kilometers a second, produces sufficient heat to melt quartz and other minerals, leaving characteristic microtextures that can prove useful for identification purposes. The same effect could be realized with cometary impacts, although these impactors with their ice and rocky centers would likely explode at altitude in the atmosphere and not reach the surface. The Tunguska explosion over Siberia in 1908, often considered to be a cometary impact, was probably a large stoney meteorite that broke up after entering the troposphere.

The SEM–EDS is particularly useful in studying ejecta from asteroid impacts, especially the shock lamellae and melt features that distinguish them from grains produced in other sedimentary environments.

12 Paleosols

Both soils and paleosols are ideal candidates for study by SEM–EDS (Mahaney, 1990a), but curiously few researchers employ these imaging and spectrographic techniques in studying morphogenesis (Catt, 1986; Birkeland, 1999; Mahaney et al., 2000b). Whether the objective is the study of finite changes in the intensity of weathering over short or long lengths of time, or the microscopic characterization of sands in buried paleosols (Mahaney et al., 2001c) to refine the reconstruction of paleoenvironments, SEM–EDS can play a vital role in imaging microstructures of various kinds. When combined with detailed particle size analysis (Mahaney et al., 2001c), the SEM–EDS becomes a powerful tool in analyzing the sand fraction to elicit information on degrees of etching and dissolution among light and heavy minerals. This applies to both soil stratigraphic units and individual soils or paleosols. Other additional important information can be gleaned from various coatings of carbonates, oxides, hydroxides, and clay minerals which cling to sand particles despite strong leaching and washing in the laboratory.

Once samples are collected, dried in the laboratory, and prepared for particle size analysis, the recovered sands may be stored for later microscopic and chemical analysis. Depending on various research objectives, it may prove beneficial to carry out mineralogical and/or chemical analyses on the sand fractions, and to determine the weathered state of phyllosilicates such as mica, or the weathered state of more resistant minerals such as quartz. If the soil or paleosol chronosequence is short-lived, study of micas or amphiboles might be particularly important from a relative-age dating point of view (Birkeland, 1999, Table 8.2). If the time spectrum is longer, say in the order of middle to late Pleistocene, quartz might be the mineral of choice. In the latter case, three-dimensional imagery by SEM may provide enough high-resolution data to quantify dissolution differences, thus providing insight into relative differences in age between pedons (Mahaney, 1992a). In still other cases where soils might indicate little difference in age between pedons in a chronosequence, study of the sand fraction might reinforce and support the soil profile evidence, or soil morphology may have reached a steady state for those age deposits and the only differences seen may be from the microscopic study of sand fractions.

In Quaternary lithostratigraphic sequences, SEM microtextural evidence has been used to study preweathering, that is, weathering presumed to have occurred prior to glacial transport (Mahaney & Kalm, 2000). Quartz grains with well-preserved dissolution etching, partially reformed during glacial transport into triangular-faceted grains with multiple crushing microfeatures, have been shown to provide evidence of subaerial weathering environments prior to glaciation (Mahaney & Kalm, 1995, 2000). Even if there is no dating precision, there is at least a semi-quantitative estimate of the strength of paleoleaching (see Mahaney, 1990a).

In this chapter the intent is to illustrate the importance of SEM–EDS in providing petrographic baseline data on the composition of the sand fraction in paleosols and the analysis of dissolution characteristics and coatings on sands in a variety of different environmental settings. Starting on Mount Kenya, where there is clear-cut information on paleosols ranging in age from early to late Quaternary, examples are drawn from other tropical alpine localities in the northern Andes, the middle latitudes of southern Ontario and central Germany, and Antarctica, to provide an overview of paleosols in different areas.

Teleki Valley, Mount Kenya

The western approach to Mount Kenya, Teleki Valley, is the most frequently studied catchment on the mountain (Gregory, 1894; Baker, 1967; Mahaney, 1990a). A left lateral moraine below Met Station (3000 m a.s.l.), first identified by Gregory (1894), is a compound moraine system containing two paleosols with a combined thickness of 5–6 m that extend into a saprolite of Middle Pleistocene age (Mahaney, 1990a). Each paleosol contains a loess–ash sequence of horizons over till. Below the upper paleosol of post-Teleki age (= post-Illinoian), an older paleosol of post-Naro Moru age overlies a saprolite presumed to be of early middle Pleistocene age. Clay mineral data and microscopic study of the sand fractions in this pedostratigraphic succession, at various sites, show important relationships between ratios of metahalloysite and gibbsite, which reveal considerable information about paleoleaching and age. For one, metahalloysite forms slowly from volcanic glass (Mahaney, 1991b) and is found only in pre-Liki (= pre-Wisconsinan in North America; pre-Weichselian in Europe) paleosols. Second, the ratio of gibbsite/metahalloysite is a leaching indicator: the higher the ratio, the greater the leaching. Close scrutiny of relict and buried paleosols shows this to be a quantifiable relationship, which is replicated from one paleosol to another (Mahaney, 1990a). Both fibrous and nodular forms of gibbsite seem to develop from successively greater leaching of SiO_2 from metahalloysite, as indicated from XRD and confirmed by SEM–EDS.

High-resolution SEM of the TV23 profile (Mahaney, 1990a) at 2950 m a.s.l., which contains both the post-Teleki and post-Naro Moru paleosols, illustrates the importance of examining a mix of grain weathering microfeatures and coatings to establish relative weathering criteria.

The Teleki-age (Illinoian) moraine crest contains the well-documented TV23 site (see Mahaney, 1990a, for location) with a relict paleosol of post-Teleki age still exposed to the subaerial atmosphere. It overlies a buried paleosol formed in loess and till of post-Naro Moru age (<780 ka); the soil formed during the mid Brunhes Chron. In the 2C2ox horizon of the relict paleosol, a mix of quartz, volcanic grains and volcanic glass (figure 12.1A) illustrates the importance of glass as a source of clay minerals and other weathering products, especially halloysite and gibbsite (see Mahaney, 1991b). In this frame almost all the glass is clay coated, whereas the quartz is relatively fresh and some grains are broken. The enlargements of individual glass grains in figure 12.1B and C show copious amounts of the nodular form of gibbsite.

Still further down the TV23 profile, quartz (upper left in figure 12.2A) is relatively minor and glass dominates in the tephra that forms, along with loess, the epipedon of the buried paleosol. The enlargement of the glass grain shown in figure 12.2B shows the fibrous form of gibbsite similar to that in other profiles studied in the same moraine system (Mahaney, 1990a). In figure 12.2C the high resolution imagery shows many fibrous strands of gibbsite dominating over metahalloysite. The ratio of gibbsite/metahalloysite provides data for an estimate of the paleoleaching that occurred in the two paleosols, the relative intensity being useful in assessing the strength of soil water movement during the post-Teleki and post-Naro Moru glaciations (Vortisch et al., 1987; Mahaney, 1990a). Aggressive leaching in the post-Naro Moru paleosol led to the removal of metahalloysite and increases in the relative amount of gibbsite present in the paleosols.

Venezuelan Andes

The Ped5, VII paleosol, formed in outwash sand and peat overlying the early Mérida till (early Wisconsinan) near Pedregal, in the north-western Venezuelan Andes (Mahaney & Kalm, 1996; Mahaney et al., 2001c), provides important information on interstadial weathering and soil formation. Infinite AMS dates of >60 ka suggest an age for the paleosol in Isotope Stage 3, or immediately following the recession of early Mérida ice at the beginning of the Pedregal Interstadial.

Postglacial soils in till, outwash, and mass wasting deposits are all Entisols (Mahaney & Kalm, 1996) with thin profiles and maximum Fe oxide and hydroxide release in the Cox horizons. Soil weathering since the last glaciation is slow, hampered by thick Ah horizons with up to 15 percent organic matter which soak up most meteoric water, inhibiting downward percolation and the resultant build-up of B horizons. There are no Spodosols (cf. podzols) (Soil Survey Staff, 1998) presently forming at the surface, but at Pedregal Spodosols are found as buried paleosols in the Ped5 section (Mahaney & Kalm, 1996). As an example of Spodosol development, one of the key features is excessive leaching of Al and Fe compounds from the E horizon and their deposition in the Bs horizon caused by acids in organic-rich surface horizons (either Ah or O; see Earl-Goulet et al. (1998) for a discussion of the processes involved).

A predominantly quartz-rich sand was examined by SEM–EDS to analyze grains for evidence of acid dissolution and Fe or Al coatings in the Bs horizon (figure 12.3). The data show the advantage of imagery supported by EDS spectra

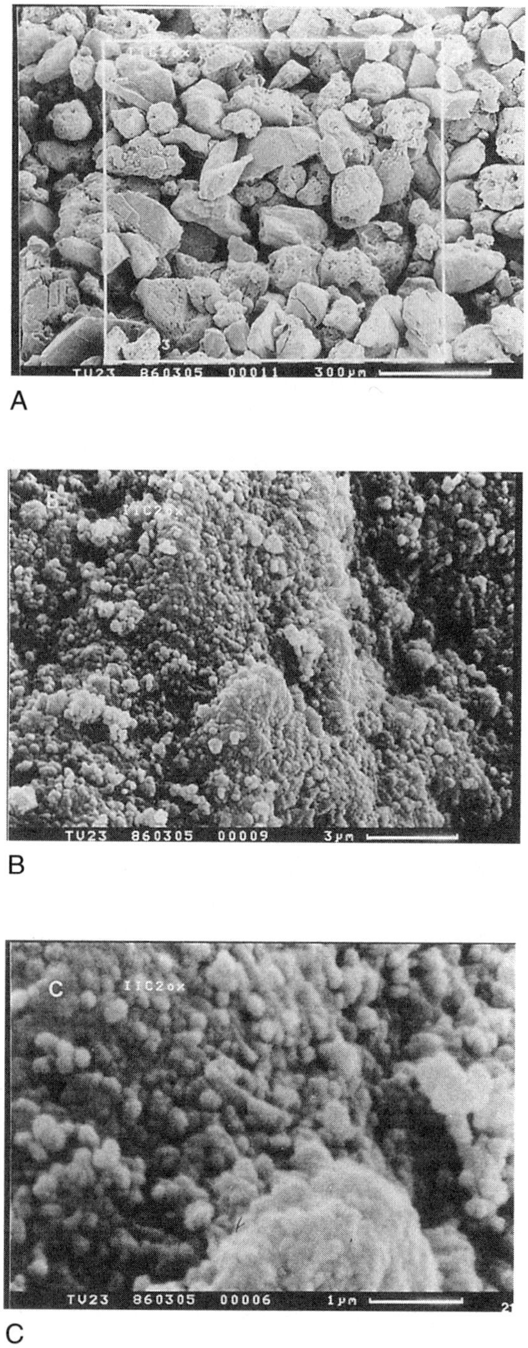

Figure 12.1 (A) Teleki Till at site TV23 with a mix of quartz, volcanic glass, and other grains. (B) Enlargement (×10) of clay coatings on volcanic glass in the 2C2ox horizon of the post-Teleki paleosol. (C) Still greater enlargement of the clay size coatings in B showing dominant nodular forms of gibbsite. Reprinted from Vortisch et al. (1987) by permission of the editor.

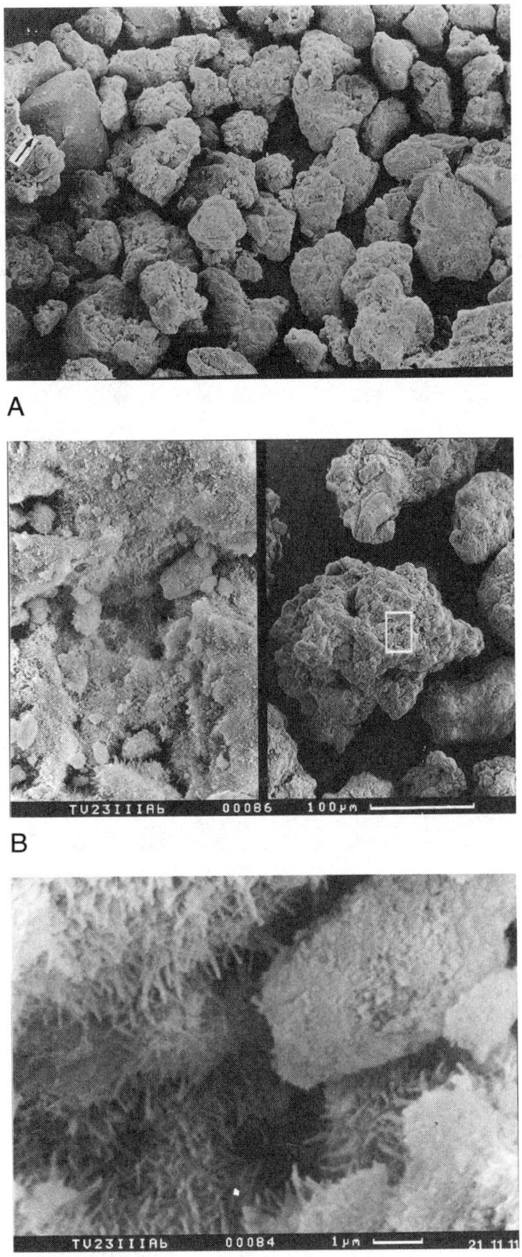

C

Figure 12.2 (A) Grains in the Ab horizon in the post-Naro Moru paleosol. Quartz is rare (arrow) with volcanic glass dominant and well coated with clay. (B) A grain of volcanic glass (right). The enlargement (×10) shows dominant fibrous forms of gibbsite. (C) Greater enlargement of the gibbsite in B showing the detail of the fibrous strands of gibbsite with metahalloysite the dominant coating on the grain in the upper right. Reprinted from Vortisch et al. (1987), by permission of the editor.

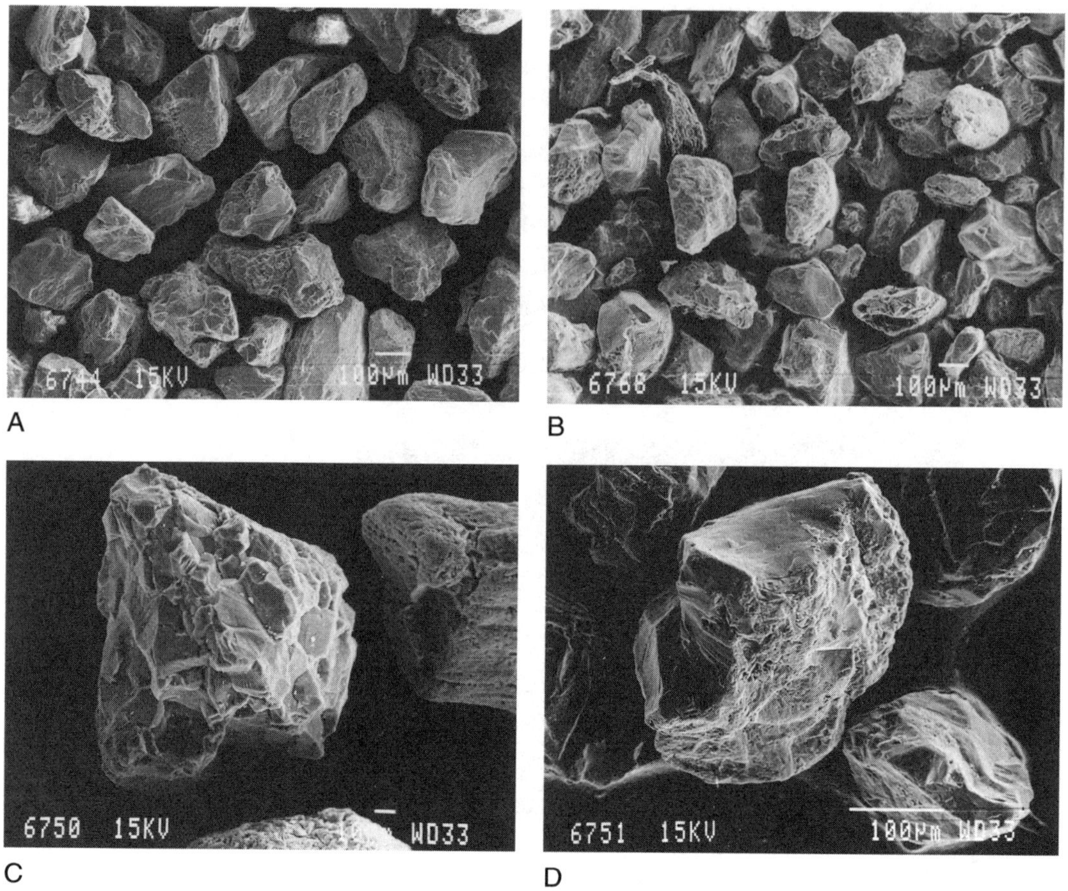

Figure 12.3 (A) Mixed grains, predominantly quartz, well washed and uncoated in the E horizon of the Ped5, VII paleosol (Spodosol), north-western Venezuelan Andes. (B) Partially coated grains of fine sand in the Bs horizon of the paleosol. (C) Fresh quartz (center) with large v-shaped wedge-like microfeatures in outwash, acid washed from overlying peat dated at >60 ka in the E horizon. (D) Quartz, with fresh fracture face (top), partial dissolution microfeatures, and thin discontinuous coating of Fe (lower grain) in the Bs horizon.

data. In this case the difference between the E horizon (acid washed) and the Bs (Fe and Al accumulation) supports the field assignment of this profile to the Spodosol soil order. The total time involved in profile genesis is not known. The age of the upper peat is AMS dated at >60,720 years BP (Isotrace, To-7250). I put the time available for pedogenesis at around 2 ka or less, judging by younger dates higher in the section (Mahaney et al., 2001c). A study of other representative quartz grains from younger profiles in the succession might well yield important information on degrees of acid washing from younger peats and the formation of Bw (color B horizons in Inceptisols) and Cox horizons of Entisols (see Mahaney & Kalm, 1996).

Rouge River, Ontario

The Rouge River, described in chapter 8, is home to a soil stratigraphic sequence ranging from the present to Late Glacial–Lake Iroquois time (later Pleistocene). Profile R43 (see figure 8.1 for details) is of special interest because it formed in landslide debris starting between 5 and 6.5 ka in sediment with imperfect drainage that gave rise to the genesis of smectite (montmorillonite) and vermiculite, both of which are found as coatings on sands in the soil. Investigations with the SEM–EDS show thin, discontinuous coatings on sand-size particles despite abundant amounts of both clay species in the clay fraction (figure 12.4). The EDS spectrum (figure 12.5) indicates higher than average Si, which reflects the thin coatings of 2:1 clay minerals.

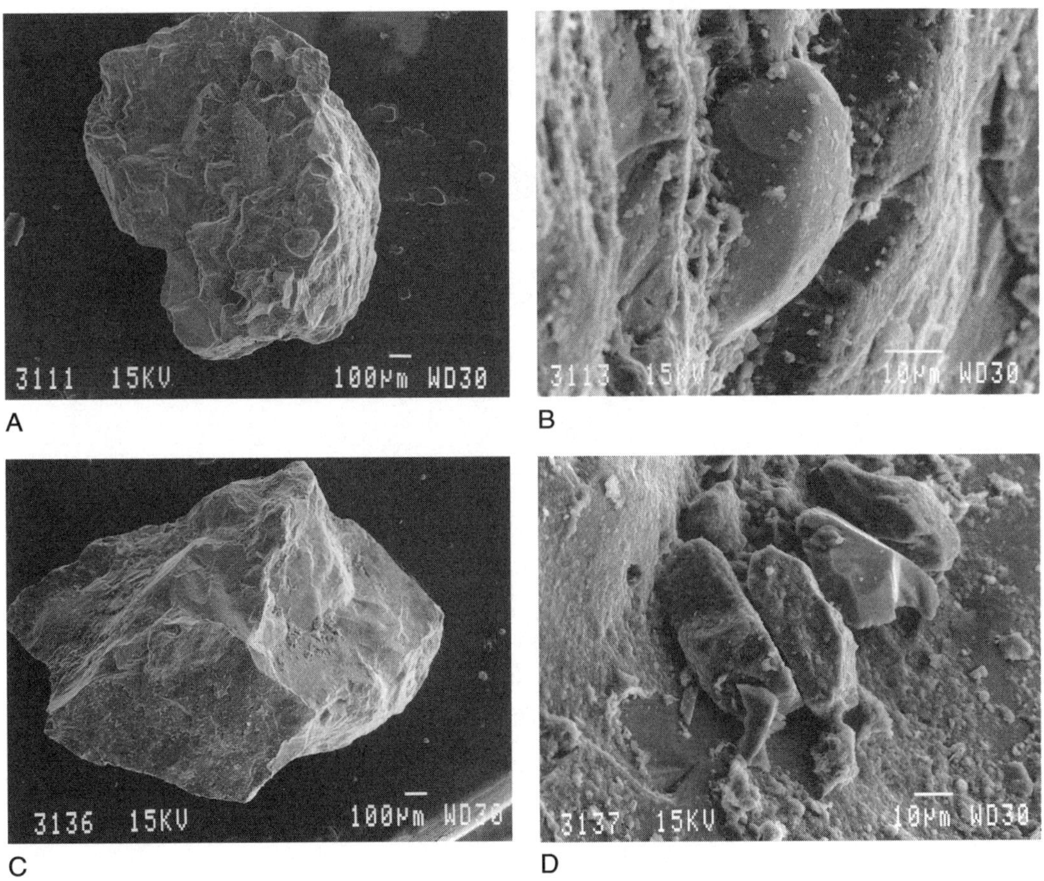

Figure 12.4 (A) Andesine grain, with small fracture faces probably produced by gravitational transport, and with a discontinuous cover of clay minerals. (B) Enlargement of the surface in A showing an adhering rounded particle of quartz. (C) Angular grain of quartz partially coated with clay. (D) Coated surface of the grain in C showing smaller particles: the fresh particle is quartz, and the coated ones are plagioclases.

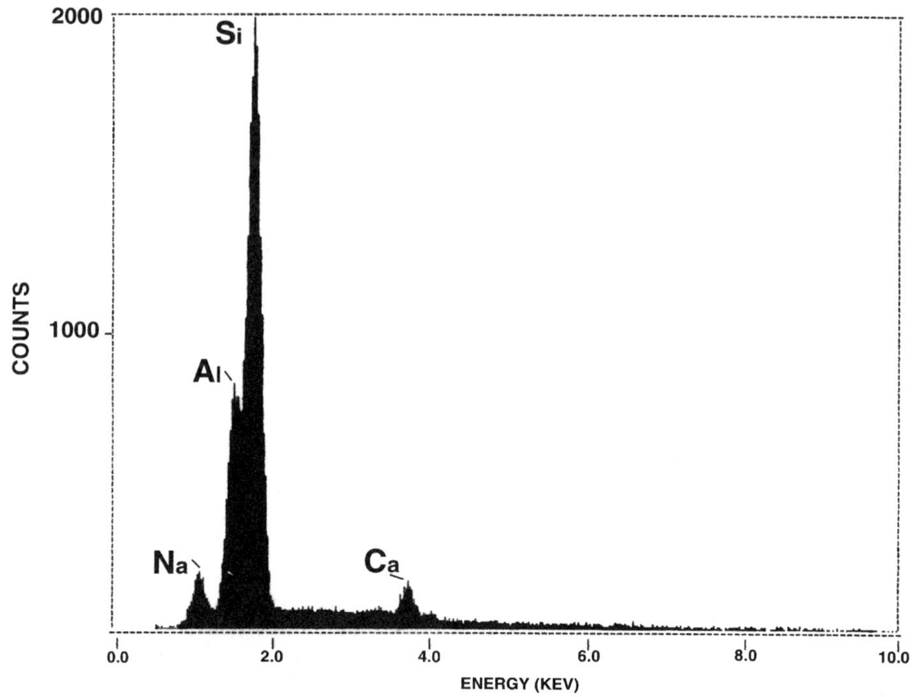

Figure 12.5 EDS spectrum of the andesine grain in figure 12.4A showing relative peak heights for different elements.

Loessic Paleosols

The Central European loess record described by numerous workers, including Kukla (1975) and Morrison (1978), remains unrivaled for its complex paleosols, many of which extend beyond the Brunhes–Matuyama boundary of 0.78 Ma. At Dreihausen, in central Germany, the loess record shows depositional–weathering episodes of Brunhes age (Mahaney et al., 1993), which represent glacial–interglacial cycles. Above each loess in the sequence, a brown earth paleosol, with an argillic horizon (Alfisol; Soil Survey Staff, 1998), formed under mixed deciduous forest during various interglacial periods.

Antarctica

Paleosols in the Antarctic have been studied by many workers (Bockheim, 1979; Campbell & Claridge, 1987) but few researchers have used the SEM–EDS to study Fe oxides and salt accumulations (Mahaney et al., 2001a). In the Antarctic Dry Valleys, pedostratigraphic successions often consist of a

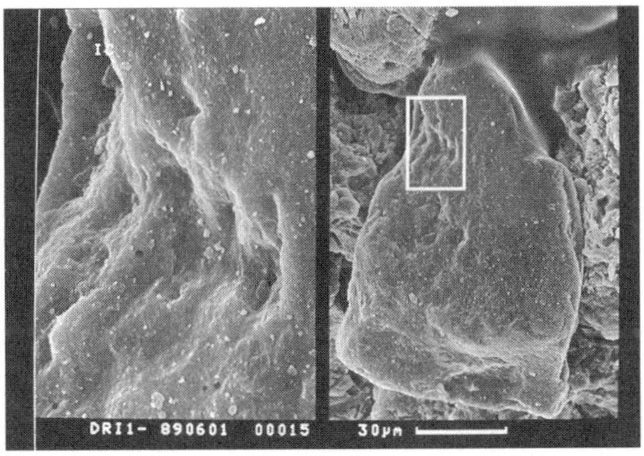

A

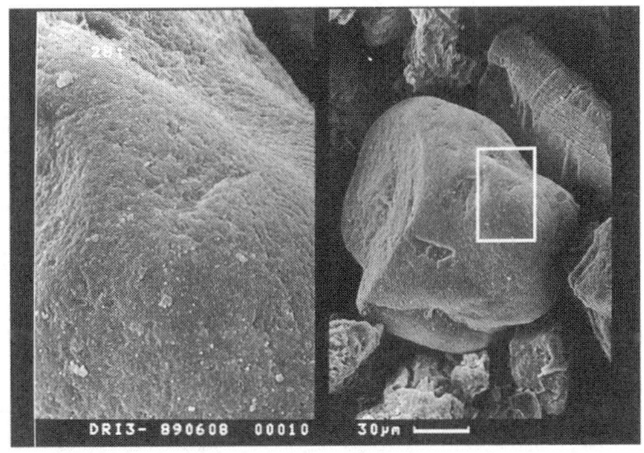

B

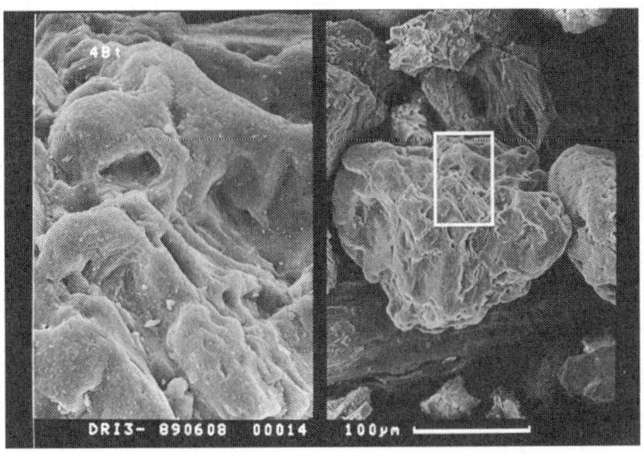

C

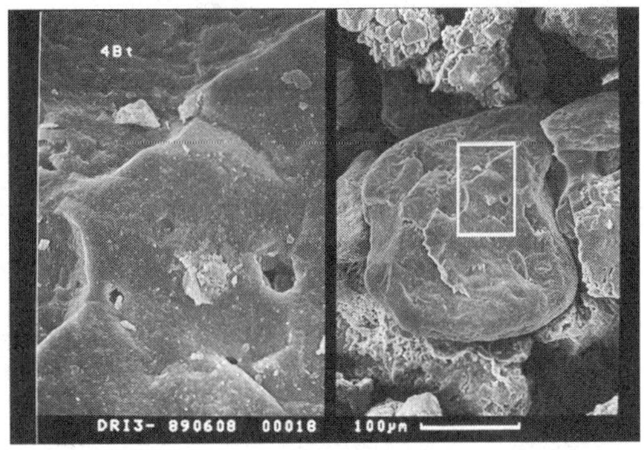

D

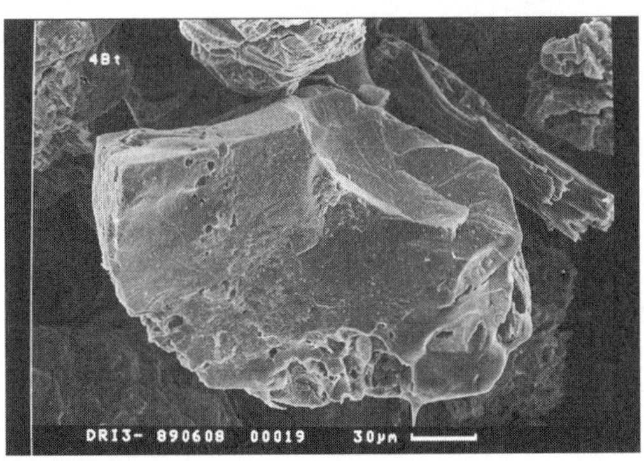

E

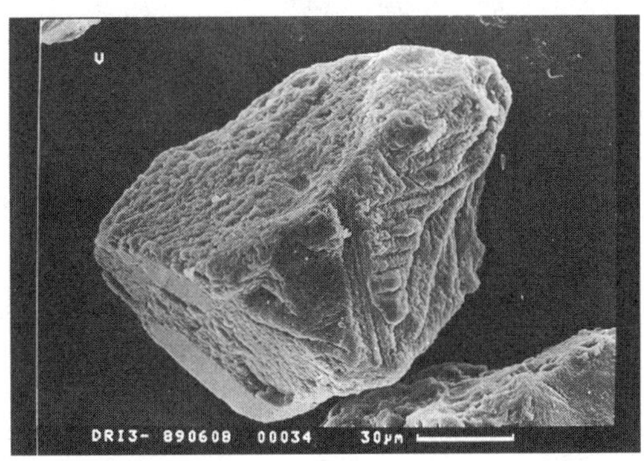

F

Figure 12.6 Weathering sequence in the DRI3 profile, central Germany (enlargement ×5). (A) Quartz in the surface Alfisol, a relict paleosol exposed at the surface during almost the whole of the last glaciation. (B) Quartz showing greater dissolution, relative to A above, in the buried Bt horizon of the Eemian paleosol (enlargement ×5). (C) Quartz in the lower paleosol (4Bt horizon) showing multiple dissolution cavities but few adhering particles (enlargement ×5). (D) Orthoclase in the 4Bt horizon with equally well advanced dissolution in the DRI3 paleosol (enlargement ×5). (E) Cr-bearing amphibole that looks remarkably like quartz with multiple dissolution cavities in the lower paleosol. (F) Well weathered and coated plagioclase in the oldest loess. Preweathered grains of this nature provide information on weathering prior to emplacement in the early middle Quaternary.

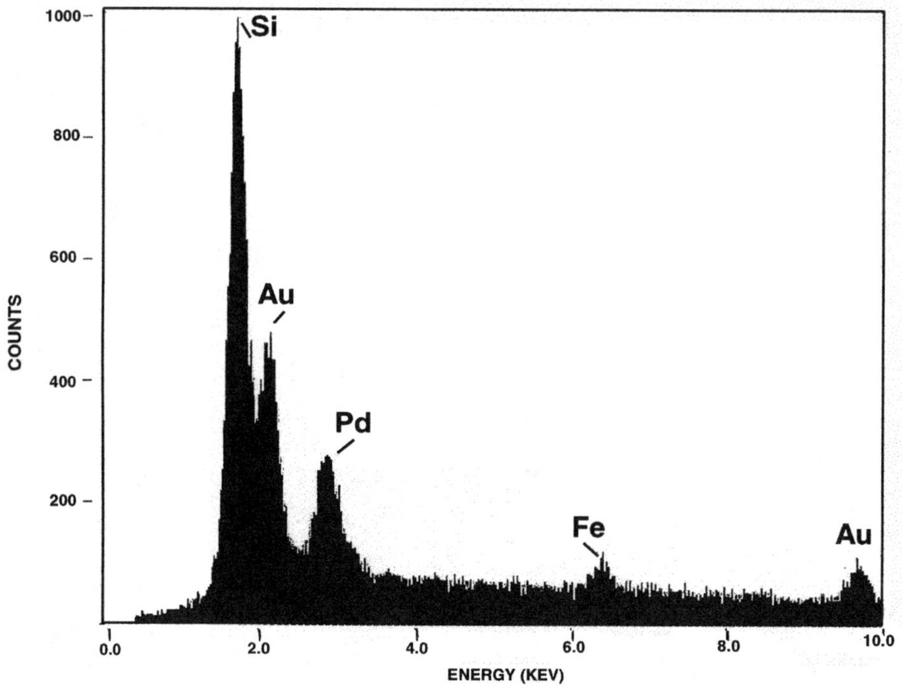

Figure 12.7 EDS spectrum of the Fe coating on the quartz grain in figure 12.6B. Gold (Au) and Pd are coatings which make the surface conductive. The Fe-peak provides a measure of the coating thickness, which is time-dependent, and would be expected to increase with time of exposure and/or intensity of weathering.

pebble–cobble pavement, underlain with an Fe-rich horizon and salt-rich horizon. The entire profile of approximately ~25 cm depth, overlies a number of buried profiles with the same sequence of horizons. Research undertaken on these profiles has attempted to answer questions regarding the amount of Fe release and the build up of weathered coatings, as well as the origin and accumulation of salts. As shown in figure 12.8, sands in Antarctic paleosols archive considerable data on the composition and distribution of both Fe oxides and salts. When combined with data on Fe extracts (see Mahaney et al., 1999, for the method, and Mahaney & Hancock, 2001, for data), the SEM–EDS can be used to image grains with variable Fe coatings, and the thickness can be compared with relative increases in oxalate and dithionite extractable Fe. The accumulation of salts, primarily $MgSO_4$ and $CaSO_4 \bullet 2H_2O$, which undergo pronounced variations in concentration (Mahaney et al., 2001a), as determined by conductivity measurements, can be imaged by SEM and their composition established by EDS. Salt efflorescences, as shown in various frames of figure 12.8, may play a role in the physical weathering of mineral grains.

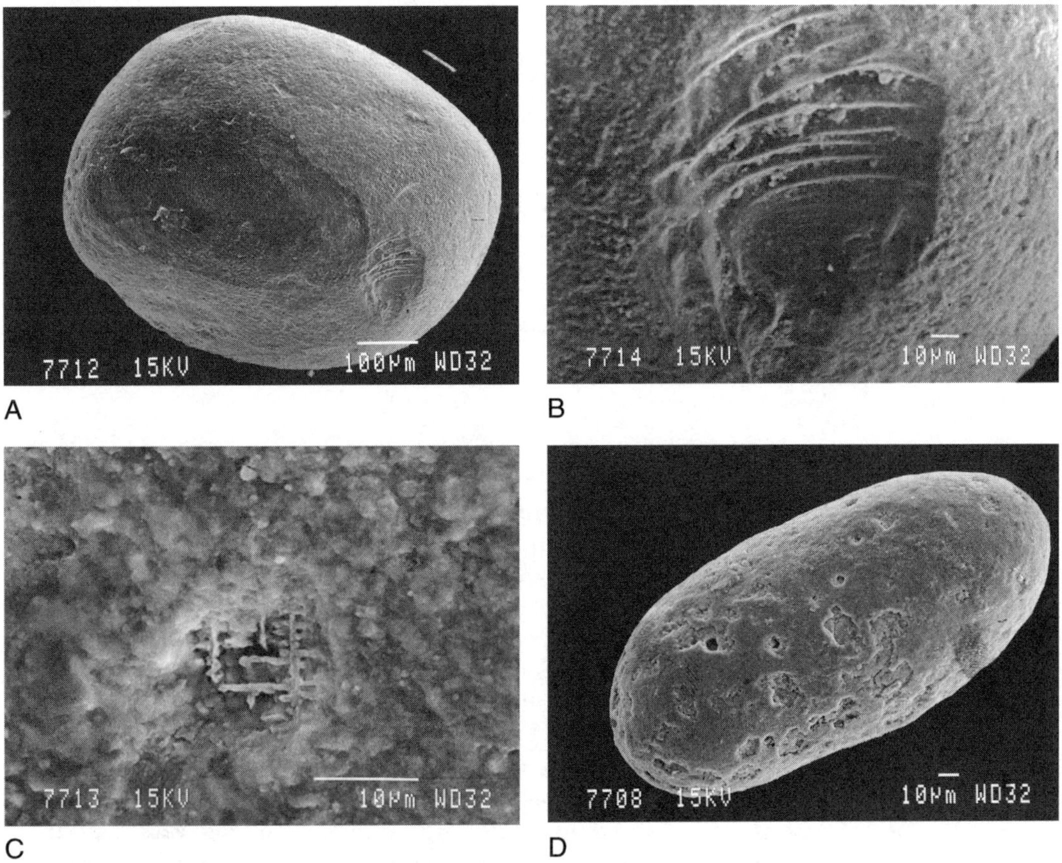

Figure 12.8 Mixed quartz and pyroxene grains in Profile 828, horizon C, New Mountain, Antarctica. (A) Round quartz grain in a salt-rich horizon of a mid-Miocene paleosol which may have an outwash origin in the early warm phase of Antarctic glaciation. (B) Enlargement of the conchoidal fracture in A with Fe coating and salt efflorescences. (C) Enlargement of salt efflorescence ($MgSO_4$) in the quartz in A. (D) Spindle-shaped, etched pyroxene coated with Fe and salt efflorescences.

Conclusions

These examples of soil and paleosol analyses show the importance of SEM imagery and EDS analysis in the study of soil morphogenesis. SEM imagery may be used to assess the state of weathering in fractions of sand, and the thickness and composition of coatings on quartz and other minerals; the result is high-resolution, three-dimensional imagery. From tropical alpine buried and surface paleosols to paleosols at higher latitudes, the imagery supports analyses by other methods such as XRD and INAA (neutron activation) and it has a multitude of uses to soil scientists and stratigraphers alike.

13 Weathering and Diagenesis

Detrital grains may be studied by SEM and EDS to determine whether grain microfeatures might prove useful in determining both clast sedimentologic and diagenetic history. Grains from clay to sand size are prime candidates for study, and information about shape, sphericity, packing structures, grain cohesion, and the chemistry of cementing agents make the SEM–EDS the instrument of choice in such investigations.

Often high-resolution imagery will shed considerable light on post-depositional alteration and modification of grains as the weight of overlying sediments leads to the expulsion of initial gases and liquids and subsequent penetration of groundwater. The fluctuation of groundwater and the invasion of meteoric water lead to the genesis of structures and coatings of various kinds. Grain assemblages can shed important information on post-deposition processes. Indeed, the cementation process and the action of pore fluids that leave reprecipitated silica, carbonate, and/or iron coatings, help to reconstruct events following deposition. Thus, the SEM–EDS can be an important instrument in the study of post-depositional modification of grains following transport by different geologic agents. In sum, analysis of coatings and dissolution microtextures (see Borger, 1993b, for a weathering intensity scale) on grain surfaces might prove as important in reconstructing diagenetic processes as fracture and abrasion microfeatures are in reconstructing transport histories. The ratio of fracture–abrasion to coating microtextures might provide a sliding scale of the slow conversion of nearly pristine grains, little affected by diagenesis, to grains and grain assemblages greatly altered by pore waters and inexorably slow diagenetic alteration.

Sand grain microfeatures that are related to diagenetic alteration have been studied with the light microscope by many workers (Delvigne, 1998). Less frequently the SEM has been used to determine environmental histories (see, e.g., Krinsley & Doornkamp, 1973; Mahaney, 1990a; Mahaney & Kalm, 1995). Using backscattered imaging of grains (Krinsley & Smalley, 1973; Krinsley et al., 1998), it has been possible to reconstruct multiple environmental histories including diagenesis, wind velocity (Krinsley & Wellendorf, 1980), weathering and diagenetic histories (Krinsley & Doornkamp, 1973; Mahaney, 1990a, 1992b; Mahaney & Rutter, 1992), metamorphic alteration (Retallack & Krinsley, 1993), and paleoclimatic reconstructions (Krinsley & Marshall, 1987; Mahaney, 1990a; Mahaney & Kalm, 1995, 2000). Additional fracture and abrasion microtextures provide important evidence of damage inflicted on sands during transport by different geologic agents, all of which is subject to modification by diagenesis. Diagenesis modifies many of the original mechanical and abrasion microtextures on sand surfaces, and must be taken into account when environmental reconstruction is carried out.

Cambrian Sandstones

Grains from Devonian and Cambrian sandstone beds in Estonia have been studied (Mahaney & Kalm, 1995, 2000) to elicit information on microtextures and composition, relative to overlying tills of Quaternary age, that might prove useful in environmental discrimination. As part of this study the Rannamöisa Section (figure 13.1) was included to gain information on sands of Lower Cambrian age that might have been incorporated into Quaternary tills. The

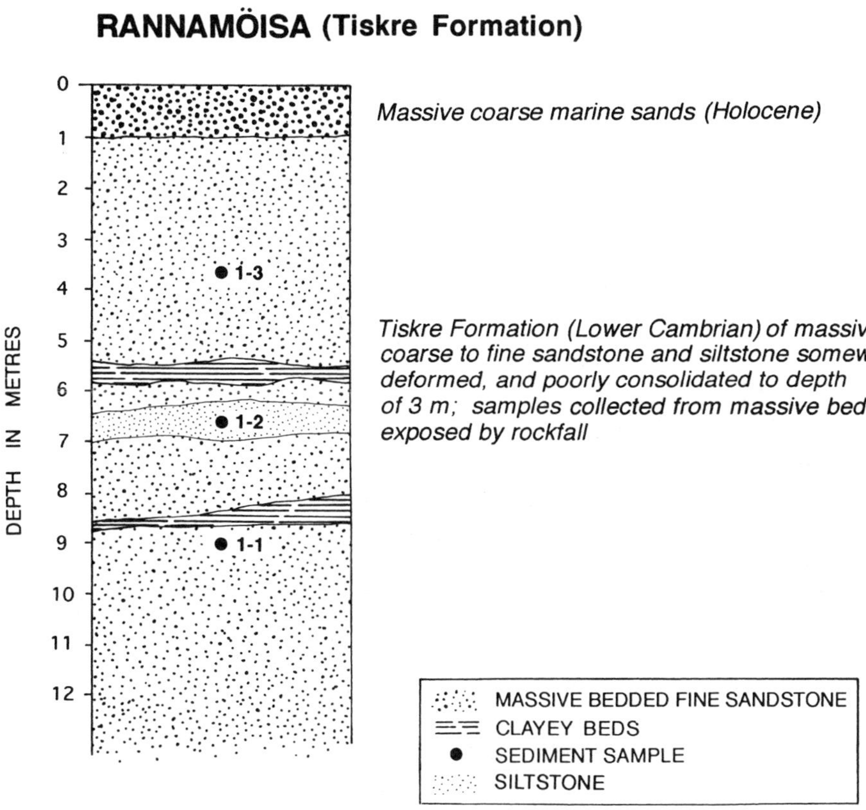

Figure 13.1 Rannamöisa Section, northern Estonian Coast, showing the location of the Cambrian sandstone samples.

coating microfeatures, composition of sands (including assemblages of sands and silts,) grain size, and grain shape in the Lower Cambrian sandstones were studied to determine if provenance, mode of transport, and Phanerozoic diagenetic history could be reconstructed.

Sections of non-cemented siliciclastic sandstones and siltstones of Lower Cambrian age (Mens & Pirrus, 1977; Kurvits et al., 2000), up to 20 m or more in thickness, crop out on the North Estonian Coast. The easily accessible Rannamõisa member of the Tiskre Formation, located along the Gulf of Finland, west of the City of Tallinn, was chosen for investigation.

The samples were collected from massive, horizontally bedded, well-sorted and very fine grained, silty quartzitic sandstones with thin interbedded siltstone and claystone between 5.5 and 8.7 m in thickness. Massive beds as thick as 0.5 m do not show any internal laminae sets. Horizontal bedding, formed by alternation of thick massive siltstone and thin clay beds is weakly developed. Coarse massive marine sands of Quaternary age about 1 m thick cap the Lower Cambrian rocks.

Several thousand grains were analyzed from three levels—the bottom, middle, and top of the section (Kurvits et al., 2000)—and 60 grains per sample (from three bulk samples) were analyzed in detail. Collected samples were disaggregated by hand, and sieved to determine the degree of sorting. In all three samples grains fell in the range of 40–70 μm (coarse silt and very fine sand) corresponding to a high degree of sorting. Where clay is present, it is bound up in the coatings or interstitial cement holding individual grains together.

The grains studied are nearly all quartz, with minor amounts of K-feldspar, and a few grains of mica indicating it is derived primarily from sedimentary rock. No heavy minerals were recovered in the coarse fraction in the laboratory separation process, although some were noted in the bulk samples. The single outstanding feature of the sands in this section is their angularity or subangularity. Unlike sands investigated in Devonian outcrops 100 km to the south (Mahaney & Kalm, 1995, 2000), only angular to subangular grains were recovered from the Rannamõisa section. This is partially caused by the smaller grain size in the Cambrian samples, compared to Devonian outcrops. According to Blatt (1970), quartz grains smaller than 125 μm show little rounding and the mean grain size in the Rannamõisa Section is ~60 μm. Also, Viiding and Konsa (1976) studied quartz grain morphology from Lower Cambrian rocks and found a decrease of angularity in the 100–50 μm fraction, and an increase in subangular to subrounded grains upward.

The fact that all grains are angular or subangular indicates that the transport distance must have been short, possibly with saltation and rolling, or the transport path did not cause essential mechanical wastage. Thus, material has been transported in suspension. Since water transport, even with fine sediment of coarse to medium silt, would leave grains with round edges and surfaces full of v-shaped percussion scars (Krinsley & Doornkamp, 1973), it is possible to rule out this geologic agent. The angularity of the grains is similar to mass wasted material (Bull et al., 1987), but without fracture faces of any kind (see Mahaney et al., 1991 and Chapter 8). Eolian transport, as discussed by Krinsley & Wellendorf (1980), is possible. Quartz and orthoclase sands in the Navajo and Entrada sandstones, of Jurassic age, in the south-western United States (next section) also exhibit extreme angularity despite initial shoreline transport, followed by wind transport, from a source in Idaho, 1000 km to the north.

There is no evidence of glacial transport in the several thousand grains studied, as there is only one grain with remnant conchoidal fractures (figure 13.2E).

From the grain size and shape of Tiskre sands, it may be assumed they were transported in eolian suspension, possibly augmented by fluvial transport. According to Einsele (1992), the 40–70 μm fraction can be transported by moderate storms as short-term suspensions in air for several tens of kilometers. Material carried in suspension may settle down in shallow water near the mainland, with storm waves transporting grains to deeper water to form massive beds. This type of structure, forming in the outer parts of the shore face, below the everyday wave base, may escape destruction. These conditions favor the formation of massive bedding, similar to beds in the Tiskre Formation, and could contribute to preserving the angularity of quartz and K-feldspar grains (figure 13.2). It has been suggested that massive beds, lacking any lamination, are formed either by very rapid deposition from suspension, or by deposition from very highly concentrated sediment dispersions (Blatt et al., 1980). This could be expected with low mechanical attrition.

There are no sediments representing the time interval between Early Cambrian and Early Late Cambrian in North Estonia in the area of the Tiskre beds. The upper contact of Tiskre sandstones with overlying beds is erosional. Assuming a long-lasting hiatus between partially eroded Lower Cambrian rocks and newly deposited Upper Cambrian sands equivalent to a period of subaerial weathering, one might expect extreme weathering (dissolution) of quartz and K-feldspar grains under a humid climate.

The extreme etching of K-feldspar and dissolution of quartz is prevalent on all grains studied (figure 13.2). Fresh grains were not encountered and extreme chemical weathering obliterated most surface microtextures making it impossible to use fracture and abrasion microfeatures in reconstructing environmental history. With respect to the K-feldspars, very deep etching follows the cleavage planes of individual minerals, with both shallow and deep microfeatures extending into some grains. Corrosion of quartz has led to the production of deep etch pits which in many cases have nearly obliterated the grain. Some of the liberated silica, from both quartz and K-feldspar, has been reprecipitated on the weathered quartz and feldspar grains as authigenic coatings (figure 13.2F).

Coatings of silica, prevalent on approximately 25 percent of all sands, is very apparent in sample 1-1 at the base of the section. Etched features are common on approximately 75 percent of all grains studied. Grain assemblages are all loosely cemented with SiO_2. Since quartz accounts for approximately 60 percent of all grains studied, the SiO_2 must originate from diagenetic alteration of quartz and K-feldspar by pore fluids, probably fluctuating with the waxing and waning of groundwater. Such fluctuations, more common in coarser arkosic sands, might also lead to rapid release of silica on the greater surface area of finer sands such as are present here. Ingress and egress of pore water could be related

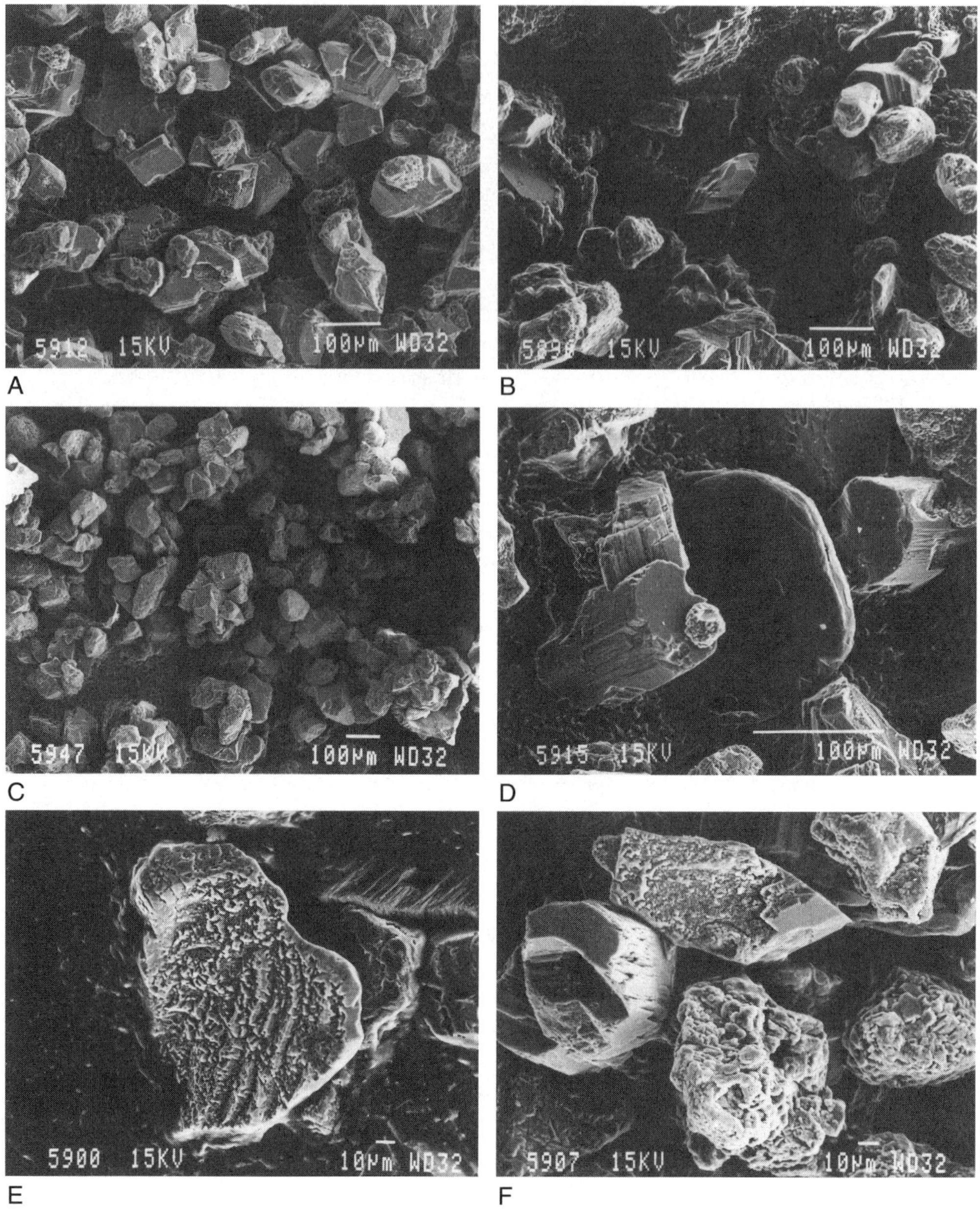

Figure 13.2 (A) General frame of sands and coarse silts in sample Rann1-2 with approximately equal proportions of fresh to weathered grains. (B) General frame of sands and coarse silts in sample Rann1-1 with a somewhat lower ratio of fresh to weathered grains. (C) Poorly cemented grain assemblages in sample Rann1-3. (D) K-feldspar (left) and mica (right) in relatively good shape with infrequent v-shaped percussion scars indicating water movement. (E) Quartz with weathered and partly coated curved grooves from an unknown agent (less than 0.1 percent of grains carry this microtexture) in sample Rann1-1. (F) Clay-coated K-feldspars with etching along cleavage planes in sample Rann1-2, which consists almost exclusively of very fine sand and coarse silt. Reprinted from Kurvits et al. (2000) by permission.

to downcutting by surface streams, lowering the water table and channeling meteoric water through the cap sandstone at the surface of the outcrop. Silica coatings are thick, and thereby probably represent time frames of tens of millions of years.

Micas, which comprise less than 1 percent of the minerals present, indicate comminution, probably from eolian or turbulent water transport. Approximately 20 coatings on various grains were studied with the EDS and all were of SiO_2 only, either amorphous or crystalline. These were most probably wreckages from diagenetic weathering of quartz and K-feldspar.

Grain micromorphological analysis of the Lower Cambrian siliciclastic rocks in the Rannamöisa section of northern Estonia indicates a clast size of 40–70 μm, including well-sorted coarse silt and very fine sand. All grains studied are angular to subangular in shape. Extreme chemical alteration of the grains did not allow exact microtextural analysis, although grain shape suggests eolian transport over short distances. The lack of sedimentary structures (massive beds lacking any lamination) supports the idea of very rapid deposition from suspension. This may have occurred during storms generating backflow carrying material from shore to an unspecified depth below wave base. The source is probably within 500 km and the erosive agent did not allow grain rounding. The huge mainland 200–300 km to the east and south of Rannamöisa, covered with Lower Cambrian clay and siltstones, was the source of the clastic sediment.

Of the several thousand grains analyzed, no fresh grains were encountered. All grains have moderate to extreme chemical etching and dissolution, microfeatures which are attributed to diagenesis. Fine sand aggregates consisting of quartz and K-feldspar, weakly indurated and cemented with SiO_2, are present in all three samples, with the most extreme etching/dissolution and reprecipitated coatings at the base of the section. These aggregates may represent drainage pathways for meteoric water moving down to the water table at various times. Perhaps they represent the most frequently occurring depth of groundwater fluctuation from their position at the base of the section over long periods of time. Frequent inundation caused by fluctuating water tables followed by flushing, would lead to rapid release of SiO_2 into pore waters, followed by precipitation with changing Eh–pH.

Jurassic Sandstones

Jurassic sands from the Entrada and Navajo formations of the south-western United States were examined by SEM–EDS to determine their source, transport history, and diagenesis. The sands provided some unusual microtextures compared to grains transported in wind systems elsewhere (see chapter 6), in terms of shape, individual relief characteristics, and coatings on particles. The sands reported here may have had a source area to the north in what is present-day Idaho (D. Netoff, personal communication, 1998). The samples were transported first in Jurassic seas, reworked southward by near-shore currents, followed by transport in a wet eolian environment with the water table close to the surface. Vast amounts of sand were fed into giant erg systems, which were spread out over vast areas of what is today western Utah, southern Arizona, and western Colorado. Because there were highlands present in these areas, some of the grains could have been transported in fluvial systems but it is impossible to estimate the percentage.

The samples are a mix of quartz, plagioclase, orthoclase, mica, and calcite. Grain shape is dominantly subangular (figure 13.3A and B) which is not surprising given the small size of most grains (average diameter 100 μm) and their similarity to the very fine sand and coarse silt described in the previous Cambrian model from the Gulf of Finland.

The prevalence of adhering particles is important in that it might be related to grinding and fragmenting during transport. Because adhering particles are coated, comminution must have preceded the invasion of pore waters, either from meteoric or subsurface sources. Even fresh, uncoated grains have up to 20 percent of their surfaces covered with adhering particles, many of which have compositions different from the host grains (figure 13.3D).

Edge rounding (figure 13.3A), which is prevalent on 30–40 percent of grains, appears to have occurred later during fluvial transport. Rounded edges are coated with the same thickness as the more angular facets of other grains, indicating that the coatings are most likely diagenetic. Because edge rounding is limited to only portions of whole grains, it is considered to be the result of short distance transport by an active rounding agent.

Particle coatings, as determined by EDS, show a range of chemistries consistent with common as well as uncommon cementing agents. There are calcite grains without coatings, feldspar and quartz with thick coatings of carbonate, and other feldspar and quartz grains that are coated with variable thicknesses of Fe and Si. In some cases, the coatings have adhering particles in them; some particles appear to have been attached to the grain surface initially and later engulfed by cement. In others, the coating covers the entire grain surface and the adhering particles are attached to the surface of the coat. This might imply a softening of the coat due to the invasion of pore fluids, during which time invading particles settled onto the larger grains.

On some grains a mix of carbonate and iron coatings occur; on still other grains clay minerals dominate. The clay mineralogy of these coatings is smectite for the most part, with a high Si:Al ratio and some Ca. Irregular surfaces on coatings probably relate to an underlying presence of small silt-sized particles of different minerals. A coating with a mix of Fe and Cu, with a Fe:Cu ratio of approximately 4:1, was observed on one grain and appeared to be an anomalous occurrence that might be explained by additional analysis.

The ratios of the major coating agents are fairly constant on all grains irrespective of their mineralogy: Ca to Fe is approximately 3:1, and Si to Fe is 1:1. This suggests the prevalence of carbonate, and despite the low number of calcite grains encountered in the samples, there is probably a history of carbonate-rich water invading the sediments as they were slowly lithified. The source of Fe is somewhat mysterious, as few heavy minerals were encountered in this survey, although many adhering particles belong to the amphibole group. Perhaps this is the source, although there does not appear to be sufficient amphibole to provide the coatings observed. Iron might have migrated from fine-grained Fe-rich beds above and below the Entrada Formation. Silica coatings are derived

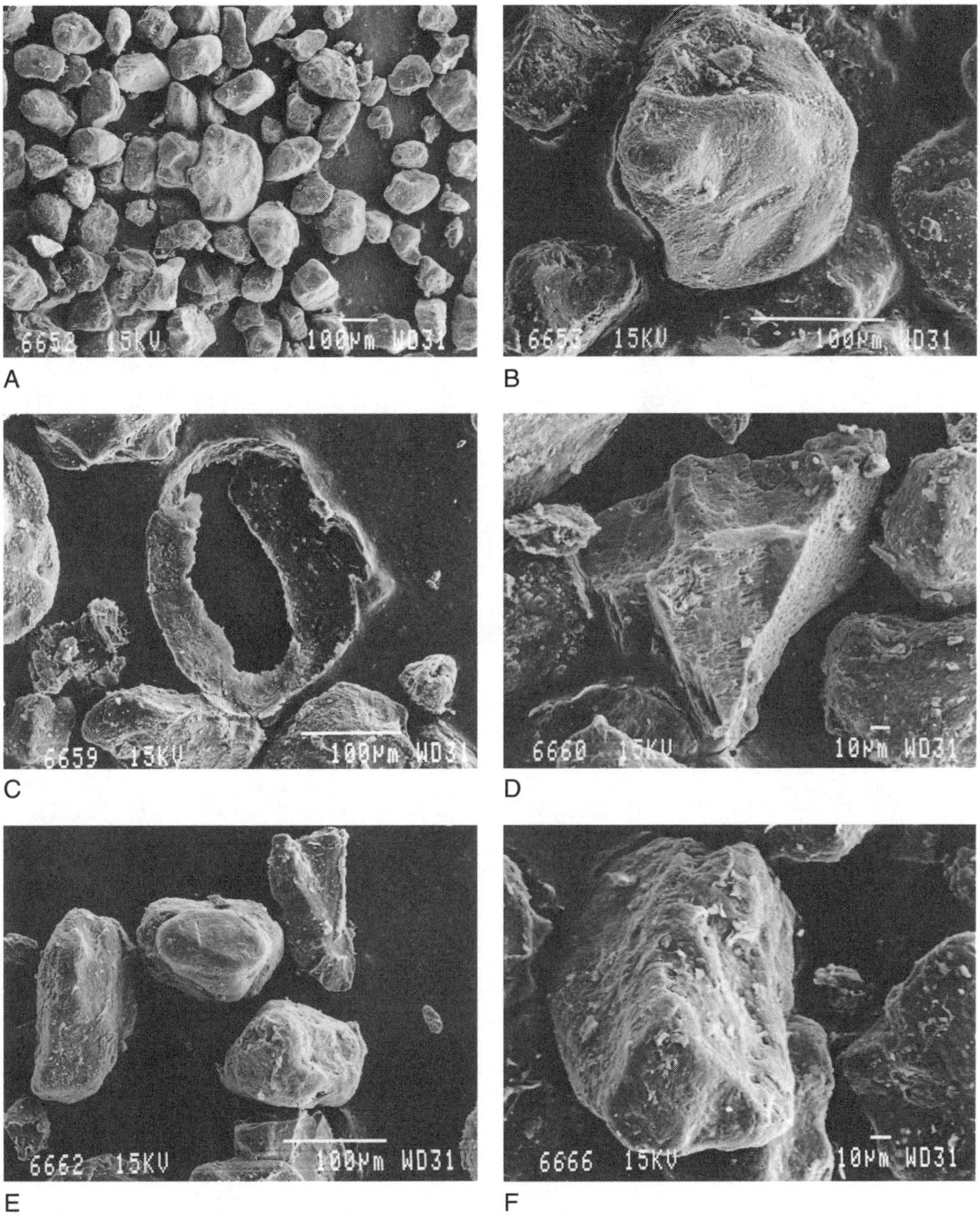

Figure 13.3 (A) General frame of disaggregated subangular grains in samples of Entrada sandstone, south-eastern Utah. (B) Quartz with rounded edges and Ca-rich clay coating. (C) Well-weathered quartz in Entrada sandstone sample CJP100XB. (D) Angular calcite grain with minor dissolution and slight carbonate coating. (E) Three K-feldspar grains ranging from subangular to subrounded with quartz to left. All grains have clay coatings of Ca and Fe, with a Ca:Fe ratio of about 3:1. (F) Quartz with Fe coating and many adhering particles in sample 1013X in the Entrada sandstone.

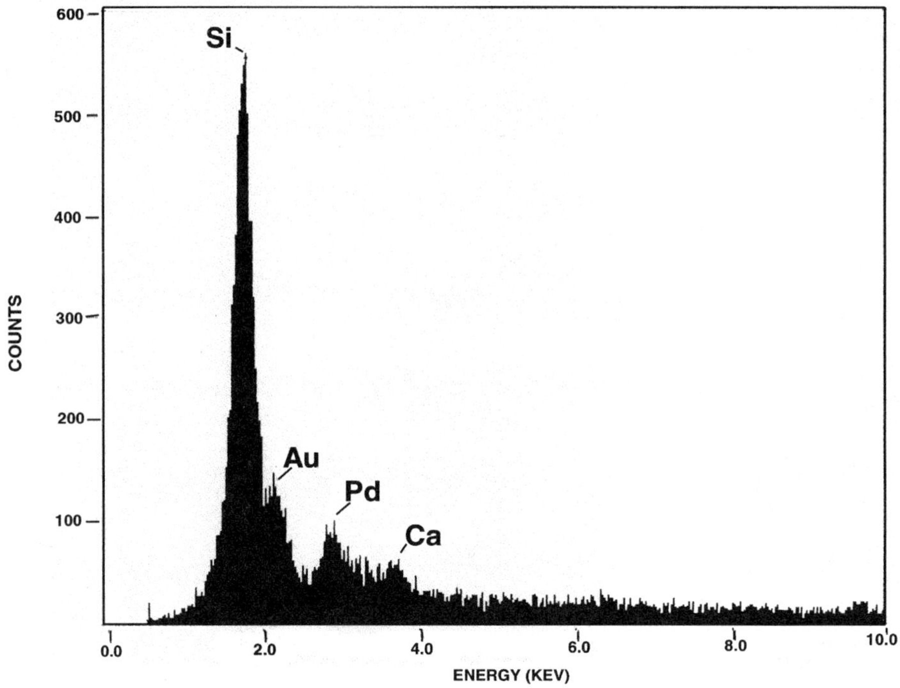

Figure 13.4 EDS spectrum of a quartz grain with a carbonate coating. Peaks of Au and Pd are from the conductive coating.

from the prevalent quartz found in all samples; apparently pore water invasions have been of sufficient magnitude and frequency to diagenetically alter grains originally transported by water and wind.

Most grains are chemically weathered, with microfeatures that appear to be the result of diagenesis. The ratio of fresh to weathered grains is approximately 1:4, with almost all weathered grains coated. The weathered grains contain a higher than usual number of adhering particles.

Marine Cretaceous

Bedrock in southern Alberta, including the Oldman Formation of Cretaceous age, and terrestrial Eocene beds, has been studied for diagenetic effects and chemical composition (Mahaney et al., 1988b) along with tills in the overlying Quaternary succession (Mahaney & Stalker, 1988). While diagenetic properties are common in the tills, which include mainly new growth of quartz, grains in the marine Cretaceous include anhydrite or gypsum on quartz and K-feldspar grains, indicating little in the way of diagenetic alteration. In some cases, entire grains of gypsum show an intact crystal structure with little alteration (figure 13.5A–C).

Most grains are euhedral (figure 13.5D and E) with sharp edges and gentle "creases," the enlargement (Fig 13.5D) of which (×5) shows microfold features where the skin of the quartz has been disrupted, presumably by collisions with other grains. Still other grains appear to be thin spalls with microplatelets; the enlargement (figure 13.5E) showing tears and fractures, some with odd-shaped lunate features. Still other grains come complete with radiating fractures presumably resulting from impact, possibly with other grains in fast-moving traction currents (figure 13.5F).

The mean grain size is very fine sand, of approximately 60 μm diameter, and is positively skewed with little variation in shape. All grains are sharp-edged and very similar in shape to the Cambrian grains described above. While there are many fresh grains (figure 13.6A) in this suite of samples, compared with those in figure 13.5, there is also a high percentage of weathered grains, including quartz (figure 13.6F). Some grains carry shallow fractures (figure 13.6B), or deep fractures that result in grain breakage (figure 13.6C and D). In figure 13.6C, the fracture pattern shows v-shaped as well as circular fractures that might result from dense, turbulent flow in water. In any case, fracture patterns are rare on these grains due to concealment from diagenetic coatings. The grain breakage shown in figure 13.6C is enlarged in D. Straight fractures might indicate sudden collision or strain release whereas convex–concave patterns may result from torque forces that attempt to bend the grain. The enlargement in figure 13.6D shows a clean separation with tears on the bottom. This may result from excessive loading, first by marine sediment, and later by cryostatic pressure from Quaternary ice, and deposition of thick (50 m) till. Quartz and other grains appear to be remarkably free of load-induced fractures, displaying mainly mild or soft diagenetic weathering, although the reasons for this situation are unknown.

With increasing load and sediment packing, intense edge fractionation of grains is to be expected. This is shown in figure 13.6E where the entire side of the grain in the center has collapsed, and in figure 13.6F where a soft weathered quartz grain has been reformed with the mild relief characteristic of many grains in this sample suite.

Many grains have coatings of nodular gibbsite (figure 13.7A) whereas others are moderately to intensely weathered, either from former subaerial exposure or from diagenesis (figure 13.7B). Gibbsite, in particular, is indicative of aggressive leaching, which means either intense flushing during pedogenesis or intense flushing with groundwater during diagenesis.

New growth of quartz is common here just as it is in the younger overlying tills and may be a morphological feature relying on dissolution processes operating on neighboring quartz grains. The presence of Ca-plagioclase grains with minor dissolution (figure 13.7D), side by side with quartz showing extreme dissolution effects, makes a simple interpretation difficult at best. Because of their susceptibility to hydrolysis, plagioclases are among the most short-lived recorders of microtextures. Their presence, therefore, suggests little in the way of invading pore water in this environment. If present, groundwater would not be fluctuating to any great degree, as the plagioclase could not stand continual flushing.

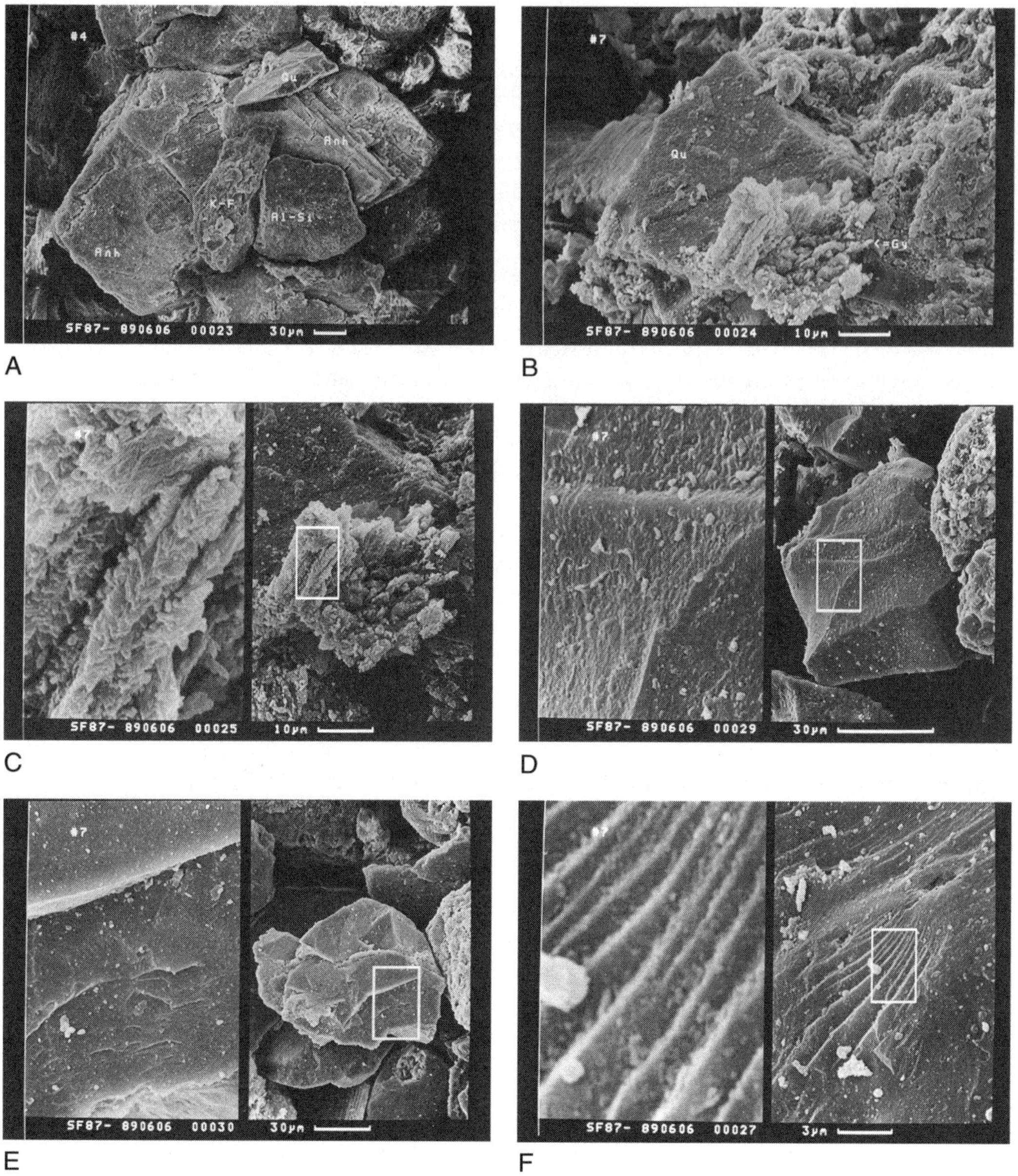

Figure 13.5. Marine Cretaceous from Wellsch Valley, Saskatchewan (see Mahaney & Stalker, 1988). (A) Mixed grains including K-feldspar, albite (Al-Si), and anhydrite (Anh) without gypsum indicating little diagenesis effect. (B) Thin layer of chert over quartz (Qu) with rosettes of gypsum (Gy). (C) Enlargement of the gypsum grain in B. The split enlargement (left) is ×5. (D) Abraded angular quartz grain showing smoothed facets. The enlargement is a split of ×5 showing few adhering particles and what could be new growth. The weathered grains to the right are plagioclases. (E) Slightly etched quartz showing tears and fractures with lunate shapes of unknown origin. The enlargement is ×5. (F) Crater cone shape formed in quartz with a radiating fracture pattern that could be a remnant of turbidity current collision. The enlargement (×5) shows some corrosion on fractures. Note the high frequency fracture lines with wavelengths of 0.2 μm.

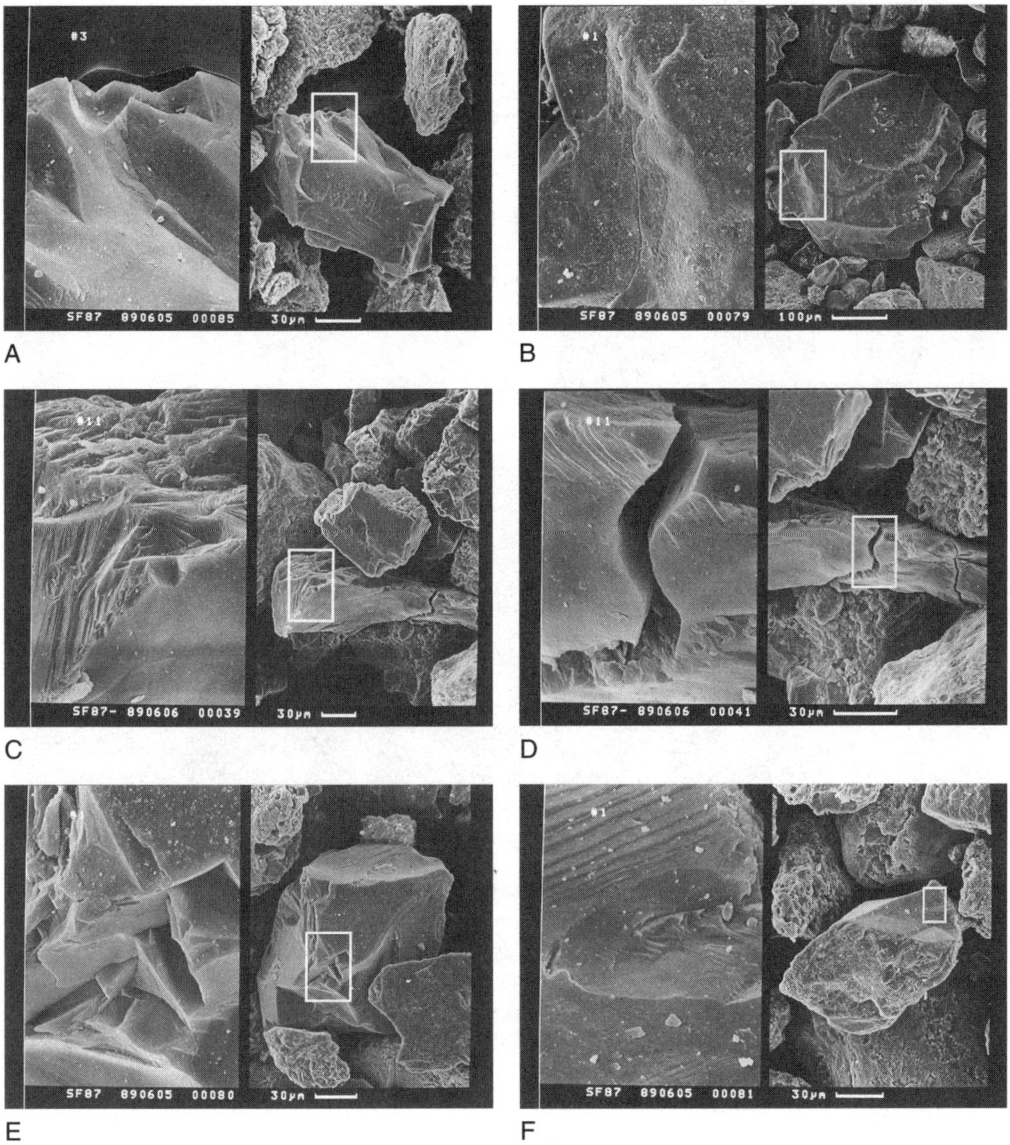

Figure 13.6 (A) Quartz with a fine coating of Si showing a "molar tooth" or sawtooth fracture (see figure 5.1) and lunate crater microtextures; the former are seen only on grains of marine origin. The enlargement is ×5. (B) For comparison, from the "Jaw Face" till above marine Cretaceous beds at Wellsch Valley, Saskatchewan, a glacial grain surrounded by mixed weathered and fresh grains. The enlargement (×5) shows a vertical crack through the center grain surface. (C) Bone-shaped quartz grain with some preweathering (upper right) and a crack through the thin limb. (D) Enlargement of a fracture in quartz. The enlargement (×5) shows tensional tear microfeatures below. (E) Regrowth of quartz partly fractured by loading of till during Quaternary glaciations in southern Saskatchewan, Canada. (F) Quartz surrounded by chemically weathered grains. The central grain is honeycomb etched with the top portion showing a cirque-like form of unknown origin. Perhaps the grain was weakened, fractured, and then split, exposing a fresh, fractured surface.

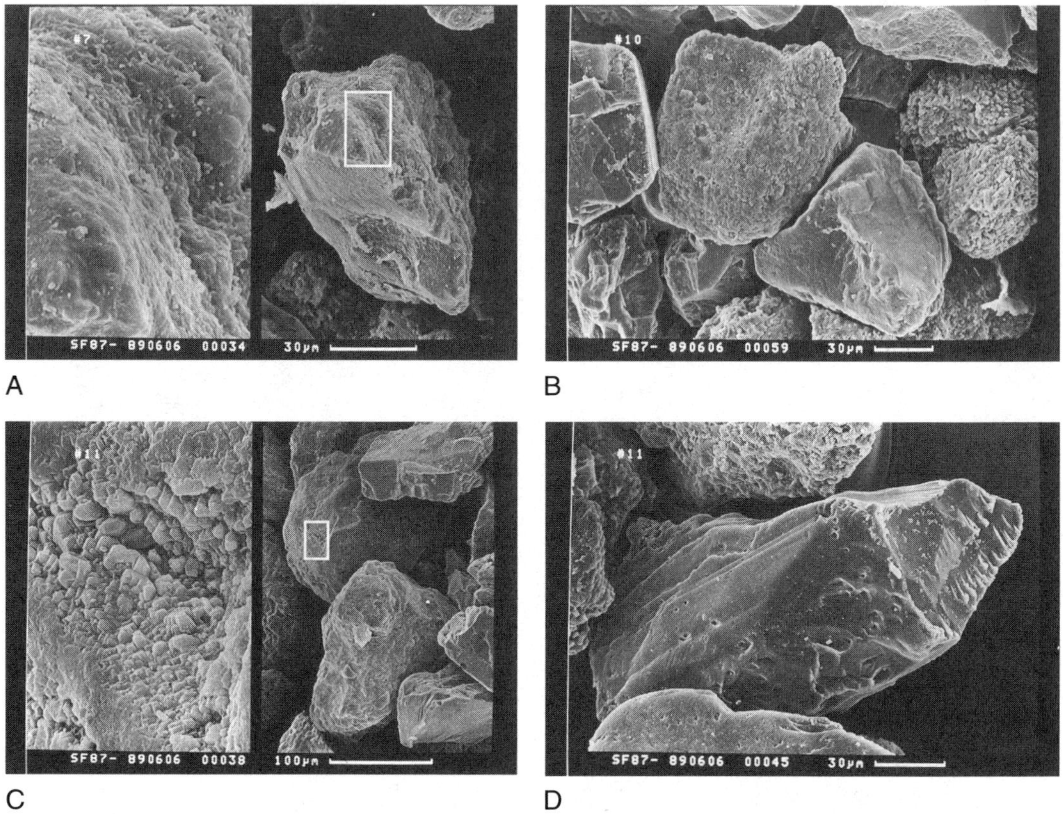

Figure 13.7 (A) Quartz grain with side furrows, all coated with Al. The enlargement is ×5. (B) The two quartz grains in the center show various stages of dissolution and are surrounded by weathered grains of unknown lithology. (C) Two quartz grains in the center. The upper grain exhibits new growth of quartz, as shown in the enlargement ×5. (D) Ca-plagioclase with minor dissolution microfeatures surrounded by weathered and coated grains.

Quaternary Examples

Examples of diagenesis in Quaternary sediments range from mild or soft chemical weathering in lithostratigraphic materials to soil parent materials close to the pedogenically altered parent material. In some cases, diagenesis came from fluctuating water tables, or from springs issuing out from sandy or gravelly aquifers. In still others it probably relates to meteoric waters issuing down through surface soils into fresh, undifferentiated deposits. In alpine basins, for example, where alternating fine and coarse beds are often inclined at steep angles (Mahaney & Kalm, 1996), groundwater may follow underground flow lines issuing out of buried soils, confounding attempts at radiocarbon dating. In still other cases diagenesis in tills, presumably from groundwater fluctuations, is detectable with XRD (Mahaney & Hancock, 1993). Indeed, clay

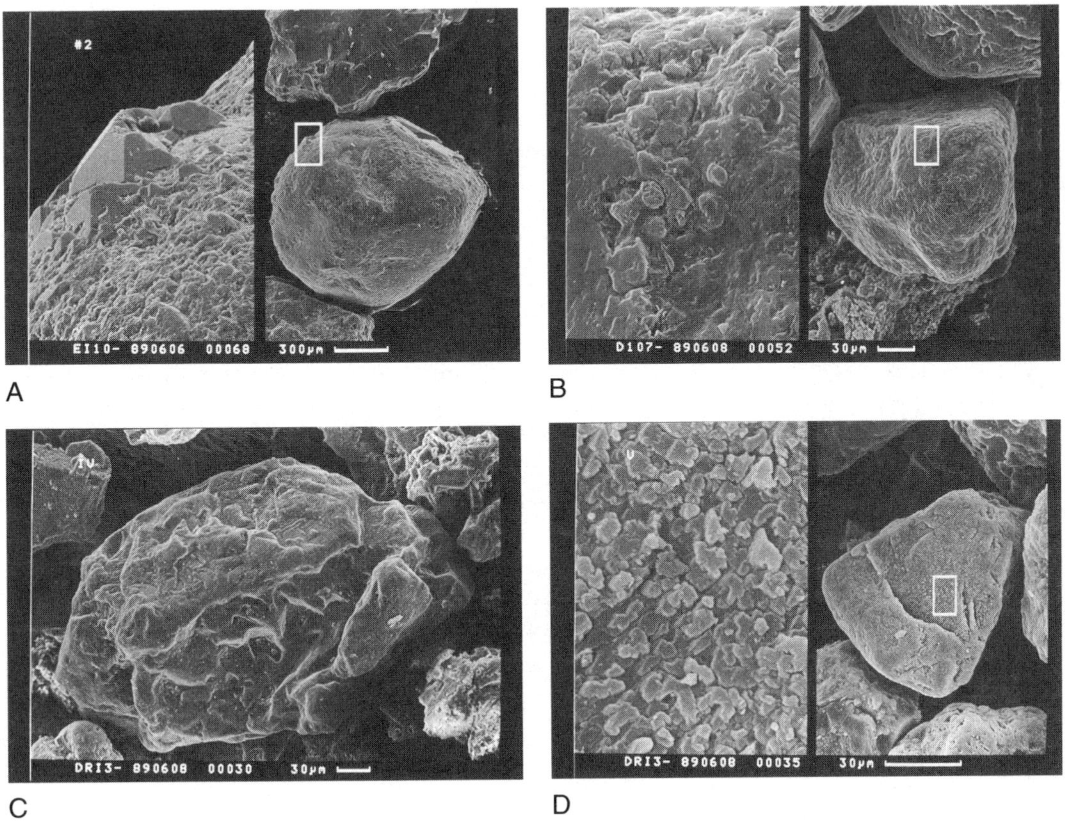

Figure 13.8 (A) Rounded coarse sand from sample E110 on Ellesmere Island showing v-shaped percussion scars over most of the grain surface, presumably from upper flow regime transport. The enlargement (×10) shows details of new growth of quartz weakly affected by dissolution. The sample is courtesy of Wes Blake (Geological Survey of Canada). (B) Grain packing in microtanks on the surface of quartz in the Nile Delta could be a transport microfeature, but most likely is linked with a dynamically stable surface into which finer grained silts could be packed during time of flood. Grains are loosely cemented with silica. The sample is courtesy of Wolfgang Andres (Frankfurt University, Germany). (C) Rounded quartz with a new growth of quartz that has also been rounded. The microfeature data would suggest the new growth of quartz at some distant locality, followed by eolian transport and rounding. The v-shaped percussion scars indicate water transport prior to emplacement. (D) Quartz in the oldest eolian bed of the Dreihausen sequence (Germany), with a middle stage of reprecipitation on a dissolution surface. Fine platelets of quartz deposited on a flat surface of very fine sand

coatings on sands, in instances like this, may provide just the proper amount of supporting information one needs to assess the degree of alteration after deposition. As with older deposits, younger Quaternary beds may exhibit similar microfeatures, including packing of grains, soft weathering, grain breakage from overloading, and other microfeatures.

As an example of new growth of quartz from diagenesis in the most extreme environments, quartz from till on Ellesmere Island is shown in figure 13.8A. This quartz grain, inherited from the underlying sandstone, is rounded with multiple v-shaped percussion cracks and dissolution etching in the mid and upper grain positions. As shown in the enlargement, a new growth of quartz has formed preferentially on a portion of the grain subjected to extreme dissolution. The major question in a situation like this is whether the new growth occurred in till or in some previous depositional setting. Because the new growth is fresh, it would seem likely to be post-depositionally formed following glaciation.

Sediment from the Nile Delta often contains eolian grains originating in the surrounding deserts; these grains show a multitude of soft to severe weathering, breakage from grain-to-grain packing, and pockets of finer grains packed into microtanks or microdepressions on larger grain surfaces (figure 13.8B). EDS analysis of packed grain features shows they are cemented largely by silica.

In the pedostratigraphy described in central Germany at Dreihausen (figure 13.8C and D) by Mahaney et al. (1993), wind-blown and volcanigenic deposits yielded quartz with new growth, formed on rounded grains, which were presumably rerounded by eolian transport. In other words, the second rerounding occurred after the formation of new growth at some other locality. In unit V of the Dreihausen succession, reprecipitated quartz appears as angular, thin platelets of silica disconnected from the grain surface but bound with silica cement. The age of this grain is probably Holsteinian, which allows plenty of time for diagenesis.

Conclusions

Certain diagenetic effects are easily observed and analyzed using the SEM–EDS, and these effects can aid in the interpretation of grain modification following deposition. Soft to severe grain alteration may be subject to speculation as to where it actually occurred, but grain packing, grain breakage, and ratios of different cementing agents adds a dimension not obtainable with other instruments. When exposed to subaerial weathering, such diagenetically altered sediments may add an important dimension to interpretations that are made about soil mineral weathering.

With the EDS, important information may be gathered about cementing agents. With high-resolution imagery it is often possible to determine the degree to which a cement pervades a sediment. Often the cementing material is mixed with varying ratios of Si:Al:Fe:Ca and these relative ratios have important implications for groundwater fluctuations and the lithology of the main mineral complex.

The widespread occurrence of regrowth of quartz begs the question as to how much fluctuating groundwater contributes to the dissolution of Si and regrowth of quartz elsewhere.

Heavy Minerals 14

Heavy and ultraheavy minerals, although minor in terms of weight percentage compared to the bulk weight of a sample, are none the less important in solving sourcing or provenance questions (Folk, 1968). In addition, they are useful in assessing stages of weathering in paleosols (Tejan-Kellu et al., 1991), degree of diagenesis in sediments (Blatt & Sutherland, 1969; Morton & Johnsson, 1993), and they may assist in separating parent materials (Codarcea, 1977) in a soil–paleosol profile (Mahaney et al. 1981; Mahaney & Halvorson, 1986). The assemblages and abundance of heavy minerals have been used to correlate beds in loess (Muller & Negendank, 1974; Codarcea, 1977; Krogh et al., 1993) and to establish provenance. Heavy minerals may have an igneous and/or metamorphic origin, or they may be detrital minerals recycled from sedimentary bedrock as well as clasts of sediments found in Quaternary sediments.

With a specific gravity higher than a standard of 2.85, heavy minerals commonly occur as minor constituents or accessory minerals in rocks, sediments, and soils. With weights usually less than 1 percent in sands and slightly higher in finer grained material, they comprise mainly magnetite, ilmenite, zircon, rutile, kyanite, garnet, tourmaline, sphene, apatite, hornblende, augite, and biotite. Gold, an ultraheavy malleable mineral, occasionally found along with quartz in tills and stream sediment, contains microforms that result from thinning, folding, and flattening (Mahaney, 1992b; Mahaney & Milner, 1998). This malleable mineral is very distinctive when compared with other heavy and light minerals.

Heavy minerals may be separated from a bulk sample by many methods including use of special sieves (sizing) and jigging, panning, vibration, magnetic separation, and heavy liquid separation. To avoid the use of heavy liquids which are generally expensive and toxic, gravimetrically separated samples can be a useful alternative: a batea with a circular design, wide flaring rims and indented center is used to wash and separate sand, silt, and clay size material, as outlined by Krumbein & Pettijohn (1938). The separation of minerals by specific gravity and use of heavy liquids is also described in the same source and in Carver (1971).

The main moderate magnetic minerals, including amphibole, pyroxene, ilmenite, and hematite, along with the weakly magnetic species of epidote, garnet, spinel, and Fe-rutile may be separated using a magnet as outlined by

Krumbein & Pettijohn (1938). All magnetic and nonmagnetic minerals have something to offer the microtexturist, in that with the SEM–EDS, it is possible to achieve high-resolution imagery and rapid identification by chemical signature. The microform of the grain may offer clues as to its transport history as well as provide information on preweathering or diagenetic modification.

The examples used here come from Quaternary sediments in Bolivia and Indonesia where information on glacial processes (Bolivia) and paleosols (Sumatra) are particularly relevant. For example, minerals such as zircon, with a hardness of 7.5, might be expected to inflict severe damage on less resistant minerals such as quartz (hardness 7.0), orthoclase (hardness 6.0), calcite (hardness 3.0), and gold (hardness 2.5) (Hurlburt, 1971). The "microstriator" effect in glacial sediment arose from a SEM study of heavy minerals in Bolivian tills by Mahaney & Milner (1998). It was based primarily on the prevalence of striated quartz and progressive changes in the shape of zircon, from "stubby" to "pencil" forms, at the base of Bolivian glaciers. The "pencil form" probably acts like a stylus, inscribing less resistant grains with longitudinal grooves and striations.

Bolivian Tills

In a routine investigation of heavy minerals in Bolivian tills (Mahaney and Milner, 1998) zircon grains were found to be survivors of glacial crushing as well as an important recorder of environmental history. Zircon (hardness 7.5) and spinel (hardness 8.0) ranked as effective recorders of crushing and scratching events—so much so that both grains may well crush and striate less resistant minerals such as quartz, affecting even less resistant minerals such as gold, and intergrowths of quartz and gold. Because zircon is more abundant than members of the spinel family, attention focused on well-worn "pencil forms" of zircon, thought to be inherited from pristine, blocky-blunt, euhedral forms that evolved into intermediate blunted, worn-down grains, part of a chain of form-changes that resulted in the genesis of a zircon "stylus." It is the stylus, with its "pencil form," together with abundant adhering particles, considered to result from glacial grinding (Smalley, 1966), that Mahaney & Milner (1998) concluded might be responsible for numerous striations on abundant quartz and less abundant gold in the tills. With a hardness of 7.5, zircon is the most prevalent heavy mineral capable of inflicting damage on quartz (hardness 7.0; Hurlburt, 1971).

As shown in the various frames of figure 14.1, zircon assumes stubby–blocky forms that eventually develop rounded to sharp edges (figure 14.1B), culminating in a perfectly formed stylus, or "pencil form" (figure 14.1C). Sometimes zircon is protected by armoring, which, when it breaks off, may damage the grain (figure 14.1B and D). Compared with kyanite (figure 14.1D), zircon may exhibit a dual striator form (figure 14.1E and F).

Various forms of the pencil form genesis are shown in the frames of figure 14.2, with pristine forms and older forms affected by dissolution (figure 14.2B). Other grains seem to have been damaged (figure 14.2E and F), possibly by collisions within the glacial mill.

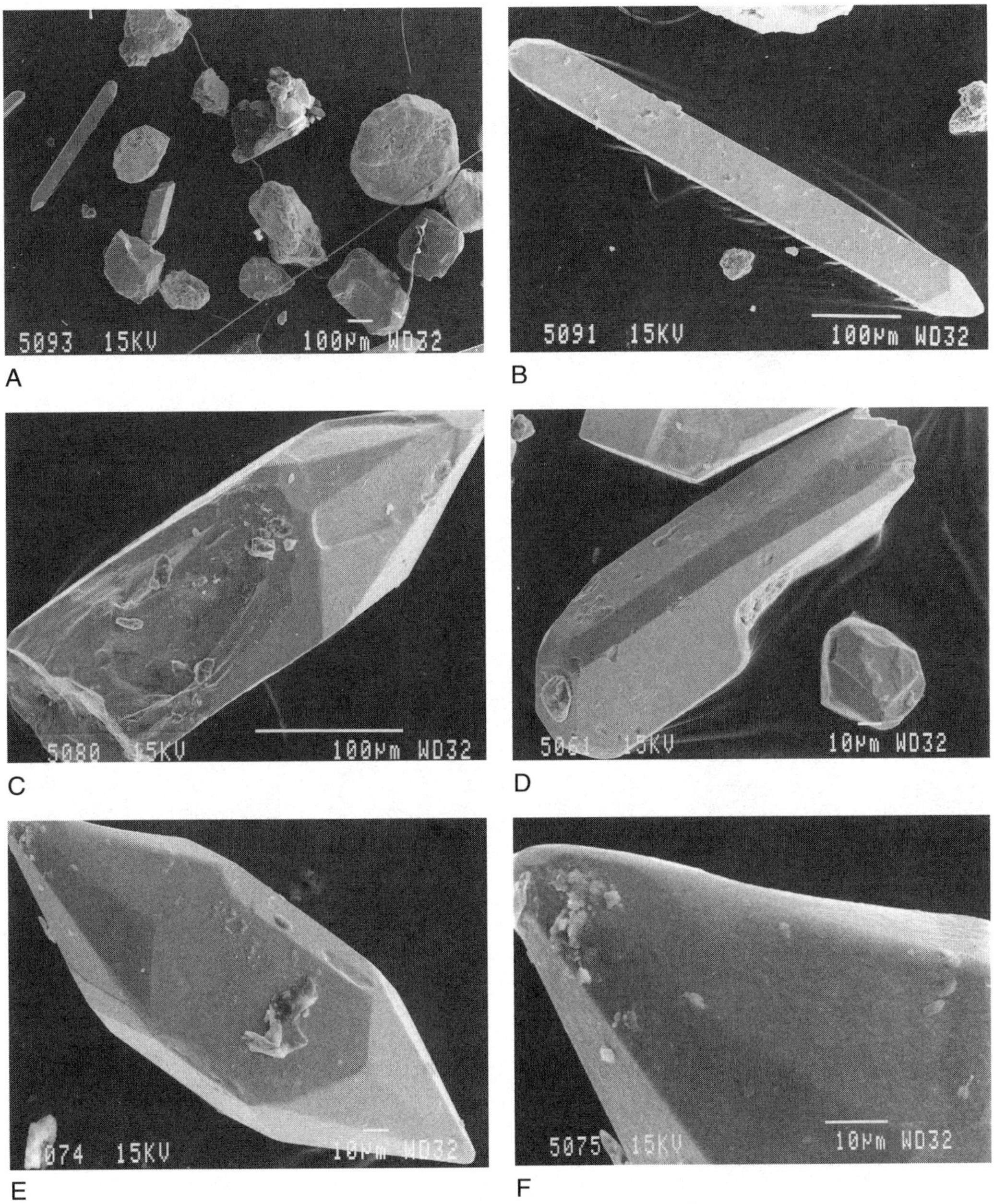

Figure 14.1. (A) General frame of equant and pencil forms of zircon in Till 27 (for till location, see Mahaney & Milner, 1998). (B) Zircon microstriator showing a sharp lower edge (striator) and a rounded upper edge which may have been protected by a gangue mineral (armoring). (C) Perfectly formed primary, euhedral zircon with a well-honed pencil point in Till 25. (D) Zircon in Till 23 with a broken edge, possibly the result of collision. The grain at the bottom is kyanite (Si, Al). (E) Dual striator zircon well abraded into a stylus at both ends in Till 25. (F) Enlargement of the upper stylus of the grain in E. Reprinted from Mahaney & Milner (1998) by permission of the publisher, Taylor & Francis AS.

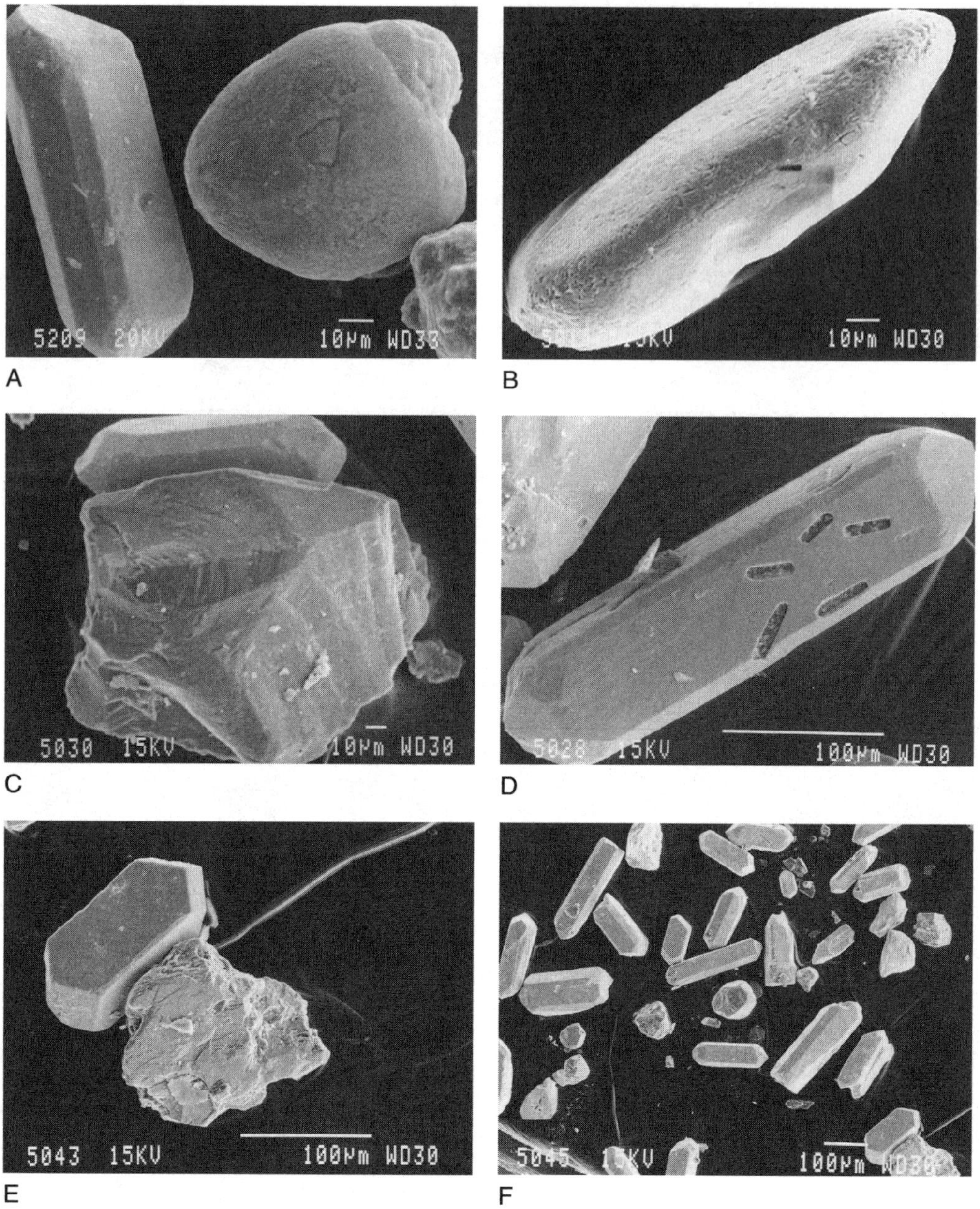

Figure 14.2 (A) Two zircons, one with a primary, pencil, or stylus form, and the other with an elliptical form with bulbous edges in Till 1. (B) Rounded edges and solution weathering on zircon in Till 17. (C) Glacially crushed quartz with primary unflawed zircon above in Till 19. (D) Zircon showing primary cavities and growth of the stylus in Till 19. (E). A well-fractured ilmenite grain next to a primary zircon grain with a stylus at one end and collision damage at the other in Till 21. (F) General frame with spinel in the middle surrounded by zircons with stylus forms and collision damage. Figure 14.2(A) reprinted from Mahaney & Milner (1998) by permission of the publisher, Taylor & Francis AS.

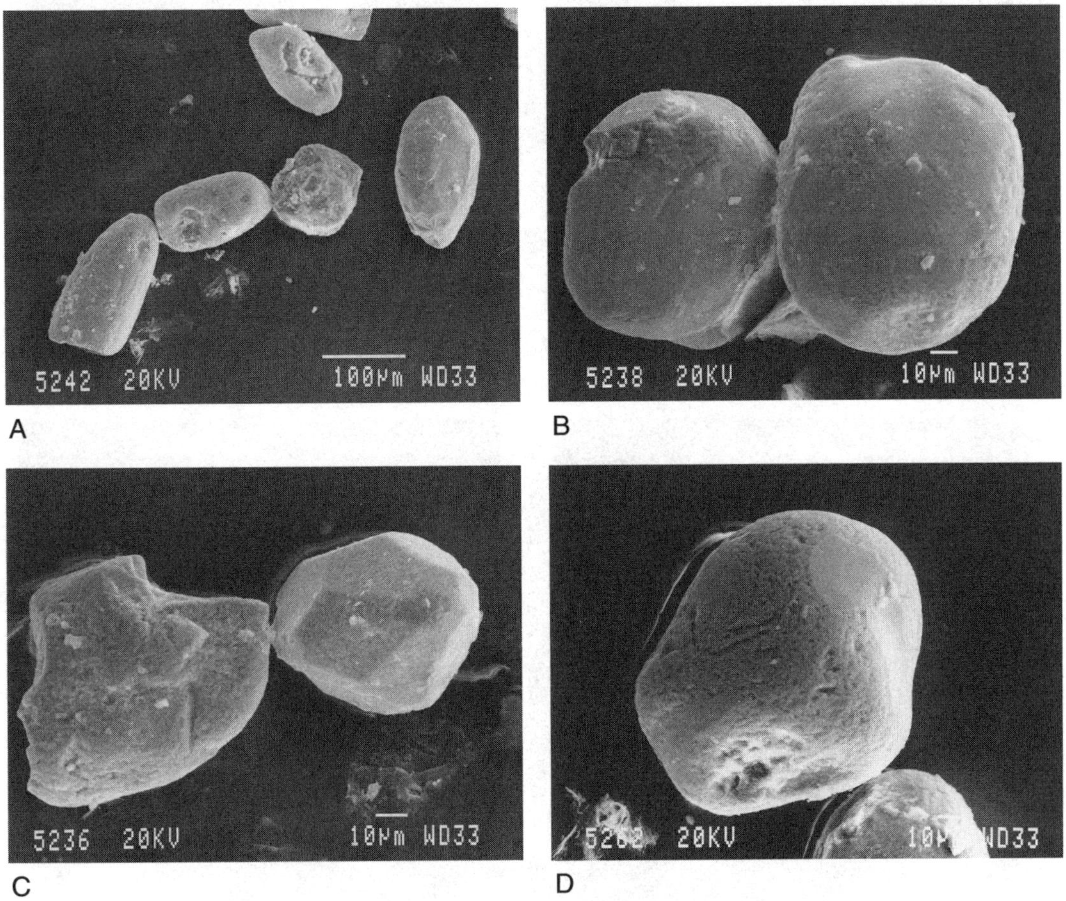

Figure 14.3 (A) Well-weathered quartz in the center flanked by zircons with rounded edges, presumably well abraded. Zircon to the far left is snapped off. (B) Worn, old zircons with variable coatings of Fe on the grain to right in Till 4. (C) Two zircons, worn with subround and angular edges. The grain to the left is coated with Fe (Si:Fe = 6:1, so reasonable Fe) in Till 4. (D) Weathered zircon with Fe–Al coatings in Till 9.

Coatings on light (figure 14.3A) and heavy minerals (figure 14.3B–D) range from oxides of Fe to combinations of Fe and Al, with variable amounts of gibbsite, the latter being indicative of aggressive leaching most probably during a previous interglacial period.

The coatings on some heavy minerals consist of various clay minerals, as indicated in figure 14.4(A). In this study illite was the dominant coating, and probably inherited from the weathering of biotite during previous interglacial periods.

Weathering of magnetic minerals, such as ilmenite, as shown in figure 14.5A and D, may be gauged against fresh grains of rutile and other grains of spinel and garnet (figure 14.5B and C) to assess relative rates of sand weathering within a paleosol.

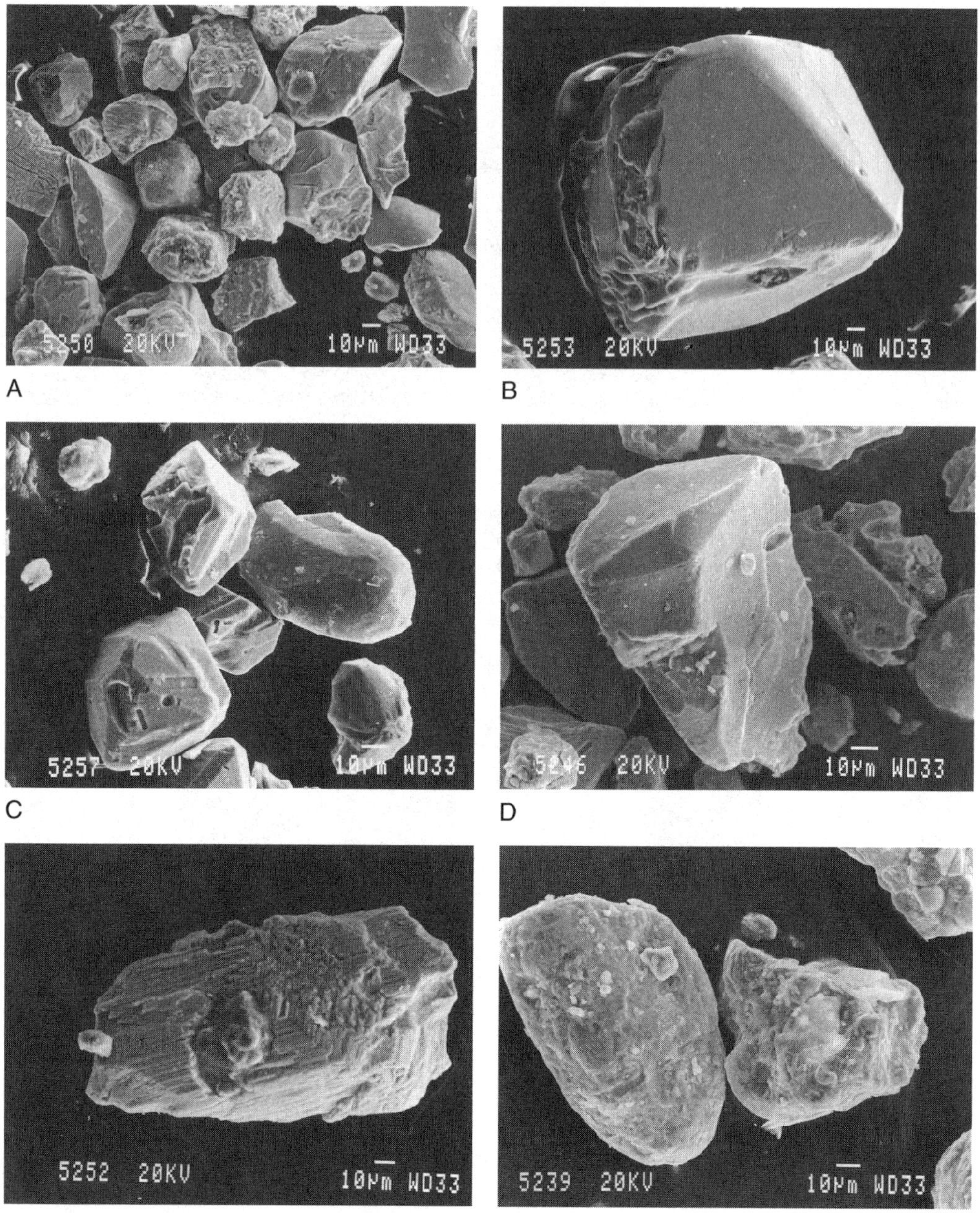

Figure 14.4 (A) Angular and subangular grains from Till 7. The grain in the center is ilmenite from Till 7 coated with clay having a Si:Al ratio of 2:l with K, so most probably illite. Other clay-coated grains are seen above and below. (B) Ilmenite with edge damage and striator in Till 7. (C) Four ilmenite grains with striators at the top and bottom. (D) Angular ilmenite with deep fractures in Till 5. (E) Rutile in Till 5 with primary form intact and edge damage from collision. (F). Two rutile grains with variable clay coatings. The grain to the left is cleaner, with some adhering particles.

Figure 14.5 (A) Slightly weathered ilmenite (left) and fresh rutile (right) in Till 6. (B) Spinel (Al–Fe–Mn with Al:Si = 0.6:1.0) with adhering particles and sharp, worn edges in Till 21. (C) Garnet with worn edges and multiple fractures in Till 17. (D) Slightly weathered ilmenite (left) with well-weathered magnetite (right) in Till 22. (E) Spinel with microstriator point, sharp edges, and multiple fractures in Till 11. (F) Fresh, angular spinel (center) with weathered ilmenite (upper right) in Till 21.

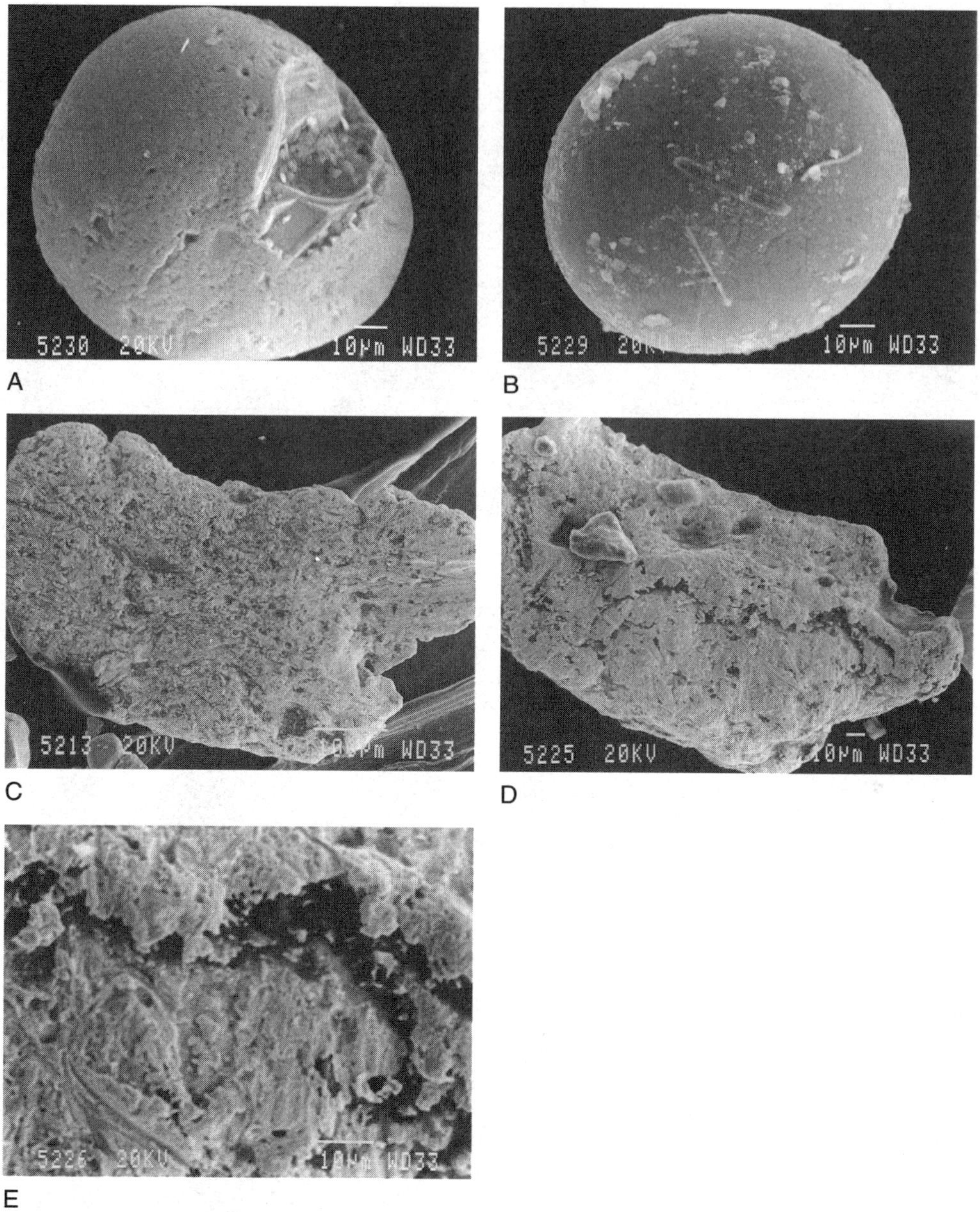

Figure 14.6 (A) Zircon, abraded and round, with a recess cavity and sawtooth fractures (very rare in glacial sediments), and multiple v-shaped percussion fractures. (B) A Ca-glass tectite grain with small zircons adhering along with other particles in Till 6. (C) Corroded gold grain in Till 2 with linear grooves faintly gouged in the grain. The EDS shows Fe and Si indicating quartz and magnetite or ilmenite embedded in the gold grain. (D) Gold grain in till with the top part folded over the bottom part in Till 6. The open suture line is a dark lobate feature. (E) Higher magnification frame of gold folded over quartz with striated curved grooves. The dark tone represents iron and/or humus in the fold. Figure 14.6(C) reprinted from Mahaney & Milner (1998) by permission of the publisher, Taylor & Francis AS.

Sawtooth fractures (see figure 5.1), very rare in glacial sediments, become something of an enigma when observed on zircon grains, as they would have to be inflicted by something harder, presumably grains such as topaz, corundum, or diamond. Whereas none of these minerals were observed in the heavy mineral fractions analyzed, a search of larger bulk samples might prove especially rewarding. Folded, flattened, and striated gold grains (figure 14.6C–E) attest to the damage inflicted by glacial crushing.

At this point it is particularly important to illustrate the importance of the energy-dispersive spectrometer as a tool in studying the various array of coatings and weathering features encountered on heavy minerals. As shown in the EDS (figure 14.7), complex coatings of Fe and Al, with Al slightly greater than Fe, can be determined. The Fe is probably obtained from the oxidation of pyrite, magnetite, and ilmenite, which are numerous within the heavy mineral concentrates. Aluminum, on the other hand, could be the product of aggressive leaching during a previous interglacial period.

Sumatran Soils

Sumatran Ultisols, in a flight of Quaternary terraces, have been studied with respect to age and weathered state as part of a geophagy project (Stambolic-

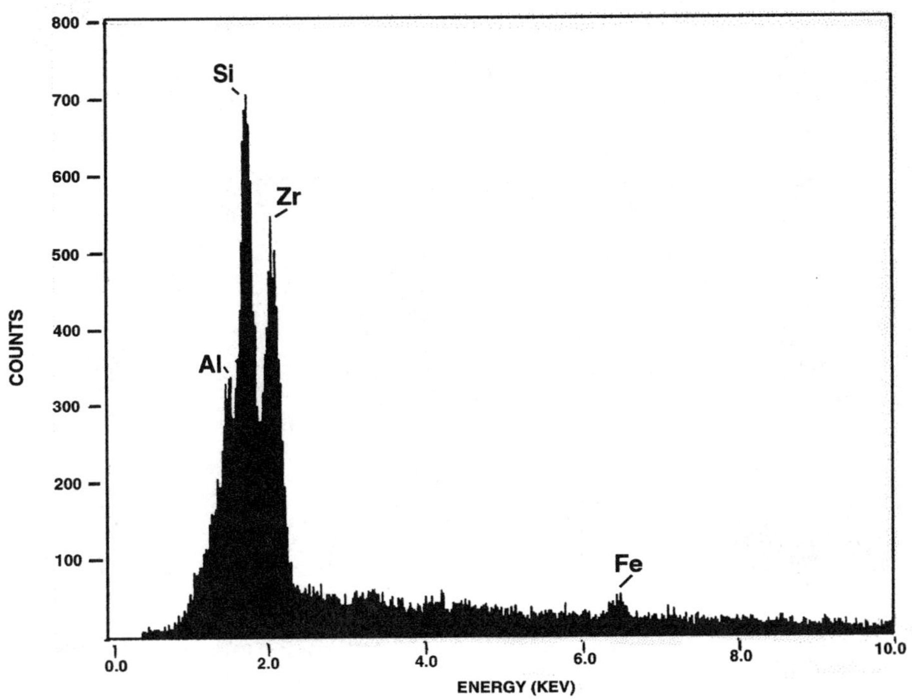

Figure 14.7. EDS spectrum of grain 5262 in figure 14.3D showing the strength of the Fe and Al coating on zircon.

Robb, 1997). The heavy mineral grains shown in figure 14.8 provide examples of the degree of weathering in B horizons, including weathering of pyroxenes (figure 14.8A) and dissolution effects on quartz and orthoclase. Hematite, shown in figure 14.8B coated with gibbsite, gives a quantitative estimate of paleoleaching and removal of 1:1 clay minerals such as halloysite. Some hematite, pseudomorphous after pyrite, lower down in the profile is coated with SiO_2 and kaolinite (figure 14.8C). The magnetite in figure 14.8D is coated with Fe-oxihydroxides and clay minerals.

In profiles where heavy mineral weathering in a soil–paleosol can be used as an age or alteration index, it is important to select grains from the surface down into the parent material. Thus, it is possible to present a true picture of grain

Figure 14.8. Mixed minerals in Suci6 (B horizon) profile, Sumatran Ultisol. The long, weathered grain is pyrite, probably a former inclusion in augite. The grains to left of center and above are quartz, the grain to the lower left is K-feldspar. (B) Hematite with a gibbsite coating in Suci7 profile, Sumatra. (C) Hematite with a Si coating in Suci4, Sumatra. (D) Magnetite with coatings in Suci4, Sumatra

alteration from biotite to magnetite and to pay close attention to coatings of various kinds including clay mineral and weathering products. As shown above, one of the most important mineral weathering ratios is that of clay minerals to gibbsite, which indicates the progressive removal of Si and the build-up of Al leading to the genesis of gibbsite. Excessive gibbsite as a weathering product comes from the stripping of Si from halloysite clays with heavy leaching, as reported by Mahaney (1990a, 1991b).

Other weathering products, including coatings of various kinds and clay minerals, can be seen and interpreted by SEM–EDS, which can provide a semi-quantitative estimate of the degree of alteration of primary minerals. One may choose the heavy mineral complex in relation to the source of Fe which is often the main coating, but other minerals are important as well, including ilmenite (providing Ti), spinel (providing Al), and rutile (providing Ti). To some degree, even the light minerals such as quartz provide coatings of Si, which often form a patina around heavy mineral grains.

Conclusions

Heavy minerals of sand and silt size may be identified and described by analysis with the SEM–EDS. Knowledge of the mineralogy and weathered state of these minerals provides important information on the source of the sediment, transport mechanisms, presence of lithological discontinuities, or addition of allochthonous materials. In soils it may be the prime means of assessing the degree of weathering that has occurred. All fractions, from coarse sand to fine silt, may be studied, but usually there are too few representative grains in the coarse and very fine fractions to constitute a representative sample. Therefore the coarse silt and fine sand fractions provide the optimum size for suitable optical analysis.

One advantage of the SEM is in studying the rounding of heavy minerals. A grain with a round outline may have little pivotability and appear as a flat plate. While difficult to see with the binocular microscope without the use of crossed polars and analysis of interference colors, with the SEM it is readily apparent in stereoscopic image. Corrosion, solution etching, and pitting are also readily apparent with the SEM and this is manifest with grains such as hornblende where weathering follows the long axis of the mineral. Decomposition of feldspars follows the cleavage for the most part, whereas in garnet, an isometric mineral, weathering and dissolution etching is random. Semi-quantitative estimates of the degree of etching may assist in studies of soil stratigraphy as well as slow chemical alteration during diagenesis.

Weathering of heavy minerals can produce clay coatings and mixed oxides of various kinds, open cavities that may fill with clay and/or other weathering products, or even create a mineral pseudomorphous after an original mineral species. In paleosols, for example, argillans (clay infills) normally studied in thin sections under the light microscope can be studied with the SEM and the chemistry of the clay can be worked out with EDS. Because sand tends to retain a portion of the clay added to it in pedogenesis, the SEM–EDS actually provides a rapid means of analysis, much more rapid in fact than preparing thin sections from micromorphology samples collected in the field. There are limitations, as the amount of organic material in the argillan cannot be worked out with the EDS.

Applied Studies

15

Various attempts have been made to use SEM surface textures on quartz and other minerals in allied disciplines such as engineering, forensic sciences, archaeology, and even geophagy, a branch of nutritional sciences. Engineering applications, for example, may include reference to the strength of substrate materials, degree of chemical dissolution of substrate minerals, and coatings of various kinds that may be related to diagenetic effects of one kind or another. Among the examples cited in other chapters of the book, engineering tests designed to determine the stress–strain relationships of quartz grains that may be encountered in glaciers of varying thickness form an important subject of enquiry by Earth scientists interested in glacial dynamics. In this chapter, special attention will be paid to the experimental crushing of quartz in the laboratory, with special reference to pressure and temperature in the glacial stress field.

Applications in the forensic sciences include the study of mainly exotic minerals or unusual microtextures found at the scene of a crime that could be traced to a specific locality, or coatings of blood or tissue on quartz or other minerals that might be studied by SEM–EDS. While some applications discussed here have been previously elucidated by Murray & Tedrow (1975), they beg additional study with the SEM–EDS.

Archaeologists interested in wear patterns on quartz and other minerals have used the SEM as a superb tool to study the morphological use-wear patterns on quartz and quartz arenite stone tools (Knutsson, 1988; Julig et al., 1992). Microstructural analysis of this kind has enabled archaeologists to distinguish the different uses of stone tools and the appearance of their wear, allowing different areas of application, as with materials such as quartz where the glassy surface prevents the use of high-power light microscopy to see use-wear patterns. These studies rely on experimental replication of quartz tools and their usefulness in determining wear-patterns on specific materials.

Geophagy, or the ingestion of natural earths as dietary supplements and/or pharmaceuticals (zoopharmacognosy), is a subject of wide interest amongst anthropologists, geographers, nutritional scientists, medical specialists, chemists, zoologists, primatologists, and microbiologists. Where attention is directed toward the analysis of ingested soils, the SEM and EDS provide excellent instrumentation to study microtextures that shed light on this interesting behavior, which incidentally includes humans. There is a long and complex history of soil ingestion, closely connected to the study of paleonutrition, which can be traced back to prehistoric times. Geophagic sites, documented by paleonutrition experts, have been investigated with the intent of identifying the chemical stimulus or stimuli that might explain this interesting behavior (Mahaney et al., 1997b). Specific coatings on sand particles of the ingested material provide insight into various chemical stimuli and motivations for their uses. For example, consumption of natural earths by the Great Ape primates, such as the mountain gorilla, has resulted in crushing of weathered regolith, the grains of which exhibit microtextures and intense fracturing as revealed by SEM (Mahaney, 1993).

Glacigenic–Geomorphological Applications

The central tenets of the theories of brittle and ductile fracture show that the opportunity of fracture is determined by increases in stress until strain produces slow to catastrophic crack propagation. Such fracturing may penetrate along preexisting flaws in minerals (Griffith cracks) and/or by the generation of new flaws which extend along or across mineral surfaces, either as shallow or deeply embedded cracks. Many of these are predetermined by the stress–strain energy field, much as vibrational energy is released through the grain (Mahaney, 1995b). In the most general terms, larger mineral grains require less stress to fracture than smaller ones, but much depends on the nature of the stress field, the strength of the minerals, and their weathered state; smaller grains require, on average, higher stress fields to induce failure. Put another way, the probability of fracture by breaking the weakest links in a mineral, increases with increasing size of fragments, as in a glacial system.

There are many ways to view fragment size in a glacial sediment. Repeated fracturing results in fragmentation, which requires increasing stress to achieve fracturing as particle size diminishes from sand to silt. Small silt-size fragments (<50 μm) continually comminuted and/or liberated from larger grains by grinding at the base of the ice are at the end of their size reduction run, and are apt to be preserved intact after their sojourn in the ice. These grains carry the final range of microtextures from glacial crushing. Other small fragments of the same size, subjected to low-pressure mechanical action by mass wasting, eolian and/or fluvial agencies long before they were added as load to a glacier, are likely to be unmodified by the mechanical forces of an active glacier. This is why the fine sand–silt fractions are important to study with the objective of cataloging the range of microtextures at close to the final stage of comminution. Therefore, some of the grains in the fine and medium silt fractions may have a microtexture record entirely unrelated to travel in the ice, making it hazardous and unreliable to study fine grain taphonomy or life history.

Interest in fractography, as the identification and interpretation of fractures on particle surfaces (Krinsley & Donahue, 1968; Krinsley & Doornkamp, 1973; Margolis & Krinsley, 1974; Marshall, 1987; Mahaney et al., 1988a; Mahaney, 1990a), has been pursued with the objective of using particular fracture markings, and their frequency of occurrence, as a means to gain insight into historical glacial records. In order to target certain microtextures as "glacial" it is necessary to address several important questions:

1. Are there characteristic mechanical damage features on glacial grains?
2. What are glacial microfeatures and how are they related to glacial dynamics?
3. Is it possible to deduce distance of transport and ice thickness from glacial grains?

Quartz was selected as the test material for engineering tests because it is abundant in deposits and its hardness ensures it to be an excellent recorder of damage inflicted on it. As a test, amorphous silica glass spheres were used for several reasons:

1. As an analog of quartz, they could be subjected to low stress in a confined space.
2. Using a spherical-shaped object would make it possible to compare the crushed residue with crushed grains.
3. They could be used to determine the final shape and number of fractured particles.

Since several previous fractographic studies were performed on till particles in the 50–500 μm range (Krinsley & Donahue, 1968), and on larger particles of 16–26.5 mm diameter (Iverson, 1990), it seemed prudent to carry out a laboratory experiment using similar size material; a fourth test used silica spheres approximately 3000 μm in diameter. (The results reported here were carried out in the laboratory of H.H. Schloessin, Department of Geophysics, University of Western Ontario, London, Ontario, in collaboration with the Geomorphology and Pedology Laboratory at York University.)

In the first three tests particles were taken from the fragments of a large single piece of optically translucent α-quartz (from Brazil) and crushed by low-velocity impact. After separating the particles into three size categories of 50–500 μm sizes, they were placed in a piston–cylinder device constructed at University of Western Ontario (figure 15.1). Used to compress a cylindrical volume of ice to 2000 psi capacity, it consists of a steel piston, with appropriate clearances, fitted into a brass cylinder. The piston–cylinder device with the ice sample inside was placed within an aluminum container, which contained a cooling agent maintained at a constant, low test temperature. Ice, dry ice, and dry ice with acetone were used to control low temperatures in the enclosing space

To control the position of quartz particles in the test cylinder, fragments were sprinkled on ice and widely spaced. The ice was grown layer by layer to further insure that particles were not touching one another and each layer was periodically subjected to a low vacuum of ~1 torr to eliminate as much air as possible. The first three tests were run on quartz fragments; the last test with a load of equal-size silica glass spheres. In tests 1 and 2 the ice temperatures were below −10 °C and thus similar to a cold glacier. Tests 3 and 4 were carried out with the temperature raised to pressure melting, simulating warm glacier conditions. The objective of the experiment was to determine the amount of fracturing that occurred in the charge before and after compression in ice. For tests 1 and 2,

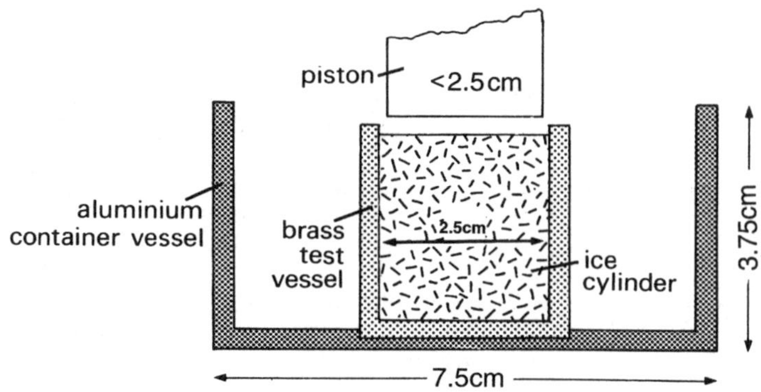

Figure 15.1 Piston–cylinder apparatus used to compress quartz fragments and silica spheres embedded in ice. The steel piston is just under 2.5 cm in diameter, the brass cylinder has an inside diameter of 25 mm, and the aluminum cooling vessel has an inside diameter of 75 mm.

comparison was made by optical microscope and SEM. In test 1, involving particles ranging from 50 to 125 μm, no change was evident either by optical or scanning electron microscopy. This would seem to confirm the theory that small and isolated particles subjected to stresses arising from contact within the ice and/or load within a glacier are least likely to fracture and most likely to survive the pressure intact.

Test 2 (figure 15.2A and B and figure 15.3), the fine sand fraction (125–250 μm), involved a load with a maximum pressure of 2000 psi or 17.5 Mpa on the ice cylinder (equaled to an ice thickness of 1.98 km) being applied for 30 min. Imagery of the recovered grains shows several new fragments with sizes much lower than the starting size. Original fragments with protrusions were subjected to considerable edge rounding and several flaky particles were likely due to physically induced spalling.

In Test 3 (figures 15.2C and 15.3), the size of charge and the cylinder pressure were approximately the same as in Test 2. In this experiment, which lasted for 40 min, the temperature was brought up to p-melting of −1.29 to 1.23 °C. The imagery of the resultant charge shows considerable change in particle size and distribution. Here several small fine fragments cling to larger ones, which survived compression. Pressure melting conditions in the test cylinder led to the formation of meltwater; ice bridges melted, allowing particles to join. Particles at the base of the brass cylinder were subjected to high hydraulic pressure causing some quartz to become embedded in the brass base. In this third test quartz grains were fractured into numerous particles being brought into mutual contact under high load. These particles exhibited unmistakable arc-shaped steps with numerous and deeply embedded conchoidal and subparallel linear fractures, with a sizeable population of fine adhering particles.

Test 4 (figures 15.2D and 15.4), the concluding test in this series, was designed to study the effect of compression on a population of well-isolated, untouched, and perfectly spherical silica glass spheres, (i.e., particles with the same minimal

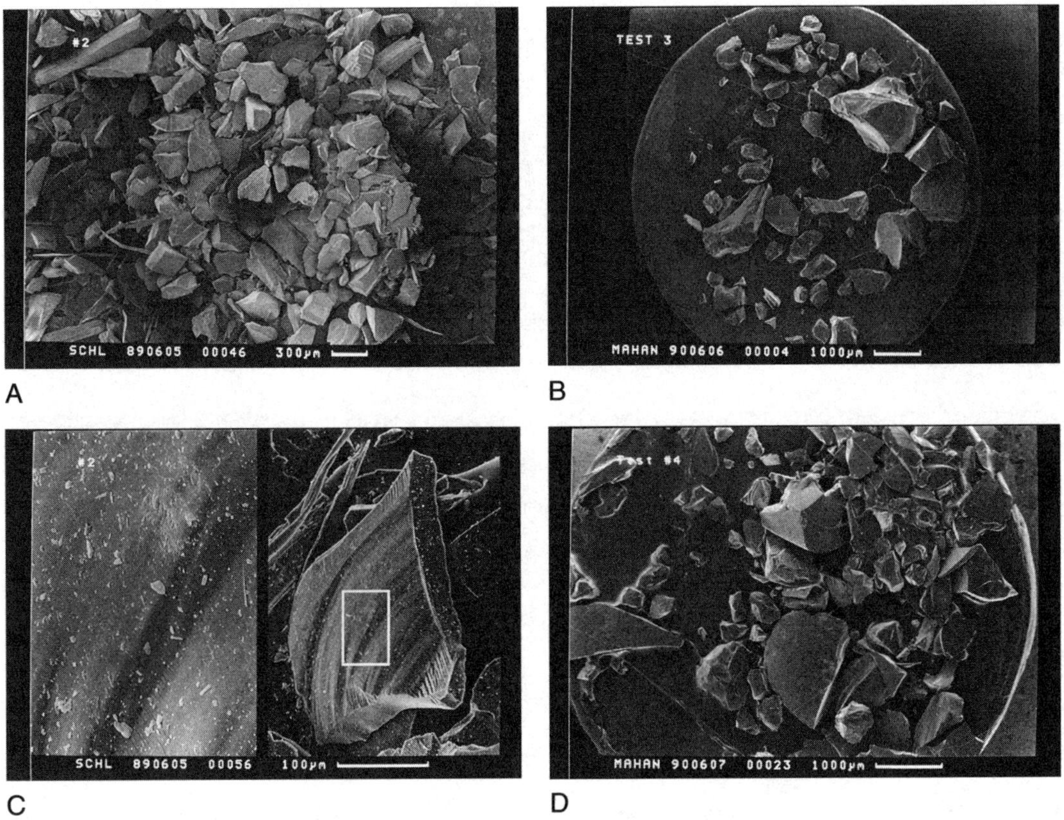

Figure 15.2 (A) Recovered charge of fragmented sands originally sieved to medium and fine sand size (125–250 μm diameter). Approximately 25–30 grains of the sample were fragmented into smaller sizes. (B) Representative quartz fragments from Test 2 of approximately 125 μm diameter, with deeply embedded conchoidal and subparallel linear fractures and sharp edges with numerous adhering particles. (C) Fragments of silica spheres, originally of 250–1000 μm size (Test 3), recovered after compression in ice at a simulated thickness of 1.98 km. (D) Recovered fragmented charge from Test 4. The original size of the silica spheres was 3.5 cm. The simulated ice thickness was 1.98 km.

stress contributing shape). These were situated to represent inhomogeneous inclusions exactly as would occur with fresh quartz released from bedrock. Because of their sphericity, they represented a set of particles with equal geometric stress concentrating factors, easily amenable to theoretical analysis. Pressure was applied at 17.5 Mpa as in the previous tests, and after a short time, the temperature was raised to p-melting. Mechanical damage on this load was severe, with four spheres totally shattered into small fragments of fine sand and silt size, five greatly reduced in diameter, and 17 escaping with only a few craters produced. Apparently the nine spheres suffering the greater damage released fine particles simulating glacial grinding, the latter originally postulated by Smalley (1966).

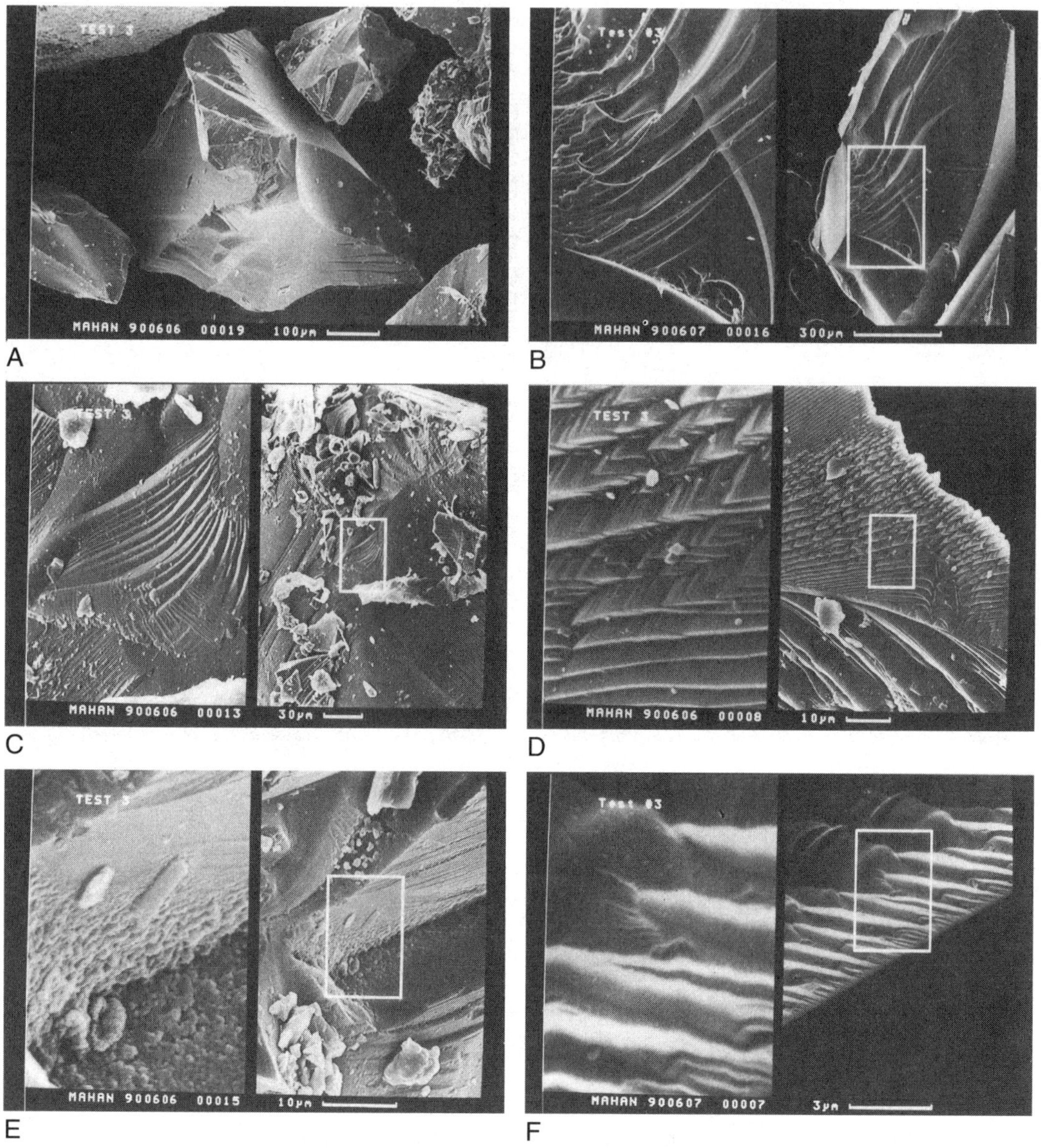

Figure 15.3 (A) Triangular-faceted quartz, well abraded and fractured. (B) Hollowed-out quartz with numerous striations merging with grooves. The enlargement (×3) shows striations merging with grooves. (C) Quartz surface with conchoidal fractures and adhering particles (enlargement ×5). (D) Edge of a quartz grain with high- and low-frequency fractures. The enlargement (×5) shows sawtooth structures common only on neotectonically released grains. (E) Small silt-size particles impressed on a quartz surface with particles above covered with a coating of Si-rich dust (enlargement ×3). (F) Ruptured grooves on a quartz surface (enlargement ×3).

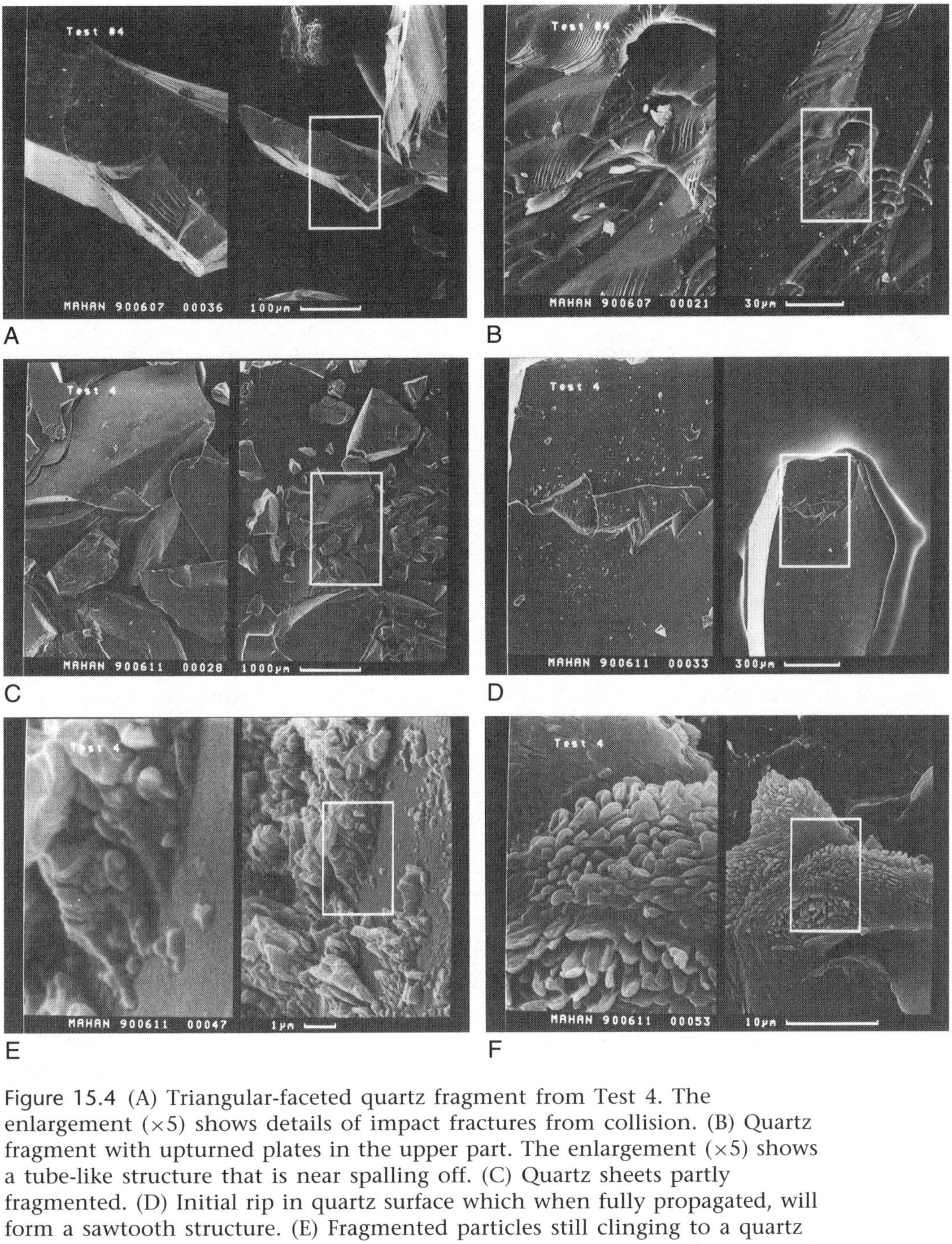

Figure 15.4 (A) Triangular-faceted quartz fragment from Test 4. The enlargement (×5) shows details of impact fractures from collision. (B) Quartz fragment with upturned plates in the upper part. The enlargement (×5) shows a tube-like structure that is near spalling off. (C) Quartz sheets partly fragmented. (D) Initial rip in quartz surface which when fully propagated, will form a sawtooth structure. (E) Fragmented particles still clinging to a quartz surface. (F) Partially shattered surface of a fragmented mechanically upturned plate.

These experiments show that coarse silt inclusions in ice are apparently unaffected by compression. Sizes ranging from fine to medium sand are affected by compression depending on temperature. Clearly, the p-melting temperature, or something close to it, produces the greatest degree of fragmentation, as shown in Test 3. Large particles of equal sphericity are subjected to considerable comminution when the temperature is lowered to the p-melting point.

Forensic Sciences

The SEM–EDS has played a significant role in many criminal cases pursued in the United States by the FBI and in Canada by the RCMP. It requires rather simple geological information to trace boulders, cobbles, pebbles, and granules from source to crime scene, but expending resolution downward, through levels of sand and silt to clay, the SEM–EDS can play a spectacular role in a forensic exercise. In tracing soil particles found on vehicles, boots, etc., used to commit crimes, identifying tooth fragments found at crime scenes, and/or tracing alibi samples claimed by a suspect to have come from a particular area, the SEM–EDS can play an important role. Moreover, salting of gold may be identified by analysis of texture and shape of grains.

In Ontario, scientists at the Centre of Forensic Sciences have developed an automated particle search (McVicar & Graves, 1997) using the SEM–EDS, to classify soils by region in the province. Compared to traditional optical microscopic methods, the automated SEM–EDS method is fast and simple. The high resolution of the system allows a statistical measure of soil similarities or differences, and determining their origin has proved successful in many criminal investigations.

The identification of teeth, as well as grooves and particles on teeth, has materially aided investigators in criminal investigations (Fairgrieve, 1994). Even tooth fragments from soil may be analyzed by SEM–EDS to reconstruct the whole tooth, which is important in matching dental records in forensic anthropology.

Salting takes many forms and the SEM–EDS can be used to identify fraudulent samples. Alluvial placer gold is commonly salted with shavings from jewelry or bar gold, which often contain parallel striations. Virgin placer gold may be salted with alluvial gold that has been previously recovered by amalgamation with mercury leaving a gray-colored coating on the salted gold. Low-grade alluvial gold in low-gradient valley sites alternating with thick, coarse gold from high-gradient alluvial environments should signal human tampering.

Forensic gold analysis could have saved a great deal of money for investors in various mining stocks. Vein gold, from the Busang Property in Indonesia, was apparently "salted" with alluvial gold. The fine-grained gold was apparently added to barren rock in a methodical manner to produce values defining a three-dimensional ore body. Detection of fraud was achieved after redrilling by referees, recovery tests, and morphology of the gold samples. As shown in figures 15.5 and 15.6, vein gold and placer gold may look similar to one another (figure 15.5A and B), but there may also be startling differences between uncorroded (figure 15.5C) and corroded (figure 15.5D) specimens based on their patinas. Human tampering (i.e., salted gold) may be identified from shaved surfaces off a metal ring which leaves shavings at the end of scratches. Indeed, the classifica-

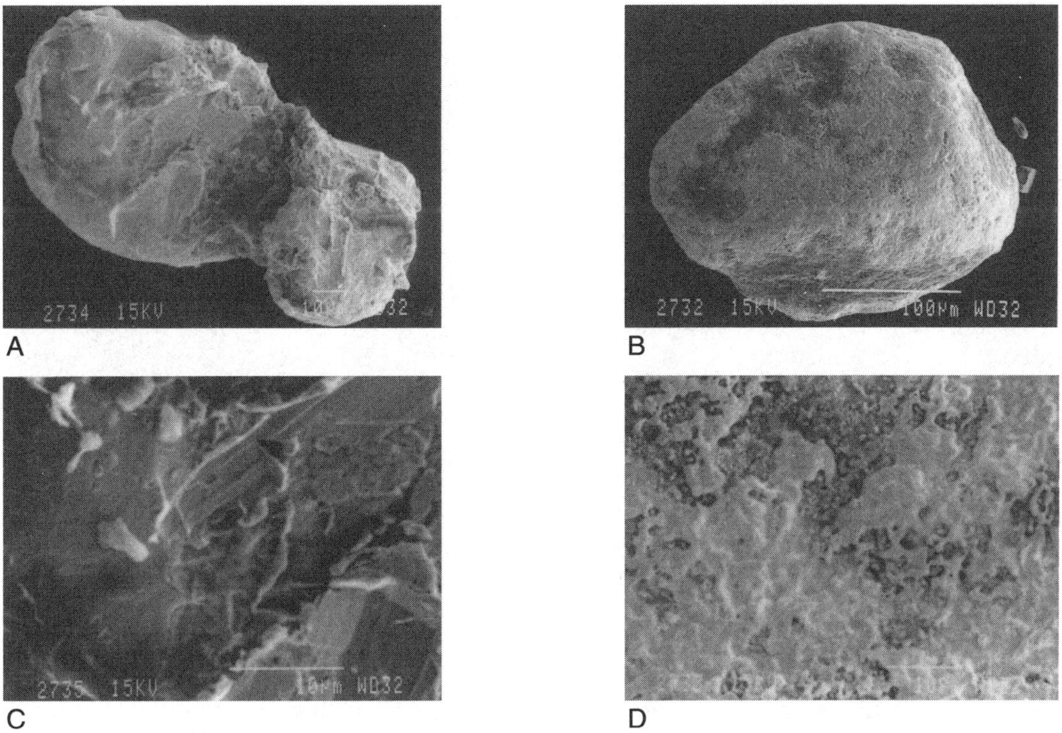

Figure 15.5 (A) Equant vein gold, not worn but very similar in morphology to alluvial gold. (B) Equant creek gold, worn and coated (dark area). (C) Fresh, uncorroded surface of the vein gold in A. Scratches show human tampering with the surface. (D) Corrosion patina (light tone) indicating a period of rest since wear occurred. Gold samples and imagery courtesy of M.W. Milner, Consultant, Toronto, Ontario.

tion scheme developed by Kinnunen (1996) and Kinnunen et al. (1997) outlines key textures that could be used to identify human tampering with natural gold. These include saw cuts, drill holes, testing facets, and burnished projections.

Vein gold, in its nascent form, has a jagged geometry (figure 15.6B), which outlines a perfect geometric mold of the vein minerals hosting it. Alluvial gold grains (figure 15.6A) are worn in a stream placer and come complete with clay coatings formed from soil in alluvium or from clay in stream sediment. It is the difference between the vein gold formed with the impression of primary accessory minerals and the stream gold, often rounded, thinned and sometimes corroded, that sets the two types apart.

Archaeology

Sand microtextures are important to archaeologists in geoarchaeological work (Julig et al., 1990), and they are used in much the same way as they are in the

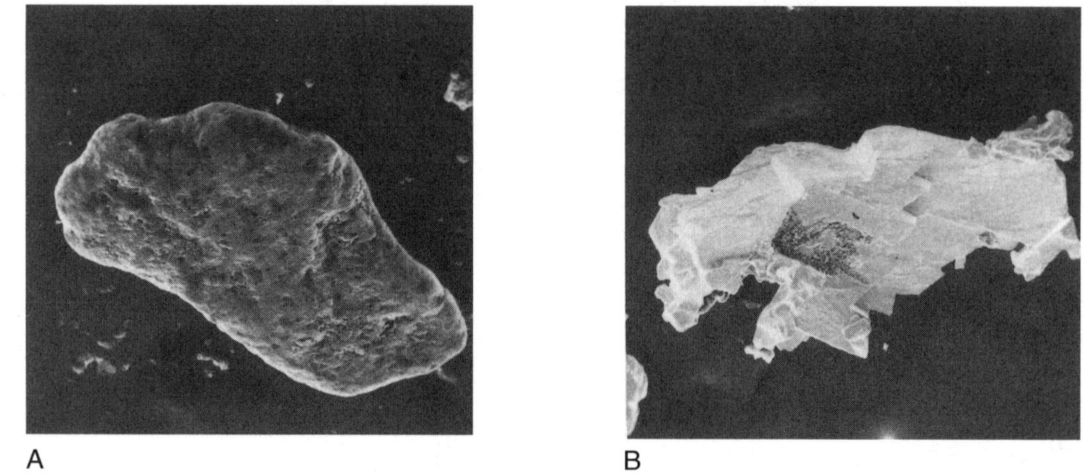

Figure 15.6 (A) Round gold from a worn stream placer with a clay coating from soil formed in alluvium. (B) Jagged gold showing a perfect mold of the vein mineral cleavage that enclosed the gold. Gold samples and imagery courtesy of M.W. Milner, Consultant, Toronto, Ontario.

Earth sciences: to establish the composition of coatings on grains, discriminate amongst sedimentary environments, and in determining site formation processes. SEM imagery is also important in the analysis and classification of quartz stone tools used in various research projects (Knutsson, 1988). The light microscope is often used in the study of natural surface textures versus use-wear pattern analysis (Kinnunen et al., 1985), but in many cases the SEM proves to be a superior instrument for studies of this kind. In addition, the SEM–EDS can be used to establish the hafting cement used to hold tool components together (Endlicher & Tillmann, 1997), and in certain instances to establish the composition of tools used to produce rock carvings (R. Dorn, personal communication, 1999). In this section the possibilities for using sand microtextures in archaeological research will be explored.

At the Sheguiandah Site on Manitoulin Island, the controversy of pre-Clovis occupation was studied by Julig et al. (1992). Briefly, the pre-Last Glacial Maximum occupation of the island, as proposed by other workers, was based on the presence of human-made artifacts in glacial diamictons which supported the hypothesis of great antiquity (pre-last glaciation) of human occupation of the New World. The main question centered about the interpretation of a glacial origin for the artifact-bearing beds lower down at the base of the section.

Quartz grains recovered from the diamictons in the lower beds were studied by SEM–EDS for their surface characteristics. These were shown to have chemically etched fractures suggesting they could have been reworked from older deposits, or even reworked during previous high lake stands. None of the grains carried fresh, unweathered glacial crushing features or abrasion (figure 15.7A–F) as displayed in chapters 5 and 9 of this book. Crescentic gouges, sometimes widely touted as representative of glaciation, were occasionally observed in

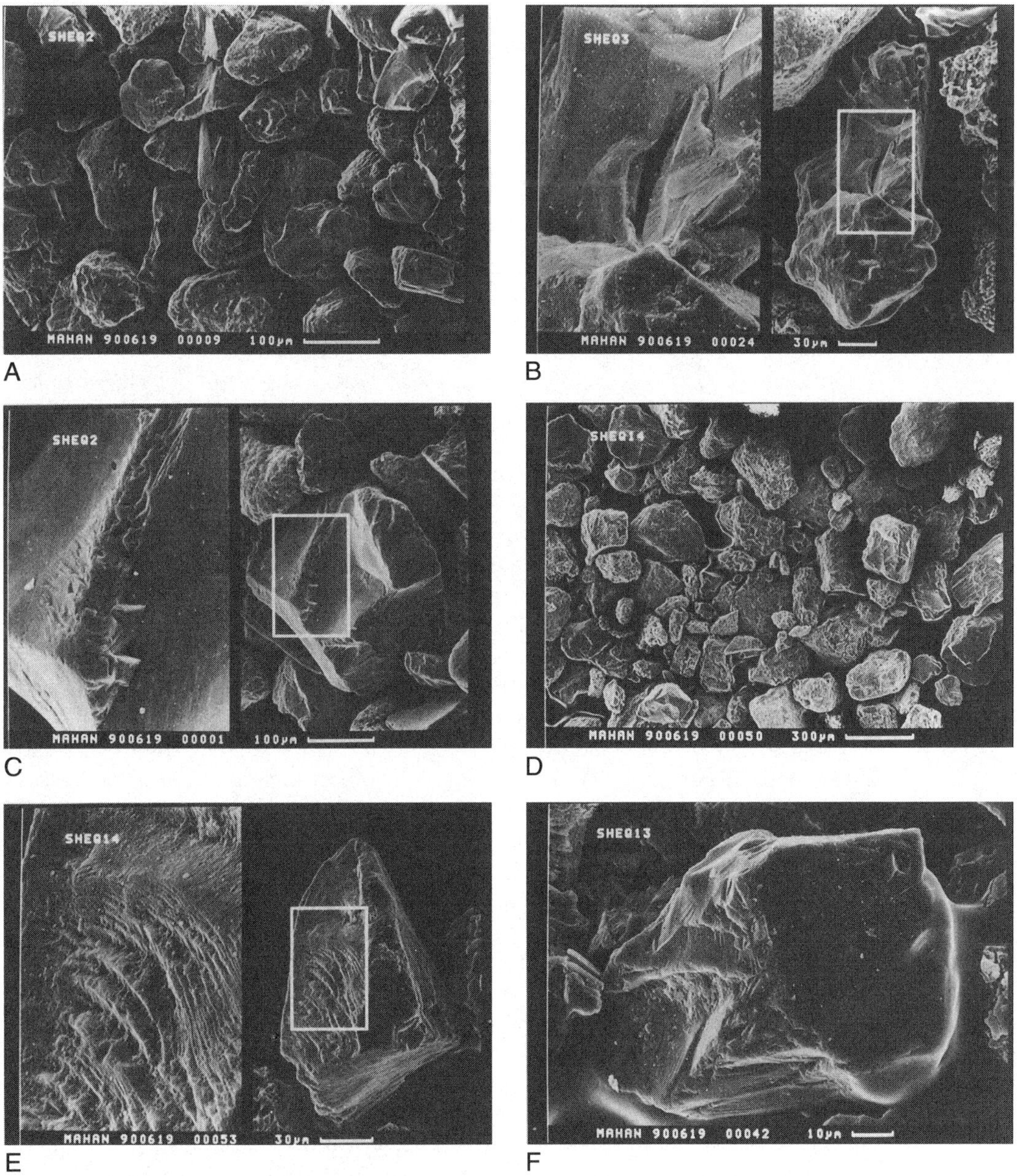

Figure 15.7 (A) General view of sands in Sheg2. (B) Deep striation on quartz with abraded and weathered subparallel and weathered conchoidal fractures in sample Sheg3. (C) Fractured and abraded quartz with sawtooth fractures which are common in neotectonic debris. (D) General view of sands in Sheg14 showing considerable weathering. (E) Quartz grain from Sheg14 with preweathered linear and conchoidal fractures. (F) Extensively crushed and preweathered quartz grain with a crescentic gouge in the upper right from sample Sheg13.

grains from the Sheguiandah diamictons, but all were well weathered and considered reworked. Other quartz grains with deep fractures or striations (figure 15.7E and F) were undoubtedly glacial grains, but from the degree of dissolution, they probably were reworked from older weathering horizons. Grains with sawtooth fractures (figure 15.7C), rather rare in glacial deposits, are common in neotectonic debris (Mahaney & Sjöberg, 1993) and in subaqueous marine debris flow deposits (Mahaney, unpublished), and might possibly be microdébitage from tool production. The somewhat fresher surfaces shown here favor the latter hypothesis. This unusual distinctive fracturing might also be produced by crushing during knapping of stone tools.

The short summary of SEM analysis at the Sheguiandah Site illustrates the importance of microscopy in geoarcheological studies. Sites like this, subjected to complex geomorphic processes, can only be studied by the joint efforts of scientists in many disciplines. In this case, reworking of quartz grains in the diamictons and granulometry studies helped to establish that the sediments are of nonglacial origin (Julig et al., 1992). In another geoarchaeological study of site formation processes at the early Holocene Paleoindian beach ridge Cummins Site (Julig et al., 1990), north of Lake Superior, SEM analysis of quartz sand grains was particularly useful in determining the origin of fine and medium sands at the base of a bog on the site. The sands were deposited by wind after a decline in the levels of Lake Minong (*c.* 9.5 ka).

Other possible analytical problems of interest to archaeologists include petroglyphs (cultural carvings in rock) and cave deposits in various areas, many occurring outside the limit of glaciation. In southern France, caves in basalt exhibit hand-hewn grooves forming part of the petroglyph. The presence of quartz powder in the rock grooves suggests the "power" tool used was quartz or quartzite (figure 15.8) (R. Dorn, personal communication, 1999). Analysis of the fractures and abrasion features on the fine-grained quartz powder left behind by grinding yielded interesting microfeatures not seen on grains subjected to natural processes. Continual grinding also produced Si dust, a coating on the basalt grooves, similar to silica precipitates described by Whalley & Langway (1980), resulting from glacial grinding. Further research into this problem might yield interesting information, allowing, for example, discrimination between naturally and culturally produced grooves or striations on particle faces.

The identification of lime plaster as the cement or binding agent holding flint sickle blades to wooden handles in 18th Dynasty culture layers of Lower Egypt has been made by Endlicher & Tillmann (1997) using the SEM. Here, the SEM was used to investigate the fabric features of the plaster and the nature of angular particles cemented by very fine-grained carbonate. As observed with the SEM, crystals of calcite cement are frequently oriented at high angles to the surfaces of the coarse angular particles, and are oriented radially into open spaces (figure 15.9), which presumably indicates a partial recrystallization of the carbonate cement. Even the ratio of calcined/uncalcined carbonate fragments in the plaster can be studied by SEM–EDS.

In the identification of glass in the sample studied (figure 15.9A), along with partial vitrification of quartz, the formation of the mineral phases tridymite and cristobalite, accompanying fritting cracks, and grain shapes caused by melting and oxidation of biotite to hematite are all indicative of a high-temperature

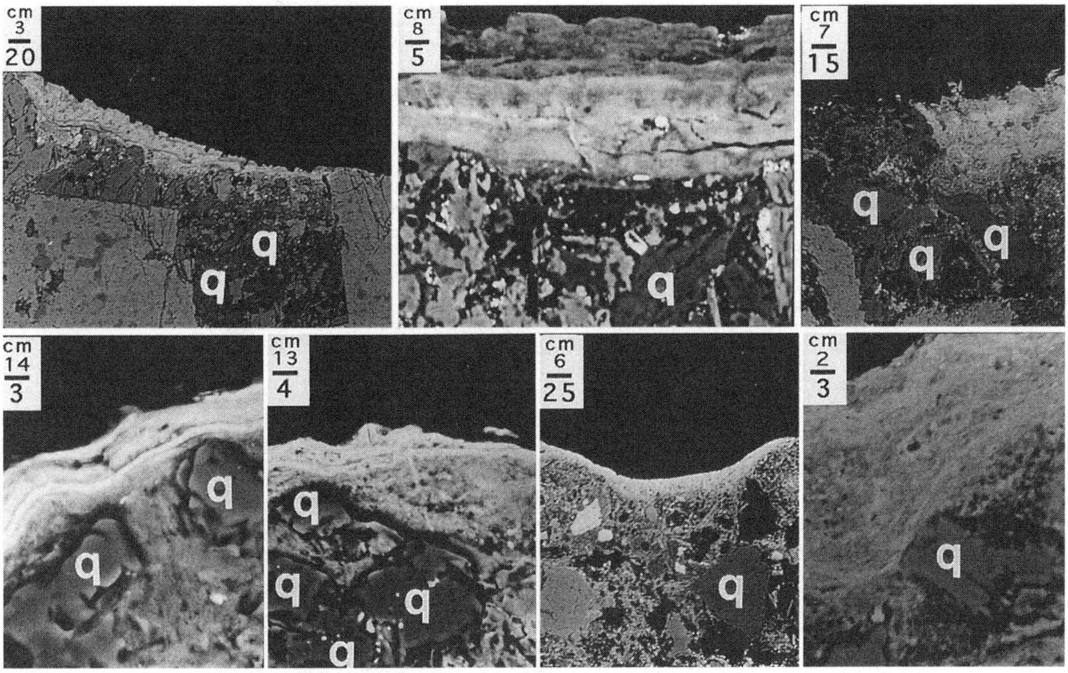

Figure 15.8 Backscattered electron micrographs of quartz grains that have been physically inserted into basalt lithologies containing no free quartz. The quartz grains (indicated by the letter q and identified chemically using an energy-dispersive X-ray detector) were found embedded within petroglyph grooves. Quartz tools were used to make the petroglyphs. Then, rock varnish slowly formed, coating the petroglyph grooves. The rock varnish is "lighter-toned" in these images, because of the higher atomic number of the manganese-iron-rich rock coating. Each of these images is given a letter–number combination that identifies the petroglyph as coming from the Conejo Mine Site in the Coso Range of eastern California. Also, each image is given a number that indicates the length of the scale bar in micrometers. For example, the image in the top left is Conejo Mine (cm) image 3, and the scale bar is 20 μm. Imagery courtesy of Ron Dorn, Arizona State University, Tempe, Arizona.

process. The mineral and chemical composition, as well as the microstructure, all indicate that the raw material came from a quartz- and feldspar-containing limestone that had been burnt and used as a plaster. These findings provide evidence for the first use of lime plaster (figure 15.9B) as a binding agent in ancient Egypt, so far known. Further archaeological conclusions are that pure lime plaster had not only been used before the 19th/20th Dynasty, but some two hundred years earlier in the 18th Dynasty.

From the examples discussed above it is clear that the use of SEM microtextures can greatly assist in archaeological investigations, either by importing criteria used in the Earth sciences or by using geoarcheological criteria. In geoarchaeology, the SEM method has not been fully exploited by researchers

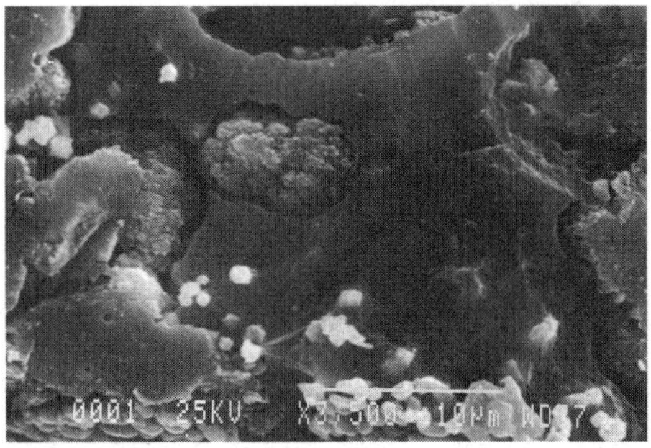

A

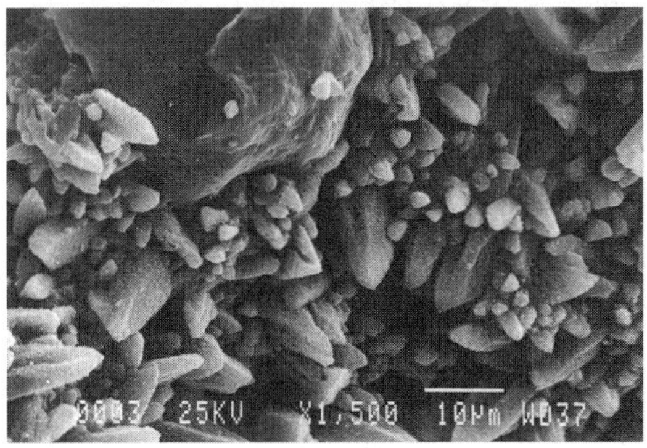

B

Figure 15.9 (A) Glass fragment with round cavities and concave embayments representing former amygdules (now partly filled up with fine-grained calcite). Figure courtesy of G. Endlicher and A. Tillman, Universität Regensburg, Germany, with permission from *Archaeometry*. (B) Calcite cement oriented and idiomorphically crystallized. Figure courtesy of G. Endlicher and A. Tillman, Universität Regensburg, Germany. Reprinted from figures 4 and 6 of Endlicher & Tillman (1997) with permission from *Archaeometry*.

(see discussion in Church, 1995), despite pioneering efforts by Moore (1978) who analyzed chert imagery, and other workers who combined SEM with other techniques such as cathodoluminescence and X-ray fluorescence (Muir et al., 1971; Evans et al., 1994). Beyond environmental reconstruction, the method allows analysis of the finest details on silt to pebble size particles.

Geophagy

Geophagy (soil ingestion) is a behavior practiced by many animals (Mahaney, 1993; Mahaney et al., 1990, 1997b). With regard to primates in general and Great Apes in particular, it is a practice widespread and documented on every continent except, obviously, the Antarctic, although it could occur among bird

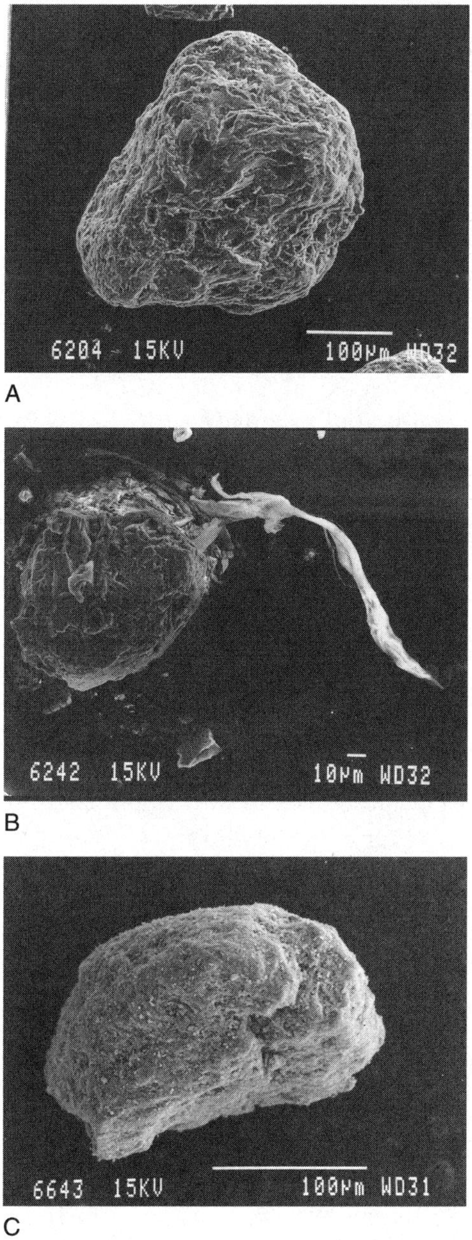

Figure 15.10 (A) Microcline with an Fe coating in soils eaten by orangutans. The coating is thicker in the depressions indicating that the craters formed before weathering. Fe coatings are common on many grains consumed by primates. The location is the Ketambe Research Area in Sumatra. (B) Rutile from natural soils eaten by orangutans with an attached root. The grain is coated with Si and Fe. The root has been replaced with Si, Al, Na, Mg, Fe, Ti, and Ca. Some elements could be important as pharmaceuticals. The location is Ketambe Research Area, Sumatra. This grain is representative of material ingested by mountain orangutans who may be medicating themselves. (C) Orthoclase coated with 1:1 clay minerals (probably halloysite) and Fe. This grain is representative of grains consumed by macaques in Japan.

nesting sites along the coast. The possible geochemical, chemical, mineralogical, and microbiological stimuli for this behavior have never been assessed to determine and validate hypotheses that might explain this behavior. The voluminous literature on this subject suggests it may involve more than a simple quest for dietary or nutritional benefits (Mahaney et al., 1997b) and may involve experimentation with natural earths that have a pharmaceutical or medicinal effect.

The ingestion of soil by primates in general and Great Apes in particular is potentially important to us, as *Homo sapiens*, especially in so far as it concerns the medicinal, chemical or mineralogical properties of the ingested material. The selection of soil as nutritional, dietary, and/or pharmaceutical/medicinal agents suggests that natural earths could be analyzed by SEM–EDS to identify critical microtextural or microstructural components important in understanding primate physiology (Mahaney et al., 1997b).

There is evidence that geophagy (Mahaney, 1993; Mahaney et al., 1997b) likewise plays a role in nutrition, dietary supplementation and pharmaceutical selection. The specific samples ingested by primates in general almost always include reddish or reddish brown clay-rich samples with appreciable quantities of 1:1 (Si:Al = 1:1) clay minerals, including combinations of kaolinite, metahalloysite, and halloysite, with small quantities of 2:1 (Si:Al = 2:1) clay minerals including smectite and illite. While there are numerous studies of the chemistry and mineralogy of the ingested earth and nearby control samples (Mahaney et al., 1997b), there are few studies that attempt to correlate ingested earth with the known geology in specific areas. The ideal instruments to study natural earths consumed by primates are the SEM and EDS as clay and other inorganic coatings on sands are considered to be primary targets for investigation.

Clay and oxide coatings on sands (figure 15.10) provide important information on the stimulus for geophagy. The sand fraction in natural earths consumed by various species of primates provides an adequate population of grains containing important information on the weathering history of the consumed material. Geophagy occurs at ancient sites with long weathering histories and with concomitant advanced etching of both the light and heavy minerals, precipitation of various oxide coatings, and adhesion of clay minerals and other weathering products. Examples of various microtextures that have provided information important in unraveling the mysteries of geophagy in various locales are reported below.

While geophagy soils are always clay-rich they usually contain enough sand to serve as collecting surfaces for coatings of various kinds. Iron is the predominant coating, although Si and Ca are common in certain localities. Clay mineral coatings are most often mixes of halloysite, metahalloysite, and/or kaolinite, sometimes accompanied by small amounts of smectite. Usually the clay mineral coating on sands can be correlated with XRD results from the clay fraction.

Scanning electron microscope study of the sand fraction of geophagic soils can be expected to provide valuable information on mineral composition, degree of dissolution, and coatings. The analysis of geophagy samples makes it possible to isolate the chemical and/or mineral materials that might explain geophagy behavior (see Mahaney (1999) for the most recent study of chimpanzees in the Mahale Mountains, Tanzania).

Conclusions

16

Sand grain microtextural analysis, as it is used to identify ancient sedimentary environments including preweathering, weathering, fractures, and abrasion, is widely applicable to investigations in allied fields such as archaeology, forensic sciences, and anthropology. Within the major geologic transporting agents, diagnostic criteria for identification, formerly widened to include many individual microtextures, is now reduced to a few criteria intensively studied in many different localities. These diagnostic environmental criteria include v-shaped percussion cracks for water; deep, straight, and curved grooves (striations) for glaciers; bulbous edges for eolian grains; and excessive grain packing in craters for mass wasting. Stretched and spindle-shaped quartz is diagnostic of tectonic processes whereas shock melting occurs only with asteroid ejecta.

All microtextures found on quartz grains are also found on other light and heavy minerals, although quartz is a well-recognized superior recorder of source rock release, transport, diagenesis, and/or pedogenesis. Other common minerals such as plagioclase, orthoclase, amphibole, and mica, all of which have one or more good mineral cleavages, may also occasionally provide meaningful data depending on the problem at hand. Less commonly, spinel and tourmaline, both chemically inert, lack prominent cleavages and are prone to exhibit conchoidal and other diagnostic surface microfeatures.

Coatings on grains of various kinds were largely overlooked in the past despite investigations undertaken by numerous workers. As shown in previous chapters, there is a wealth of information to be gleaned by studying coatings on grains, especially glacial grains where the intensity of coatings and their relationship to glacial crushing effects yield information on preweathering, followed by crushing and renewed weathering either during diagenesis or pedogenesis. First recognized by Vortisch et al. (1987), preweathering has been observed in many different environments (Mahaney, 1990a; Mahaney & Kalm, 1995, 2000; Mahaney et al., 2000a). On the one hand, if one analyzes the early Quaternary till in a sequence, it may well prove possible to determine the degree of weathering that occurred prior to the first glaciation (i.e., preweathering) and compare this with weathering that occurred following glacial crushing. On the other hand, it is possible that severe glacial crushing might all but obliterate older etching and dissolution microfeatures, leaving only fresh surfaces as a record of one or several glaciations.

Experimental crushing, which yields important information on ice thickness and degree of crushing, builds on earlier work by Mahaney et al. (1988a) and shows the severe damage inflicted on grains at the base of thick glaciers. The degree of crushing is modified somewhat by the temperature in the basal till. If the glacier is frozen to base, little grain-to-grain contact occurs, as discussed in chapters 9 and 15. However, if the temperature is close to the p-melting temperature, grain-to-grain contact increases manyfold and crushing reaches the point where many grains are nearly obliterated. Such processes lead to a reformation of grains in miniature, eventually producing the typical triple-faceted sand-sized clast carrying multiple fractures deeply embedded into the surface.

Likewise, abrasion may be so severe as to completely remove the entire fracture record from the grain surface. Also, in all the tills examined thus far, some grains are encountered that bear no hallmark of passing through the glacial system. Rather, they seem to pass unscathed through glacial transport without making contact with other grains and bearing their original microfeatures obtained in other nonglacial environments.

Source rock release is one of the most important processes to focus on in microtextural studies, but has seen the least activity over the last four decades. Outside of the work by Brown (1973) and Molén (1992), few studies have concentrated on studying grains weathered directly from bedrock, partly because of difficulties in identifying source rock for many studies. However, as demonstrated here, granitic source rock either with or without glacial influences may be easily sampled and studied. When weathered at source, even at high altitude, quartz grains at source rock sites are identified principally by the presence of fracture faces and the presence of cracked surfaces. Once entering a mass wasting system, the main microfeature, that of fracture faces, seems to endure transit to the valley bottom. Cracks, which may occur on 5–10 percent of grains at source, increase markedly down-system in talus, to around 30–40 percent, most probably as a result of boulders toppling on and impacting fine grains, causing complete fracture. Most probably fracture faces observed in fluvial and glacial sands originate at source as a result of physical release, either from snapping due to excessive weight of a chemically corroded fragment, or from ice crystal growth in a confined space where the tensile strength of the rock can be overcome.

While it is perfectly understandable that glacial, eolian, and fluvial transport mechanisms should receive the greatest attention among microtexturists, additional work is needed with mass wasting to see if the conclusions reached here will hold up with observations over a wider area. For example, most previous work with source rock release and mass wasting are from glaciated areas in Britain, Ontario, the Venezuelan Andes, and Cape Breton, Nova Scotia. The only nonglacial samples come from the Oregon coastal mountains. There is a real need to concentrate on source rock release and the effect of mass wastage transport in areas unaffected by the movement of ice, and complications caused by ice impinging on bedrock outcrops. Only rigorous testing of outcrops in unglaciated regions will confirm or deny the assertions presented in chapter 8.

The *principle of equifinality* (Mahaney & Kalm, 2000), which proposes that similar microtextures might result from the action of different geologic agents, needs to be revised and restated to exclude multiple striations, arc-shaped steps and *deep* conchoidal and subparallel linear fractures as these are the hallmark of glaciation. In addition, v-shaped percussion cracks can be eliminated as they are found only on grains transported by streams, and there is much evidence to suggest they not only deepen with increasing velocity but also occur more frequently with distance of transport. Bulbous edges, the hallmark of eolian transport, should, also be excluded. While in theory, fracture faces might be produced by almost any geologic agent, their high frequency of occurrence in mass wasted systems implies this agent can be identified by their presence in large numbers.

While many other microtextures listed in chapter 5 can be found on grains from different sedimentary environments, grains deposited by glaciers carry the most extensive array of fracture and abrasion microfeatures. While anywhere from 10 to 20 percent of grains in glaciers may escape contact with other grains, and hence sojourn in the ice without suffering physical damage, the other 80 to 90 percent of glacial grains will exit the system with fractured and abraded surfaces. These grains record damage that, to a large extent, depends on the thickness of the ice and the distance of transport with temperatures close to p-melting. While the distance of transport is nearly impossible to compute with total accuracy, it would seem that long-distance transport close to p-melting will yield a triangular-faceted quartz grain totally reformed from its original shape and size, and with the greatest damage inflicted on it. No other geological agent is capable of this transformation and none have the damaging effect that glaciers can inflict on minerals entrained in the ice.

Also, grain reworking as part of the rock cycle is best exhibited by glaciers that continually recycle older weathered grains from one glaciation to another. This is most clearly demonstrated by the increase in preweathering that has been observed in a few studies (see Mahaney & Kalm, 1995) where a succession of tills have been analyzed from progressively older to younger glaciations. In such studies the preweathering effects have noticeably increased up-section into younger late Quaternary tills.

There are many applications for microtextural analysis including the normal correlation of grains with different sedimentary environments and analysis of weathering in paleosols. When coupled with the EDS, the SEM becomes a powerful tool to study quartz and feldspar dissolution and the composition of coatings of various kinds. To this end, it provides useful information on relative age and may even be used to reconstruct leaching histories in paleosols of different age, as explained in *Ice on the Equator* (Mahaney, 1990a). The composition of clay coatings yields useful data on Si:Al ratios which help type clay species useful in understanding rock decomposition, hydrolysis, and the genesis of different clay minerals. In this latter example it would do well to couple the SEM–EDS with XRD analysis of the $<2\,\mu m$ fraction.

In the wider arena of archaeology, forensic sciences, and geophagy studies, the SEM–EDS has proven a valuable tool useful in probing the composition of cave paintings, the use of quartz tools in engraving cave features, and in matching soils from a crime victim with a source area. In geophagy, scientists are often interested in linking an ingested soil with soils in local stratigraphic columns and in working up details on the composition of natural earths eaten by one species or another. It is the chemical composition that yields important clues as to the stimulus or stimuli motivating this behavior and may involve relative concentrations of macroelements such as Fe, K, and Ca, or trace elements such as S, U, and Th. Using the SEM–EDS to study eaten soils compared with control soils has been proven to provide geophagists with a valuable analytical tool.

References

Anderson, R.S., Sorenson, M., & Willetts, B.B. 1991. A review of recent progress in our understanding of aeolian sediment transport. *Acta Mechanica Supplementum 1*. Springer-Verlag, New York, 1–19.

Assallay, A.M., Rogers, C.D.F., Smalley, I.J., & Jefferson, I.F. 1998. Silt: 2–62 μm, 9–4 ϕ. *Earth Sci. Rev.*, **45**: 61–88.

Bagnold, R.A. 1941. *The Physics of Windblown Sand and Desert Dunes*. Methuen, London.

Baker, B.H. 1967. Geology of the Mount Kenya Area, Geological Report no. 79, Kenya Geological Survey, Nairobi.

Baker, V.R., Strom, R.G., Dohm, J.M., Gulick, V.C., Kargel, J.S., Komatsu, G., Ori, G.G., & Rice, J.W. Jr. 2000. Mars' Oceanus Borealis, ancient glaciers and the MEGAOUTFLO hypothesis. Lunar Planet Sci. Conf. (CD-ROM), XXXI, 1863.

Barcilon, V., & MacAyeal, D.R. 1993. Steady flow of a viscous ice stream across a no slip/free transition at the bed. *J. Glaciol.*, **39**: 167–185.

Barlusseau, J-P., & Kounkou, G. 1989. Origin of surface textures on quartz grains in sediments of the upper Cayar deep-sea fan (Senegal–Gambia abyssal plain). *Oceanologica Acta*, **12**: 117–129.

Barndorff-Nelsen, O.E., & Willetts, B.B. 1991. *Aeolian Grain Transport 1—Mechanics*. Springer-Verlag, New York.

Biederman, E. 1962. Destruction of shoreline environments in New Jersey. *J. Sed. Petr.*, **32**: 181–200.

Birkeland, P.W. 1999. *Soils and Geomorphology*. Oxford University Press, New York.

Blatt, H., 1970. Determination of mean sediment thickness in the crust: a sedimentologic method. *Geol. Soc. Am. Bull.* **81**: 255–262.

Blatt, H., & Sutherland, B. 1969. Intrastratal solution and nanophage heavy minerals in shales. *J. Sed. Petr.*, **39**: 591–600.

Blatt, H., Middleton, G., & Murray, R. 1980. *Origin of Sedimentary Rocks*, 2nd ed., Prentice Hall, Englewood Cliffs, NJ.

Bloss, D., & Gibbs, G.V. 1963. Cleavage in quartz. *Am. Min.* **48**: 821–838.

Bockheim, J.G. 1979. Relative age and origin of soils in eastern Wright Valley, Antarctica. *Soil Sci.*, **128**: 142–152.

Boggs, S. Jr. 1995. *Principles of Sedimentology and Stratigraphy*. Prentice Hall, Englewood Cliffs, NJ.

Bomback, J.L. 1973. Stereoscopic techniques for improved X-ray analysis of rough SEM specimens. *Scanning Electron Microscopy*. IITRI, Chicago, 97–105.

Bond, G. 1954. Surface textures of sand grains from the Victoria Falls area. *J. Sed Petr.*, **24**: 191–195.

Borger, H. 1993a. Quarzkoranalyse mittels Rasterelektronenmikroskop und Dünnschliff unter besonderer Berücksichtigung tropischer Verwitterungsresiduen. *Zeitschrift für Geomorphologie*, **37**: 351–375.

Borger, H. 1993b. Penetration of solutions into quartz grains—a scale for weathering intensity. In Ford, D., McCann, B., & Vajoczki, S. (eds.) *Abstracts Volume*, Third International Geomorphology Conference, McMaster University, Hamilton, Ontario, August 23–28, 1993: 106.

Borger, H. 2000. *Mikromorphologie und Paläoenvironment*. Gebrüder Borntraeger, Berlin.

Breed, C.S., Grolier, M.J., & McCauley, J.F. 1979. Morphology and distribution of common "sand" dunes on Mars: comparison with Earth. *J. Geophys. Res.*, **84**(B14): 8183–8204.

Brown, J.E. 1973. Depositional histories of sand grains from surface textures. *Nature*, **242**: 396–398.

Bull, P. 1978. A statistical approach to scanning electron microscope analysis of cave sediments. In Whalley, W. (ed.) *Scanning Electron Microscopy in the Study of Sediments*. Geo Abstracts, Norwich, England, 212–226.

Bull, P.A. 1981. Environmental reconstruction by scanning electron microscopy. *Prog. in Phys. Geog.*, **5**: 368–397.

Bull, P.A., Culver, S.J., & Gardner, R. 1980. Chattermark trails as paleoenvironmental indicators. *Geology*, **8**: 318–322.

Bull, P.A., Goudie, A.S., Price-Williams, D., & Watson, A. 1987. Colluvium: a Scanning Electron Microscopy analysis of a neglected sediment type. In Marshall, J.R. (ed.) *Clastic Particles*. Van Nostrand Reinhold, New York, 16–35.

Campbell, I.B., & Claridge, G.G.C. 1987. *Antarctica: Soils, Weathering Processes and Environment*. Elsevier, Amsterdam.

Canada Soil Survey Committee 1977. *Soils of Canada*, v. 1, Agriculture Canada, Ottawa.

Carr, M. 1996. *Water on Mars*. Oxford University Press, Oxford.

Carver, R.E. 1971. *Procedures in Sedimentary Petrology*. Wiley-Interscience, New York.

Catt, J.A. 1986. *Soils and Quaternary Geology: A Handbook for Field Scientists*. Clarendon Press, Oxford.

Church, T. 1995. Lithic resource studies: a sourcebook for archaeologists. Special publication no. 3, Lithic Technology, Department of Anthropology, University of Tulsa, Oklahoma.

Codarcea, V. 1977. Percentage distribution of heavy minerals in the loess profiles at Paks and Mohac (grain size 0.30–0.16; 0.16–0.06 mm). *Foeldrajzi-Kaezl*, **101**: 138–143.

Cordier, P., & Gratz, A. 1995. TEM study of shock metamorphism in quartz from the Sedan nuclear test site. *Earth Planet. Sci. Lett.*, **129**, 163–170.

Coudé-Gaussen, G., & Balescu, S. 1987. Étude comparée de loess periglaciaires et peridésertiques: premiers resultats d'un examen des grains de quartz au microscope électronique a balayage. *Catena, Supplement*, **9**: 129–144.

Cremer, M., & Legigan, P. 1989. 2. Morphology and surface texture of quartz grains from OPD site 645, Baffin Bay. *Proc. Ocean Drilling Program, Scientific Results*, **105**, 21–30.

Cutts, J.A. 1973. Nature and origin of layered deposits of the martian polar regions. *J. Geophys. Res.*, **78**: 4231–4249.

Cutts, J., & Smith, R.S.U. 1973. Eolian deposits and dunes on Mars. *J. Geophys. Res.* **78**: 4139–4154.

De Boer, D.H., & Crosby, G. 1995. Evaluating the potential of SEM/EDS analysis for fingerprinting suspended sediment derived from two contrasting topsoils. *Catena,* **24**: 243–258.

Delvigne, J.E. 1998. *Atlas of Micromorphology of Mineral Alteration and Weathering. The Canadian Mineralogist, Special Publication 3.* Mineralogical Association of Canada, Ottawa.

Dial, A.L., Jr. 1984. Geologic map of the Mare Boreum area of Mars. USGS Misc. Inv. Ser. Map I-2357 (1:5,000,000).

Dial, A.L., Jr., & Dohm, J.M. 1994. Geologic map of science study area 4, Chasma Boreale region of Mars. USGS Misc. Inv. Ser. Map I-2357 (1:5,000,000).

Dowdeswell, J.A. 1982. Scanning electron micrographs of quartz sand grains from cold environments examined using Fourier shape analysis. *J. Sed. Petr.,* **52**: 1315–1323.

Earl-Goulet, J., Mahaney, W.C., Kalm, V., Sanmugadas, K., Hancock, R.G.V., & Milner, M.W. 1998. Middle-Holocene timberline fluctuation: influence on the genesis of podzols (Spodosols), Norra Strofjället Massif, northern Sweden. *The Holocene,* **8**: 705–718.

Edgett, K., & Malin, M. 1998. Activity of Mars eolian dunes—observations of a low albedo dune field at spacial resolution by Mars Global Surveyor camera. *GSA Annual Meeting,* Toronto, Ontario, Abstract.

Einsele, G. 1992. *Sedimentary Basins: Evolution, Facies and Sedimentary Budget.* Springer Verlag, Berlin.

Endlicher, G., & Tillmann, A. 1997. Lime plaster as an adhesive for hafting Eighteenth-Dynasty flint sickles from Tell El Dab'a, Eastern Nile Delta (Egypt). *Archaeometry,* **39**: 333–342.

Engel, C.G., & Sharp, R.P. 1958. Chemical data on desert varnish. *Geol. Soc. Am. Bull.,* **69**: 487–518.

Evans, J., Hogg, A.J.C., Hopkins, M.S., & Howarth, R.J. 1994. Quantification of quartz cements using combined SEM, CL, and image analysis. *J. Sed. Res.,* **A64**: 334–338.

Fairgrieve, S.I. 1994. SEM analysis of incinerated teeth as an aid to positive identification. *J. Forensic Sci.,* **30**: 557–565.

Fecher, K. 1985. The determination of surface orientation—a simple aid in quantitative analysis on rough surfaces. *Scanning,* **7**: 297–302.

Fisher, R.A., & Yates, F. 1967. *Statistical Tables for Biological, Agricultural and Medical Research,* 6th ed. Oliver and Boyd, Edinburgh.

Folk, R.L. 1968. *Petrology of the Sedimentary Rocks.* Hemphill Press, Austin, TX.

Folk, R.L. 1975. Glacial deposits identified by chattermark tracks in detrital grains. *Geology,* **3**: 475–479.

Frondel, C. 1962. *The System of Mineralogy, Vol. II, Silica Minerals.* Wiley, New York.

Ghosh, P. 1997. Geomorphology and palaeoclimatology of some Upper Cretaceous palaeosols in central India. *Sed. Geol.,* **110**: 25–49.

Glennie, K. 1970. *Desert Sedimentary Environments.* Developments in Sedimentology 14, Elsevier, Amsterdam.

Goldstein, J., Newbury, D., Echlin, P., Joy, D., Fiori, C., & Lifshin, E. 1981. *Scanning Electron Microscopy and X-ray Microanalysis.* Plenum Press, New York.

Goltrant, O., Cordier, P., & Doukhan, J-C. 1991. Planar deformation features in shocked quartz: a transmission electron microscopy investigation. *Earth Planet. Sci. Lett.,* **106**: 103–115.

Granh, Y., Nolvak, J., & Paris, F. 1996. Precise chitinozoian dating of Ordovician impact events in Baltoscandia. *J. Micropalaeont.*, **15**: 21–35.

Gratz, A.J., Fishler, D., & Bohor, B. 1996. Distinguishing shocked from tectonically deformed quartz by the use of the SEM and chemical etching. *Earth Planet. Sci. Lett.*, **142**: 513–521.

Greeley, R. 1979. Silt-clay aggregates on Mars. *J. Geophys. Res.* **84**, 6248–6254.

Greeley, R., & Iverson, J. 1984. *Wind as a Geological Process*. Cambridge University Press, New York.

Greeley, R., & Williams, S.H. 1994. Dust deposits on Mars: the "parna" analog. *Icarus*, **110**: 165–177.

Greeley, R., Williams, S.H., White, B.R., Pollack, J., Marshall, J.R., & Krinsley, D. 1984. Abrasion by aeolian particles: Earth and Mars. NASA Contract Rept. 3788: 1–50.

Greeley, R., Kraft, M., Sullivan, R., Wilson, G., Bridges, N., Herkenhoff, K., Kuzmin, R., Malin, M., & Ward, W. 1999. Aeolian features and processes at the Mars Pathfinder landing site. *J. Geophys. Res.*, **104**: 8573–8584.

Gregory, J.W. 1894. The glacial geology of Mount Kenya. *Quart. J. Geol. Soc.*, **50**: 515–530.

Hall, M.G., & Lloyd, G.E. 1981. The SEM examination of geological samples with a semiconductor back-scattered electron detector. *Am. Mineral.*, **66**: 362–368.

Hartmann, W.K. 1999. *Moons and Planets*. Wadsworth, London.

Helland, P.E., & Diffendal, R.F. Jr. 1993. Probable glacial climatic conditions in source areas during deposition of parts of the Ash Hollow Formation, Ogallala Group (Late Tertiary), of western Nebraska. *Am. J. Sci.*, **293**: 744–757.

Helland, P., & Holmes, M. 1995. Indications of the earliest occurrence of Late Miocene glaciation of Southern Greenland from surface texture analysis of quartz grains from ODP Site 918. *G.S.A. Abstracts with Programs*, North Central/South Central Sections, 57.

Helland, P., & Holmes, M. 1997. Surface textural analysis of quartz sand grains from ODP Site 918 off the southeast coast of Greenland suggests glaciation of southern Greenland. *Paleo-3*, **135**: 109–121.

Hildebrand, A. R. 1993. The Cretaceous/Tertiary Boundary Impact (or the dinosaurs didn't have a chance). *J. Royal Astron. Soc. Canada*, **87**: 77–117.

Hillefors, A. 1970. Deep weathered rock material and sand grains under the scanning electron microscope. *Swedish Geography Yearbook*, **46**: 138–164.

Hurlburt, C. 1971. *Dana's Manual of Mineralogy*. Wiley, New York.

Iverson, N.R. 1990. Laboratory simulations of glacial abrasion: comparison with theory. *J. Glaciol.*, **36**: 304–314.

Izett, G. A. 1990. The Cretaceous/Tertiary Boundary Interval, Raton Basin, Colorado and New Mexico, and its content of shock-metamorphosed minerals: evidence relevant to the K/T boundary impact–extinction theory. *Geol. Soc. Am. Special Paper 249*.

Julig, P., McAndrews, J.A., & Mahaney, W.C. 1990. Geoarchaeological interpretation of upper Great Lakes beach environments: an example from Cummins Paleoindian Site, Thunder Bay, Ontario. *Archaeological Geology of North America*, Geological Society of America, Centennial Special Volume Four, GSA, Boulder, CO, 21–50.

Julig, P., Storck, P., Beukens, R., & Mahaney, W.C. 1992. Reinvestigations of Shequiandah Site geochronology and geoarchaeology. *Current Res. Pleistocene*, **9**: 25–26.

Kaldi, J., Krinsley, D., & Lawson, D. 1978. Experimentally produced eolian surface textures on quartz sand grains from various environments. In Whalley,

W. (ed.) *Scanning Electron Microscopy in the Study of Sediments*. Geo Abstracts, Norwich, 261–274.

Kanaori, Y. 1980. Surface textures of quartz grains from fault gouges (Pt. II). Formational age and fracturing mode. Repts of the Central Electricity Board, Japan. Res. Rept. 381007, 1–35.

Kanaori, Y., Kawaguchi, I., & Mizutani, S. 1980. Surface textures of grains from fault gouges (Pt. III). Repts of the Central Electricity Board, Japan. Res. Rept. 381008, 1–21.

Kanaori, Y., Tanaka, L.K., & Miykoshi, K. 1985. Further studies on the use of quartz grains from fault gouges to establish the age of faulting. *Eng. Geol.*, **21**: 175–194.

Kanaori, Y., Yairi, K., & Ishida, T. 1991. Grain boundary microcracking of granitic rocks from the northeastern region of the Atotsugawa Fault, central Japan: SEM backscatter electron images. *Eng. Geol.*, **30**: 221–235.

Kenig, K. 1995. Surface microtextures of quartz grains from marine and land deposits of the southern Baltic. In Cato, I., & Kingberg, F. (eds.) *Proc. Fourth Marine Conference—the Baltic*, Uppsala. Sveriges Geologiska Undersokining Ser. Ca **86**: 75–86.

Kieffer, H.H., Jaksoky, B.M., & Snyder, C.W. 1992. The planet Mars: from antiquity to the present. In *Mars*. University of Arizona Press, Tucson, AZ, 1–33.

Kinnunen, K.A. 1996. Classification scheme for surface textures of gold nuggets from Finnish Lapland. *Bull. Geol. Soc. Finland*, **68**(2): 18–33.

Kinnunen, K.A., Tynni, R., Hokkanen, K., & Taavitsainen, J. P. 1985. Flint raw materials of prehistoric Finland: rock types, surface textures and microfossils. *Geol. Surv. Finland, Bull.* **334**.

Kinnunen, K.A., Johanson, B., Terho, M., & Puranen, R. 1997. Nondestructive analysis of morphology, chemical composition and physical properties of large gold nuggets from Finnish Lapland. *Geol. Surv. Finland, Special Paper 23*, 29–35.

Kirismäe, K., & Kalm, V. 1995. Local stratigraphic scheme of the Pleistocene in Estonia: an overview of the development. In Meidla, T., Jõeleht, A., Kalm, V., & Kirs, J. (eds.) *Geology of Livonia*. Estonian Geological Soc., University of Tartu, 95–106.

Kleesment, A. 1994. Subdivision of the Aruküla Stage on the basis of lithological and mineralogical criteria. *Proc. Estonian Acad. Sci., Geol.*, **43**: 57–68.

Kleesment, A. 1995. Lithological characteristics of the upper terrigenous Devonian complex in Estonia. *Proc. Estonian Acad. Sci. Geol.*, **44**: 221–233.

Kleesment, A. 1997. Devonian sedimentation basin. In Raukas, A., & Teedumäe, A. (eds.) *Geology and Mineral Resources of Estonia*. Estonian Academy Publishers, Tallinn, 205–208.

Knutsson, K. 1988. *Patterns of Tool Use: Scanning Electron Microscopy of Experimental Quartz Tools*. Societas Archaeologica Upsaliensis.

Kondratiene, O. 1993. *The Catalogue of Stratotypes for the Quaternary of the Baltic Region*. Baltic Stratigraphic Association and Lithuanian Geological Institute, Vilnius.

Krinsley, D., 1981. *Grain Surface Textural Analysis*. Kolb, C., & Clarke, E. (eds.) File Report QR 02.4.5. Institute for Environmental Studies, Baton Rouge, LA 12–17, B1–B26.

Krinsley, D., & Donahue, J. 1968. Environmental interpretation of sand grain surface textures by electron microscopy. *Geol. Soc. Am. Bull.*, **79**: 743–748.

Krinsley, D., & Doornkamp, J.C. 1973. *Atlas of Sand Grain Surface Textures*. Cambridge University Press, Cambridge.

Krinsley, D., & Greeley, R. 1986. Individual particles and martian aeolian action—a review *Sedimentology*, **47**: 167–189.

Krinsley, D., & Hyde, P. 1971. Cathodoluminescence studies of sediments. *Proc. Fourth Ann. SEM Symposium*, IIT Res. Institute, Chicago, 409–416.

Krinsley, D., & Leach, R. 1979. Simulated Martian abrasion of glassy basalt and augite. Repts. Planet. Geo. Program, NASA Tech. Mem. 80339, 311–312.

Krinsley, D., & Leach, R. 1981. Properties of electrostatic aggregates and their possible presence on Mars. *Precamb. Res.*, **14**, 167–178.

Krinsley, D., & Margolis, S. 1969. Scanning electron microscopy: a new method for studying sand grain surface textures. *Trans. NY Acad. Sci.*, **31**: 457–477.

Krinsley, D., & Marshall, J.R. 1987. Sand grain textural analysis: an assessment. In Marshall, J.R. (ed.) *Clastic Particles*. Van Nostrand Reinhold, New York, 1–15.

Krinsley, D., & McCoy, F. 1977. Significance and origin of surface textures on broken sand grains in deep-sea sediments. *Sedimentology*, **24**: 857–862.

Krinsley, D., & McCoy, F. 1978. Aeolian quartz sand and silt. In Whalley, W.B. (ed.) *Scanning Electron Microscopy in the Study of Sediments*. Geo Abstracts, Norwich, England, 249–260.

Krinsley, D., & Smalley, I. 1973. The shape and nature of small sedimentary quartz particles. *Science*, **180**: 1277–1279.

Krinsley, D., & Takahashi, T. 1962. The surface textures of sand grains: an application of electron microscopy. *Science*, **135**: 923–925.

Krinsley, D., & Tovey, N.K. 1978. Cathodoluminescence in quartz sand grains. *SEM* (1978), **1**: 887–894.

Krinsley, D., & Trusty, P. 1986. Sand grain surface textures. In Sieveking, G. De G., & Hart, M.B. (eds.) *The Scientific Study of Flint and Chert*. Proc. Fourth International Flnt Symp., Cambridge University Press, Cambridge, 201–207.

Krinsley, D., & Wellendorf, W. 1980. Wind velocities determined from the surface textures of sand grains. *Nature*, **283**, 372–373.

Krinsley, D., Friend, P., & Klimentides, R. 1976. Eolian transport textures on the surface of sand grains of early Triassic age. *Geol. Soc. Am. Bull.*, **87**: 130–132.

Krinsley, D., Greeley, R., & Pollack, J. 1979. Abrasion of wind-blown particles on Mars—erosion of quartz and basaltic sand under simulated martian conditions. *Icarus*, **39**: 364–384.

Krinsley, D., Pye, K., Boggs, S., Jr., & Tovey, N.K. 1998. *Backscattered Scanning Electron Microscopy and Image Analysis of Sediments and Sedimentary Rocks*. Cambridge University Press, Cambridge.

Krogh, T.E., Kamo, S.L., Bohor, B.F., Hildebrand, A.H., & Sharpton, V. 1993. Precise U–Pb ages of single zircons prove a Chicxulub source for Colorado and Haitian distal ejecta. *GSA, Abstracts with Program*, **25**: 222.

Krumbein, W.C., & Pettijohn, F.J. 1938. *Manual of Sedimentary Petrography*. Appleton Century-Crofts, New York.

Kuenen, P.H. 1960. Experimental abrasion 4. Eolian action. *J. Geol.*, **68**: 427–449.

Kuenen, P.H., & Perdok, W. 1962. Experimental abrasion 5. Frosting and defrosting of quartz grains. *J. Geol.*, **70**: 648–658.

Kukla, G.J. 1975. Loess stratigraphy of central Europe. In Butzer, K.W., & Isaac, G.L. (eds.) *After the Australopithecines. Stratigraphy, Ecology and Culture Change in the Middle Pleistocene*. Mouton, The Hague, 99–188.

Kurvits, T., Mahaney, W.C., & Kalm, V. 2000. Grain micromorphology in the Rannamoisa Section, Lower Cambrian, Estonia. *Proc. Estonian Acad. Sci. Geol.*, **49**: 17–27.

Landsberg, P. 1972. *World Survey of Climatology: Climates of Africa*, **10**. Elsevier, Amsterdam.

Larson, E.E., & Walker, T.R. 1975. Development of chemical remanent magnetization during early stages of red-bed formation in Late Cenozoic sediments, Baja, California. *Geol. Soc. Am. Bull.*, **86**: 639–650.

Le Ribault, L. 1975. *L'exoscopie, Methode et Applications*. Notes et Memoires, No. 12. Campagnie Francaise de Petroles, Paris.

Lewis, C.A., & Hanvey, P.M. 1993. The remains of rock glaciers in Bottelnek, East Cape Drakensberg, South Africa. *Trans. Roy. Soc. South Africa*, **48**: 265–289.

Lindé, K. 1978. Experimental aeolian abrasion of different sand sized materials: some preliminary results. In Whalley, W.B. (ed.) *Scanning Electron Microscopy in the Study of Sediments*. Geo Abstracts, Norwich, 242–253.

Lindé, K. 1984. Scanning electron microscope studies of different sands and silts. *Acta Universitatis Upsaliensis*. University of Upsala, Upsala, Sweden.

Lindé, K., & Mycielska-Dowgiallo, E. 1980. Some experimentally produced microtextures on grain surfaces of quartz sand. *Geografisk. Ann.*, **62A**: 171–184.

Lindstrom, L. 1975. Surface and bond forming properties of quartz and silicate minerals and their application in mineral processing techniques. *Acta Polytech. Scandinavica*, Chem including Metallurgy Series, **75**, 1–149.

Lindström, E. 1988. Are roches moutonnées merely preglacial forms? *Geografisk. Ann.*, **4**: 323–331.

Little, I., Armitage, T., & Gilkes, R. 1978. Weathering of quartz in dune sands under subtropical conditions in eastern Australia. *Geoderma*, **20**: 225–237.

Loen, J.S. 1994. Origin of placer gold nuggets and history of formation of glacial gold placers, Gold Creek, Granite County, Montana. *Econ. Geol.*, **89**: 91–104.

Lucci, F., & Casa, G. 1968. Surface textures of desert quartz grains: a new attempt to explain the origin of desert frosting. *Giornale di Geologia*, Series 2, **36**, 751–776.

Mahaney, W.C. 1978. Late Quaternary soil stratigraphy in the Wind River Mountains, western Wyoming. In Mahaney, W.C. (ed.) *Quaternary Soils*. Geo-Abstracts, Norwich, 223–264.

Mahaney, W.C. 1990a. *Ice on the Equator*. Wm. Caxton, Ellison Bay, WI.

Mahaney, W.C. 1990b. Macrofabrics and quartz microstructures confirm glacial origin of Sunnybrook drift in the Lake Ontario Basin. *Geology*, **18**: 145–148.

Mahaney, W.C. 1990c. Glacially-crushed quartz grains in late Quaternary deposits in the Virunga Mountains, Rwanda—indicators of wind transport from the north. *Boreas*, **19:** 81–89.

Mahaney, W.C. 1991a. Late Pleistocene and Holocene glacial chronology at Chamonix and Argentière, French Alps. *Zeitschrift für Geomorphologie*, **35**: 225–237.

Mahaney, W.C. 1991b. Distributions of halloysite–metahalloysite and gibbsite in tropical mountain paleosols: relationship to Quaternary paleoclimate. *Palaeo-3*, **88**: 219–230.

Mahaney, W.C. 1992a, Weathering and geochronology of a Quaternary paleosol sequence in lower Teleki Valley, Mount Kenya. *Catena*, **19**: 99–118.

Mahaney, W.C. 1992b. Microtextures on quartz and gold grains transported by glaciers. In Hérail, G., & Fornari, M. (eds.) *Actes du Symposium International sur les gisements alluviaux d'or*, La Paz, Bolivia, 3–5 June 1991. CNRS, La Paz, 315–325.

Mahaney, W.C. 1993. Scanning electron microscopy of earth mined, ground and eaten by mountain gorillas in the Virunga Mountains. *Primates*, **34**: 311–319.

Mahaney, W.C. 1995a. Glacial crushing, weathering and diagenetic histories of quartz grains inferred from scanning electron microscopy. In Menzies, J. (ed.) *Glacial Environments—Processes, Sediments and Landforms.* Butterworth-Heinemann, Oxford, 487–506.

Mahaney, W.C. 1995b. Pleistocene and Holocene glacier thicknesses, transport histories and dynamics inferred from SEM microtextures on quartz particles. *Boreas,* **24**: 293–304.

Mahaney, W.C. 1999. Paleoclimate and paleonutrition–paleozoopharmacognosy, a timely connection. *Frankfurter Geowissenschaftliche Arbeiten,* Series v. **25**: 123–134.

Mahaney, W.C., & Andres, W. 1991. Glacially crushed quartz grains in loess as indicators of long-distance transport from major European ice centers during the Pleistocene. *Boreas,* **20**: 231–239.

Mahaney, W.C., & Andres, W. 1996. Scanning electron microscopy of quartz sand from the north-central Saharan desert of Algeria. *Zeitschrift für Geomorphologie* N.F., Suppl-BD., **103**, 179–192.

Mahaney, W.C., & Ermuth, F. 1975. *The Effects of Agriculture and Urbanization on the Natural Environment.* Geographical Monographs, York University, 152 pp.

Mahaney, W.C., & Hancock, R.G.V. 1993. Late Quaternary stratigraphy and geochemistry of the R47 Section, Rouge River Basin, south-central Ontario, Canada: correlation with Scarborough Bluffs. *J. Quat. Sci.,* **98**, 167–178.

Mahaney, W.C., & Hancock, R.G.V. 2001. Fe–Al extracts in Antarctic paleosols: some estimates of age and translocation. *Third International Conference on Cryopedology.* Copenhagen, Denmark, August 20–24, 2001. International Union of Soil Science & International Permafrost Assoc. Abstracts Vol.

Mahaney, W.C., & Halvorson, D. 1986. Rates of mineral weathering in Rocky Mountain glacial sequences: evidence from Western Wyoming. In Colman, S.M., & Dethier, R. (eds.) *Rates of Chemical Weathering of Rocks and Minerals.* Academic Press, New York, 147–167.

Mahaney, W.C., & Kalm, V. 1995. Scanning electron microscopy of Pleistocene tills in Estonia. *Boreas,* **24**: 13–29.

Mahaney, W.C., & Kalm, V. (eds.) 1996. *Field Guide for the International Conference on Quaternary Glaciation and Paleoclimate in the Andes Mountains,* June 21–July 1, 1996. Quaternary Surveys, Toronto.

Mahaney, W.C., & Kalm, V. 2000. Comparative SEM study of oriented till blocks, glacial grains, and Devonian sands in Estonia and Latvia. *Boreas,* **29**: 35–51.

Mahaney, W.C., & Milner, M. 1998. Zircon microstriators in the sand of auriferous Andean tills: fine tools and noble metals. *Boreas,* **27**: 140–152.

Mahaney, W.C., & Rutter, N.W. 1992. Relative ages of the moraines of the Dalijia Shan, northwestern China. *Catena,* **19**: 179–191.

Mahaney, W.C., & Sanmugadas, K. 1986. Soil development as a function of time in the Rouge River Basin, south-central Ontario. *Géographie Physique et Quaternaire,* **40**: 207–216.

Mahaney, W.C., & Sjöberg, R. 1993. Scanning electron microscopy of quartz grains from two granite caves and a gorge system in Bohuslan, southwestern Sweden. *Zeitschrift für Geomorphologie,* **37**: 337–348.

Mahaney, W.C., & Stalker, A. MacS. 1988. Stratigraphy of the North Cliff Section, Wellsch Valley site, Saskatchewan. *Can. J. Earth Sci.,* **25**: 206–214.

Mahaney, W.C., Fahey, B.D., & Lloyd, D.T. 1981. Late Quaternary deposits, soils and chronology, Hell Roaring Valley, Cascade Range, Washington. *Arctic and Alpine Res.,* **13**: 339–356.

Mahaney, W.C., Piegat, J., Halvorson, D.L., & Sanmugadas, K. 1984. Evaluation of dating methods used to assign ages in the Wind River and Teton Ranges, western Wyoming. In Mahaney, W.C. (ed.) *Quaternary Chronologies*, Geobooks, Norwich, England, 225–241.

Mahaney, W.C., Vortisch, W., & Julig, P. 1988a. Relative differences between glacially crushed quartz transported by Mountain and continental ice—some examples from North America and East Africa. *Am. J. Sci*, **288**: 810–826.

Mahaney, W.C., Hancock, R.G.V., & Stalker, A. MacS. 1988b. Neutron activation analysis of tills in the North Cliff Section, Wellsch Valley, Saskatchewan. *Sed. Geol.*, **55**: 185–196.

Mahaney, W.C., Terasmae, J., Melville, H., & Boyer, M.G. 1989. Very late Holocene overbank sedimentation and radiocarbon ages of buried soils in the Rouge River Basin, south-central Ontario. *Acta Geologica Hungarica*, **32**: 325–348.

Mahaney, W.C., Hancock, R.G.V., & Watts, D.P. 1990. Geophagia by mountain gorillas (*Gorilla gorilla beringei*) in the Virunga Mountains, Rwanda. *Primates*, **31**: 113–120.

Mahaney, W.C., Vaikmae, R., & Vares, K. 1991. Scanning electron microscopy of quartz grains in supraglacial debris, Adishy Glacier, Caucasus Mountains, U.S.S.R. *Boreas*, **20**: 395–404.

Mahaney, W.C., Hancock, R.G.V., & Zhang, L. 1992. Stratigraphy and geochemistry of the Jiuzhoutai loess sequence, northwestern China. *J Southeast Asian Earth Sci.*, **7**: 215–221.

Mahaney, W.C., Andres, W., & Barendregt, R.W. 1993. Quaternary paleosol stratigraphy and paleomagnetic record near Dreihausen, central Germany. *Catena*, **20**: 161–177.

Mahaney, W.C., Claridge, G., & Campbell, I. 1996a. Microtextures of quartz grains in tills from Antarctica. *Paleo-3*, **121**: 89–103.

Mahaney, W.C., Sanmugadas, K., & Hancock, R.G.V. 1996b. Physical and geochemical analysis of a late glacial/Little Ice Age pedostratigraphic complex in the Zillertal Alps, Austria. *Zeitschrift für Geomorphologie*, **40**: 447–460.

Mahaney, W.C., Barendregt, R.W., & Vortisch, W. 1997a. Relative age of loess and till in two Quaternary paleosols in Gorges valley, Mount Kenya, East Africa. *J. Quat. Sci.*, **12**: 61–72.

Mahaney, W.C., Wrangham, R., Hancock, R.G.V., Aufreiter, S., Milner, M.W., & Pier, H. 1997b. Analysis of geophagy soils in Kibale Forest, Uganda. *Primates*, **38**: 159–175.

Mahaney, W.C., Sanmugadas, K., & Hancock, R.G.V. 1999. Extractable Fe and Al in soils in the middle Teton chronosequence, W. Wyoming, USA. *Zeitschrift für Geomorphologie*, **43**: 393–407.

Mahaney, W.C., Milner, M.W., Voros, J., Kalm, V., Hütt, G., Bezada, M., Hancock, R.G.V., Aufreiter, S. 2000a. Stratotype for the Mérida Glaciation at Pueblo Llano in the northern Venezuelan Andes. *J. South Am. Earth Sci.*, **13**: 761–774.

Mahaney, W.C., Milner, M.W., Sanmugadas, K., Kalm, V., Bezada, M., & Hancock, R.G.V. 2000b. Late Quaternary deglaciation and Neoglaciation of the Humboldt Massif, northern Venezuelan Andes. *Zeitschrift für Geomorphologie*, Suppl.-Bd **122**: 209–226.

Mahaney, W.C., Dohm, J., Baker, V.R., Newsom, H., Malloch, D., Hancock, R.G.V., Campbell, I., Sheppard, D., & Milner, M.W. 2001a. Morphogenesis of Antarctic paleosols: Martian analog. *Icarus*, **154:** 113–130.

Mahaney, W.C., Stewart, A., Kalm, V. 2001b. Quantification of SEM microtextures useful in sedimentary environmental discrimination. *Boreas*, **30**: 165–171.

Mahaney, W.C., Russell, S.E., Milner, M.W., Kalm, V., Bezada, M, Hancock, R.G.V., & Beukens, R. 2001c. Paleopedology of middle Wisconsin/Weichselian paleosols in the Mérida Andes, Venezuela. *Geoderma* **104**: 215–237.

Mainguet, M., & Moreau, C. 1982. Aspects of the aeolian dynamic in the northern circumpolar sand seas of Mars. NASA Tech Mem. 85127, 193–196.

Margolis, S., & Krinsley, D. 1971. Submicroscopic frosting on eolian and subaqueous quartz sand grains. *Geol. Soc. Am. Bull.*, **82**, 3395–3406.

Margolis, S., & Krinsley, D. 1974. Processes of formation and environmental occurrence of microfeatures on detrital quartz grains. *Am. J. Sci*, **274**: 449–464.

Marshall, J.R. 1979. The simulation of sand grain surface textures using a pneumatic-gun apparatus and quartz plates. Ph.D. thesis, University College, London.

Marshall, J.R. 1987. *Clastic Particles*. Van Nostrand Reinhold, New York.

Mazzullo, J. 1987. Origin shapes in the St. Peter Sandstone: determination by Fourier shape analysis and scanning electron microscopy. In Marshall, J.R. (ed.) *Clastic Particles*. Van Nostrand Reinhold, New York, 302–313.

Mazzullo, J., & Ritter, C. 1991. Influence of sediment source on the shapes and surface textures of glacial quartz sand grains. *Geology*, **19**: 384–388.

McKee, T., & Krinsley, D. 1978. Clay sized quartz produced by crushing. In Sturgis, J. (ed.) *Ninth International Congress on Electrical Microscopy*, Toronto, Canada. **1**: 486–487.

McKee, T.R., Greeley, R., & Krinsley, D. 1979. Simulated eolian erosion of quartz. In Baley, G.W. (ed.) *37th Ann. Proc. Electr. Microscopy Soc. Am.* San Antonio, TX, 624–625.

McVicar, M., & Graves, W.J. 1997. Forensic comparison of soils by automated Scanning Electron Microscopy. *Can. Soc. Forensic Sci. J.*, **30**: 241–261.

Mens, K., & Pirrus, E. 1977. *Stratotypes of the Cambrian Formations of Estonia*. Valgas, Tallinn (in Russian with English summary).

Middleton, G.V., & Kassera, C.A. 1987. Variation in density of v-shaped impact pits on quartz grains with size of grains, intertidal sands, Bay of Fundy. *J. Sed. Petrol.*, **57**: 88–93.

Miller, R., & Olson, E. 1955. The statistical stability of quantitative properties as a fundamental criterion for the study of environments. *J. Geol.*, **63**: 376–387.

Molén, M. 1992. SEM-microtextures: attempting to solve the problems of the origin of diamictons and diamictites. MSc Thesis, York University, N. York, Ontario.

Moore, J.H., Hatton, R.E., Scott, R.F., Spitzer, C.R., & Shorthill, R.W. 1977. Surface materials of the Viking Lander sites. *J. Geophys. Res.*, **82**: 4497–4523.

Moore, P.N. 1978. Scanning Electron Microscopy examination of some archaeologically significant cherts of the Western Tennessee River Valley. Unpublished MA thesis, Memphis State University, Memphis, TN.

Morrison, D. 1995. Target Earth. *Astronomy*, October, 34–41.

Morrison, R.B. 1978. Quaternary soil stratigraphy—concepts, methods and problems. In Mahaney, W.C. (ed.) *Quaternary Soils*. Geo-Abstracts, Norwich, England, 77–108.

Morton, A.C., & Johnsson, M.J. 1993. Factors influencing the composition of detrital heavy mineral suites in Holocene sands of the Apure River drainage basin, Venezuela. *Geological Society of America Special Paper 284*, 171–185.

Mueller, G. 1967. *Methods in Sedimentary Petrology*, Hafner Publishing, New York.

Muir, M.D., Grant, P.R., Hubbard, G., & Mundell, J. 1971. Cathodoluminescence spectra. In *Scanning Electron Microscopy—1971*. IIT Research Institute, Chicago.

Muller, M.J., & Negendank, J.F.W. 1974. Investigations of heavy minerals in the sediments of the Mosel. *Geologische Rundschau*, **63**: 998–1035.

Murray, R., & Tedrow, J.C. 1992. *Forensic Geology*. Prentice Hall, Englewood Cliffs, N.J.

Nelson, R.L. 1954. Glacial geology of the Frying Pan River Drainage, Colorado. *J. Geol.*, **62**: 325–343.

Oberbeck, V.R., Marshall, J.R., & Aggarwal, H. 1993. Impacts, tillites and the breakup of Gondwanaland. *J. Geol*, **101**: 1–19.

Öpik, A. 1927. Die Inseln Odensholm und Rogö. Ein beitrag zur Geologie von NW-Estland. *Publications of the Geological Institution of the University of Tartu*, **9**: 1–69.

Orr, E.D. 1981, Beach and glacial particle morphogeny, Montauk Peninsula, Long Island, N.Y. MA Thesis, University of Texas at Austin, Texas.

Osborne, R.H., Bomer III, E.J., Wang, Y-C., & Lu, Y. 1993. Application of a tumbler experiment using granodiorite grus to examine the character of quartz-grain fracture in high gradient streams. *Geological Society of America Special Paper 284*, 211–234.

Pascoe, K.J. 1961. *An Introduction to the Properties of Engineering Materials*. Blackie, London.

Passchier, C.W., & Trouw, R.A.J. 1998. *Microtectonics*. Springer-Verlag, Berlin.

Passchier, S., Uścinowicz, S., & Laban, C. 1997. Sediment supply and transport directions in the Gulf of Gdańsk as observed from SEM analysis of quartz grain surface textures. *Prace Państwowego Instytuto Geologicznego*, Warsaw.

Perttunen, M., & Hirvas, H. 1982. An attempt to use the roundness of quartz grains for till stratigraphy. *Bull. Geol. Soc. Finland*, **54**: 25–33.

Petersen, R.G., & Calvin, L.D. 1986. Sampling. In Klute, A. (ed.) *Methods of Soil Analysis*. American Society of Agronomy, Madison, WI., 33–51.

Pettijohn, F. 1975. *Sedimentary Rocks*. Harper, New York.

Piotrowski, J.A. 1992. Till facies and depositional environments of the upper sedimentary complex from the Stohler Cliff, Schleswig-Holstein, North Germany. *Zeitschrift für Geomorphologie*, N.F. Suppl.-Bd: 37–54.

Piotrowski, J.A. 1996. Dynamik und subglaziale hydrogeologie der weichselzeitlichen eiskappe in Zentral-Schleswig-Holstein. Berichte Geol.-Paläont. Inst. Universität Kiel, Nr. 79.

Poirier, J.P. 1985. *Creep of Crystals: High Temperature Deformation Processes in Metals, Ceramics and Minerals*. Cambridge University Press, Cambridge.

Porter, J. 1962. Electron microscopy of sand surface textures. *J. Sed. Petr.* **32**: 124–135.

Powers, L., Brueckner, H., & Krinsley, D. 1979. Rb–Sr ages from weathered and stream transported quartz grains from the Harvey Peak Granite, Black Hills, South Dakota. *Geochim. et Cosmochim. Acta*, **43**: 137–146.

Puura, V. 2000. Scenarios of cratering and crater survival in Precambrian shield and shallow Palaeozoic platform areas in the Baltic Sea Region. In Plado, J., & Pesonen, L.J. (eds.) *Meteorite Impacts in Precambrian Shields*. Impact 4th ESF Workshop, Lappajärvi–Karikkoselkä–Sääksjärvi, Finland.

Pye, K. 1983. Formation of quartz silt during humid tropical weathering of dune sands. *Sed. Geol.*, **34**: 267–282.

Pye, K., & Mazzullo, J. 1994. Effects of tropical weathering on quartz grain shape: an example from Northeastern Australia. *J. Sed. Res.*, **A64**: 500–507.

Pye, K., & Sperling, C. 1983. Experimental investigation of silt formation by static breakage processes: the effect of temperature, moisture and salt on quartz dune sand and granitic regolith. *Sedimentology*, **30**: 49–62.

Pye, K., & Tsoar, H. 1990. *Aeolian Sand and Sand Dunes*. Unwin Hyman, Boston, MA.

Raukas, A. 1995. properties, origin and stratigraphy of Estonian tills. In Ehlers, J., Kozarski, S., & Gibbard, P. (eds.) *Glacial Deposits in North-East Europe*. Balkema, Rotterdam, 93–101.

Raukas, A., & Kajak, K. 1995. Quaternary stratigraphy in Estonia. *Proc. Estonian Acad. Sci. Geol.*, **44**: 149–162.

Raukas, A., & Kajak, K. 1997. Quaternary cover. In Raukas, A., & Teedumäe, A. (eds.) *Geology and Mineral Resources of Estonia*. Estonia Academy of Sciences Publisher, Tallinn, 125–136.

Raup, D.M. 1986. *The Nemesis Affair*. W.W. Norton, New York.

Reed, S.J.B. 1997. *Electron Microprobe Analysis*. Cambridge University Press, Cambridge.

Retallack, G., & Krinsley, D. 1993. Metamorphic alteration of fine structure in a Precambrian (2.2 GA) paleosol from South Africa. *Precambrian Res.*, **63**: 27–41.

Roaldset, E. 1984. Tertiary (Miocene–Pliocene) interbasalt sediments, NW-W-Iceland, *Jökull*, **33**: 39–56.

Robinson, D.A., & Williams, R.B.G. 2000. Accelerated weathering of a sandstone in the High Atlas Mountains of Mococco by an epilithic lichen. *Zeitschrift für Geomorphologie*, **44**: 513–528.

Rogerson, R.J., & Hudson, H.M. 1983. Quartz surface microtextures and grain size characteristics of Quaternary sediments in the porcupine strand area of coastal Labrador, Newfoundland, Canada. *Can. J. Earth Sci.*, **20**: 377–387.

Sagan, C. 1973. Sandstorms and erosion on Mars. *J. Geophys. Res*, **78**: 4155–4161.

Schieber, J., Krinsley, D., & Tennison, E. 2002. Provenance studies of fine-grained sediments with scanned cathodoluminescence of quartz: potential applications in planetary exploration. Mars Symposium, March 11–16, Houston, Texas, Lunar Planetary Inst., Extended Abstracts, in press.

Setlow, L.M., & Karpovich, R.P. 1972. "Glacial" microtextures on quartz and heavy mineral sand grains from the littoral environment. *J. Sed. Petrol.*, **42**: 864–975.

Seyedolali, A., Krinsley, D., Boggs, S., Jr., O'Hara, P., Dypvik, H., & Goles, G. 1997. Provenance interpretation of quartz by scanning electron microscope–cathodoluminescence fabric analysis. *Geology*, **25**: 787.

Shennan, S. 1988. *Quantifying Archaeology*. Edinburgh University Press, Edinburgh.

Shepard, F.P., & Young, R. 1961. Distinguishing between beach and dune sands. *J. Sed. Petrol.*, **31**: 196–214.

Sheridan, M.F., & Marshall, J.R. 1987. Comparative charts for quantification analysis of grain-textural elements on pyroclasts. In Marshall, J.R. (ed.) *Clastic Particles*. Van Nostrand Reinhold, New York, 98–121.

Sheridan, M.F., & Kortemeier, C.P. 1987. Quantitative analysis of surface features on pyroclasts: an example from the Monte Guardia Sequence, Lipari. In Marshall, J.R. (ed.) *Clastic Particles*. Van Nostrand Reinhold, New York, 122–133.

Sjöberg, R. 1987. Caves as indicators of neotectonics in Sweden. *Zeitschrift für Geomorphologie*, **63**: 141–148.

Sjöberg, R. 1989. Hunnebo Klovar i norra. Bohuslän. *Grottan*, **1**: 26–29.

Slayter, E., & Slayter, H. 1992. *Light and Electron Microscopy*. Cambridge University Press, Cambridge.

Smalley, I.J. 1966. The properties of glacial loess and the formation of loess deposits. *J. Sed. Petrol.*, **36**: 669–676.

Smart, P., & Tovey, N.K. 1988. *Electron Microscopy of Soils and Sediments, Vol. 1. Techniques*. Oxford University Press, Oxford.

Soil Survey Staff. 1998. *Keys to Soil Taxonomy*, USDA, Washington, DC.

Sorby, H.C. 1880. On the structure and origin of non-calcareous stratified rocks. *Quart. J. Geol. Soc.*, 46–92.

Stambolic-Robb, A. 1997. Geophagy amongst free-ranging Sumatran orang-utans (*Pongo pygmaeus abelii*) of Gunung Leuser National Park and ex-captive Bornean orang-utans (*Pongo pygmaeus pygmaeus*) of Sungai Wain Forest, Indonesia. MSc. Thesis, York University, N. York, Ontario.

Stoeffler, D., & Langenhorst, F. 1994. Shock metamorphism of quartz in nature and experiment: 1. Basic observation and theory. *Meteoritics*, **29**: 155–181.

Suuroja, K., & Suuroja, S. 2000. Neugrund structure—the newly discovered submarine Early Cambrian impact crater. *Lecture Notes in Earth Sci.*, **91**: 417–443.

Suuroja, K., & Saadre, T. 1995. The gneissbreccia erratic boulders from north-western Estonia as witnesses to an unknown impact structure. *Bull. Geol. Surv. Estonia*, **5/1**: 26–28 (in Estonian).

Suuroja, K., Saadre, T., & Kask, J. 1999. Geology of Osmussaar Island. *Estonian Maritima*, **4**: 39–63.

Tanaka, K.L., & Scott, D.H. 1987. Geologic map of the polar regions of Mars. USGS Misc. Inv. Ser. Map I-1802-C (1:15,000,000).

Tejan-Kellu, M.S., Fitzpatrick, R.W., & Chittleborough, D.J. 1991. Scanning electron microscope study of zircons and rutiles, from a podzol chronosequence at Ycooloola, Queensland, Australia. *Catena*, **18**: 11–30.

Torsvik, T., Smethurs, M., van der Voo, R., Trench, A., Abrahamsen, N., & Halvorsen, E. 1992. Baltica. A synopsis of Vendian-permian palaeomagnetic data and their paleotectonic implications. *Earth Sci. Rev.*, **33**: 133–152.

Tovey, N.K., & Krinsley, D. 1980. A cathodoluminescent study of quartz sand grains. *J. Micros.*, **120**: 279–289.

Tsoar, H., Greeley, R., & Peterfreund, A.R. 1979. Mars: the north polar sand sea and related wind patterns. *J. Geophys Res.*, **84**: 8167–8180.

Tulaczyk, S., Kamb, B., Scherer, R.P., & Engelhardt, H.F. 1998. Sedimentary processes at the base of a west Antarctic ice stream: constraints from textural and compositional properties of subglacial debris. *Sedimentology*, **68**: 487–496.

Viiding, H., & Konsa, M. 1976. Data on typomorphic varieties of mineral in terrigenous deposits. Narbutas, V.V. (ed.) *Mokslas*, Vilnius, 60–67 (in Russian).

Viiding, H., Kajak, K., Mark-Kurik, E., & Kurus, V. 1981. Aruküla Formation. In Sorokin, V. et al. (eds.) *Devonian and Carboniferous in the Peribaltic*. Zinatne, Riga, 112–126 (in Russian).

Vortisch, W., Mahaney, W.C., & Fecher, K. 1987. Lithology and weathering in a paleosol sequence on Mount Kenya. *Geologica et Paleontologica*, **15**: 67–91.

Watt, I.M. 1985. *The Principles and Practice of Electron Microscopy*. Cambridge University Press, Cambridge.

Waugh, B. 1965. Preliminary electron microscope study of the development of authigenic silica in the Penrith Sandstone. *Proc. York. Geol. Soc.*, **35**: 59–69.

Waugh, B. 1970. Formation of quartz overgrowths in the Penrith Sandstone as revealed by scanning electron microscopy. *Sedimentology*, **14**: 309–320.

Wellendorf, W., & Krinsley, D. 1980. The relation between the crystallography of quartz and upturned aeolian plates. *Sedimentology*, **27**: 447–453.

Whalley, W.B. 1978. An SEM examination of quartz grains from subglacial and associated environments. *Scanning Electron Microscopy*, **1**: 353–360.

Whalley, W.B., & Krinsley, D.H. 1974. Scanning electron microscope study of surface textures of quartz grains from glacial environments. *Sedimentology,* **21**: 87–105.

Whalley, W.B., & Langway, Jr., C.C. 1980. A scanning electron microscope examination of subglacial quartz grains from Camp Century Core, Greenland—a preliminary study. *J. Glaciol,* **25**: 125–132.

Williams, L. 1947. Surface textures of sedimentary particles and grains. Ph.D. dissertation, University of Chicago, Chicago.

Xu, Y.B. 1985. A straining test within an SEM and its application. *Scanning,* **8**: 34–39.

Index